NATURE, KNOWLEDGE AND NEGATION

CURRENT PERSPECTIVES IN SOCIAL THEORY

Series Editor: Harry F. Dahms

Volume 1:	1980 Edited by Scott G. McNall and Garry N. Howe
Volume 2:	1981 Edited by Scott G. McNall and Garry N. Howe
Volume 3:	1982 Edited by Scott G. McNall
Volume 4:	1983 Edited by Scott G. McNall
Volume 5:	1984 Edited by Scott G. McNall
Volume 6:	1985 Edited by Scott G. McNall
Volume 7:	1986 Edited by John Wilson
Volume 8:	1987 Edited by John Wilson
Volume 9:	1989 Edited by John Wilson
Volume 10:	1990 Edited by John Wilson
Volume 11:	1991 Edited by Ben Agger
Volume 12:	1992 Edited by Ben Agger
Volume 13:	1993 Edited by Ben Agger
Volume 14:	1994 Edited by Ben Agger
Supplement 1:	*Recent Developments in the Theory of Social Structure*, 1994 Edited by J. David Knottnerus and Christopher Prendergast
Volume 15:	1995 Edited by Ben Agger
Volume 16:	1996 Edited by Jennifer M. Lehmann
Volume 17:	1997 Edited by Jennifer M. Lehmann
Volume 18:	1998 Edited by Jennifer M. Lehmann
Volume 19:	1999 Edited by Jennifer M. Lehmann
Volume 20:	2000 Edited by Jennifer M. Lehmann
Volume 21:	*Bringing Capitalism Back for Critique by Social Theory*, 2001 Edited by Jennifer M. Lehmann
Volume 22:	*Critical Theory: Diverse Objects, Diverse Subjects*, 2003 Edited by Jennifer M. Lehmann
Volume 23:	*Social Theory as Politics in Knowledge*, 2005 Edited by Jennifer M. Lehmann
Volume 24:	*Globalization between the Cold War and Neo-Imperialism*, 2006 Edited by Jennifer M. Lehmann and Harry F. Dahms
Volume 25:	*No Social Science without Critical Theory*, 2008 Edited by Harry F. Dahms

CURRENT PERSPECTIVES IN SOCIAL THEORY VOLUME 26

NATURE, KNOWLEDGE AND NEGATION

EDITED BY

HARRY F. DAHMS
Department of Sociology, University of Tennessee, Knoxville, USA

Emerald

United Kingdom – North America – Japan
India – Malaysia – China

Emerald Group Publishing Limited
Howard House, Wagon Lane, Bingley BD16 1WA, UK

First edition 2009

British Library Cataloguing in Publication Data
A catalogue record for this book is available from the British Library

ISBN: 978-1-84950-605-2
ISSN: 0278-1204

CONTENTS

PART II: KNOWLEDGE

PART III: NEGATION

LIST OF CONTRIBUTORS

Robert J. Antonio	Department of Sociology, University of Kansas, Lawrence, KS, USA
J. I. (Hans) Bakker	Department of Sociology and Anthropology, University of Guelph, Wellington, Canada
John Barry	Institute for a Sustainable World, Queen's University Belfast, Belfast, Northern Ireland
James Block	Department of Political Science, DePaul University, Chicago, IL, USA
Steven P. Dandaneau	Chancellor's Honors & Haslam Scholars Programs, University of Tennessee, Knoxville, TN, USA
Stephen Gennaro	Children's Studies, Division of Humanities, York University, Toronto, Ontario, Canada
Daniel M. Harrison	Department of Political and Social Sciences, Lander University, Greenwood, SC, USA
Lawrence Hazelrigg	Department of Sociology, Florida State University, Tallahassee, FL, USA
Douglas Kellner	Philosophy of Education Chair, Social Sciences and Comparative Education, University of California-Los Angeles, Los Angeles, CA, USA
Lauren Langman	Department of Sociology, Loyola University of Chicago, Chicago, IL, USA

Timothy W. Luke	Political Science and Program Chair, Government and International Affairs, School of Public and International Affair, Department of Political Science, Virginia Tech, Blacksburg, VA, USA
Patricia Mooney Nickel	School of Social and Cultural Studies, Victoria University of Wellington, Wellington, New Zealand
Vanessa Walilko	Independent Scholar, Evanston, IL, USA
Michael E. Zimmerman	Center for Humanities and the Arts, University of Colorado, Boulder, CO, USA

INTRODUCTION

As the title of this volume suggests, its unifying theme is the interdependence between nature, knowledge, and negation. How we aspire to achieve knowledge is intrinsically related to how we conceive of and exist in and with nature. In turn, how we think about and relate to nature is a function of the kinds of knowledge we pursue and the purposes we ascribe to the pursuit of knowledge. Yet, efforts related to facilitating and accumulating knowledge are intrinsically entwined with concrete socio-historical contexts, as are modes of existing in nature. Depending on the values and priorities that shape as well as represent social life in a particular society, its inhabitants are likely to regard nature as a dimension of reality that must be protected, nurtured, dominated, subjugated, or exploited. If most members of a given society would be fully cognizant of the perimeter circumscribing the relationships between that society and nature, the need for social theory should be limited, and the importance of critical theory close to negligible. As social philosophers and theorists of society have been trying to verbalize for more than two centuries, rigorous awareness of the character of prevailing society–nature (or culture–nature) links in *modern societies* has required focused and steady intellectual effort, as in societies of this type, the mode of relating to nature is not only highly dynamic, but also – under the aegis of industrialization – constructive and destructive at the same time. According to much of the history of social theory (its actual history, as opposed to the history of misinterpretations of social theories), from Hegel to Marx, to Weber, to the Frankfurt School theorists, to Foucault, Beck, and many others, how a society "sees" its relationship to nature is not likely to be conducive to grasping the real character of the relationship. Rather, how modern societies are contingent on a kind of conditioning that prevents human beings from facing the facts of what makes modern society possible shapes how humans perceive nature. Those perceptions are more likely than not integral features of modern society that are difficult to access without the help of social as well as critical theory, due to the kind of gravity they exert on human existence (Lemert, 2007). Put differently, without the willingness to "negate" the programming of individual as well as group consciousness that modern societies require to maintain stability and to function, "reality" will remain hopelessly elusive, and the ability to face facts a skill that is as difficult to sustain as the recognition of and commitment to a truly meaningful conduct of life. Without the practice of *determined negation* in Adorno's sense becoming integral to the life of humanity, socially compelling notions of "reality" and "meaning" will remain fleeting at best.[1]

From various angles, the contributions included in this volume address how in modern society, the study of nature, the pursuit of knowledge, and the practice of negation are contingent on one another, in terms of both their present condition and their future possibility. As an ongoing practice rather than a canonical body of thought, social theory since its inception in the late eighteenth and early nineteenth centuries has been concerned with how the very existence of modern society at the same time depends on and requires the continuous transformation of the relationship between *forms of social life* and *nature*, at all levels of human civilization, from the experiences and identities of individual human beings to ever more complex and contradictory forms of social, political, and economic organization, and related opportunities for social solidarity or lack thereof. Our nature as social beings, that is, as actually existing human beings who are – and exist by being – "hooked" (as it were) into actual social ties and actually existing societal circumstances in concrete ways both facilitate and limit opportunities to grasp how and who we are. The riddle of classical as well as more recent social theorists endeavored to illuminate, if not to solve, pertains to the challenge of understanding how, in a context shaped by change that is characterized by more or less recognizably self-reproducing structural principles and cultural patterns, the persistence of modern society is dependent on a process of redefining both the meaning of nature and its relationship to humanity – along a continuum that reaches from the quality of human existence at one end to the sheer number of people living in particular countries, regions, continents, and, ultimately, on planet Earth, on the other.

The very possibility of modern society has been dependent on the ever more expansive, refined, and efficient extraction of a diverse set of resources, to maintain and protect social stability and social order, as both are manifest in particular societies – in social orders in their specificity, in time, and in space (Dahms, 2008, 2009). In other words, "modern society" denotes a form of social organization that is sustained by the extraction of a multitude of resources from nature, culture, and human beings that goes hand in hand with return compensation only to the degree that it is necessary to secure the continuous functioning of the societal totality, in ways that reconstitute and facilitate the further proliferation of the dominant patterns. These patterns are characteristic of the structure of distributing wealth that corresponds to the extraction of resources, access to those resources, control over the transformation of those resources, and the distribution of the benefits resulting from the process – the ongoing functioning of the societal totality – in the form of profits, dividends, salaries, wages, governmental transfer payments, and efforts to foster cultural activities to the extent to which sustaining them seems to be the necessary precondition for the survival of the species, and to prevent the rapid destruction of the natural environment, from the level of local conditions to the entire planet.

In the present historical context – the early twenty-first century – we are reminded daily of the urgent need to adjust conceptual, methodological, and especially theoretical tools that for the most part were conceived and developed during the previous two centuries, to the societal realities that are entwined with the ongoing dynamics between industrialization, the spread of market economies, and the rise and relative decline of nation-states. We have shaped and, to an increasing extent, created current realities for ourselves and our descendants, without being in the position of assessing with certainty which of these realities are beneficial to us, in what ways, to what degree, and what price we pay for sustaining and actively supporting them. Indeed, despite a continuously expanding body of work dedicated to clarifying this so-called age of "globalization" and its benefits for humanity, a large segment of the social-theoretical discourse is testimony to continuing efforts to construct a clear-cut balance sheet of the pros and cons of "modernity" – from modernization to economic growth and "development," technological progress, political democracy, secularization, individual autonomy, and the freedom to believe whatever we want to believe. Yet, it seems, the greater the determination to arrive at a definite and indisputable assessment of where we are, of where humanity finds itself today, and of what lies ahead – that is, of what can and needs to be done to increase the likelihood of qualitative improvements of life on Earth – the less successful the outcome appears to be and the more contentious the responses.

Clearly the most conspicuous and perhaps the most disputed case is the debate about climate change, its causes, and consequences. Despite a far-reaching – though not perfect – consensus about the disturbing link between industrialization and climate change among many scientists, outside of academia and research centers, views about the responsibility of (parts of?) humanity for changes in the global climate and future developments differ widely, if not wildly. The chapter by Robert Antonio, which opens the section on *nature*, confronts the issue of climate change head on, citing extensively from the ever more expansive related literature, calling social theorists to the task of considering, and ultimately tackling an issue whose importance for the future of the human species (as well as many, if not most other species) may not be possible to overstate. The chapter constitutes Antonio's first systematic contribution to a theory of the social significance of climate change as related to the global growth imperative. In this endeavor, he brings to bear the fruits of an extensive body of work that is located in the tradition of theorizing the problematic nature of modern society as it is grounded in the works of Marx and Weber, and combinations of both, as most evident in the tradition of Weberian Marxism (Antonio, 2000, 2002, 2005a, 2005b; Antonio & Glassman, 1985). Antonio's chapter is followed by two critical responses, from Michael Zimmerman and John Barry. Zimmerman, the author of works such as *Contesting Earth's Future: Radical Ecology and Postmodernity* (Zimmerman, 1994), a classic of sorts in environmental philosophy, contends that we must remain focused on the task of developing and pursuing strategies that are more likely than not to lead to tangible results, thus setting the stage for similarly oriented subsequent efforts. By contrast, Barry, the author of *Environment and Social Theory* (Barry, 1999), argues from a critical-theoretical perspective that stresses the importance of thinking in terms of radical change and expands on several of Antonio's points, while also providing a number of criticisms. Zimmerman's and Barry's assessments of Antonio's contribution are followed by his reply to their critiques in which he makes explicit implications for the role and responsibilities of social theorists that had been latent in the opening chapter. The section on *nature* concludes with one chapter by Timothy Luke, author of works such as *Ecocritique* (Luke, 1997), on the need to develop "planetarian accountancy," and by Daniel Harrison's attempt to conceptualize risk and danger in consumer societies.

The second section, dedicated to the issue of *knowledge*, presents two chapters addressing dimensions of the debate about public sociology, after Lawrence Hazelrigg sets the stage by revisiting one of Jürgen Habermas' most important books, *Knowledge and Human Interests* (Habermas [1968]1972) – the work that first drew attention to his mode of theorizing and critical theory in the English-speaking world, especially in the United States. Habermas features prominently among the most important social theorists of the modern age, in Hazelrigg's (1989a, 1989b, 1995) truly magisterial treatment of how the program of the social sciences reproduced and reinforced precisely the circumstances in which they were and are embedded. Rather than thematizing with critical intent the vast societal reality built upon the capitalist mode of production, to use Marx's phrase (including especially the link between the system of production and the production of life itself), the social sciences have not merely been active participants in the latter; they only would be able to truly illuminate conditions of modern social life after having scrutinized the connections between their practices and the conditions of their existence in society. Inspired by Ben Agger's critical perspective on public sociology, Patricia Mooney Nickel presents the most problematic and self-defeating features of Michael Burawoy's popular advocacy of public sociology as a political project – as a necessary precondition for re-energizing our willingness and ability to consider truly alternative yet real possibilities to be pursued, a willingness and ability the current state of affairs keeps consistently undercutting – a fact whose implications Burawoy neglects to confront. In turn, J. L. (Hans) Bakker regards the fragmentation of sociological discourse as a major impediment to translating knowledge about social life and society into a meaningful and transformative praxis. Arguing from a vantage point informed by Charles S. Peirce's semiotic epistemology, Bakker contends that "objectivity" in any form needs to be recognized as a social construct resulting from mutually agreed-upon semiotic sign systems.

The common theme of the third section is the need to remain committed to the practice of *negation*, as a theoretically indispensable move in a world where the gap between appearance and reality appears to be reaching proportions that no longer can be framed adequately in terms of cognitive dissonance or contradictions (Marcuse, 1968). Lauren Langman's opening chapter examines the history of the study of subjectivity and self, concentrating on the contributions of Theodor W. Adorno. Langman, who recently co-edited *The Evolution of Alienation: Trauma, Promise, and the Millennium* (Langman & Kalekin-Fishman, 2006), here contents that the distinguishing feature of critical theory is that it offers a normatively grounded critique of domination that communicates utopian hope for the "good life," allowing human existence and life experience beyond power, exploitation, destruction of nature, and gratification based on the accumulation of consumer goods. Langman's chapter is followed by the first of what will be three studies of the FBI's surveillance of Herbert Marcuse. Co-authors Stephen Gennaro and Douglas Kellner – who has published several books on and relating to Marcuse (esp. Kellner, 1998, 2001, 2004, 2006; Kellner, Lewis, & Pierce, 2008) – tell the story of how Marcuse was depicted in FBI documents that were constructed between 1966 and 1972 as a Marxist, an inspiration to New Left and counter-cultural activists, a revolutionary, an influential author, public speaker, and teacher – and perceived as a threat to U.S. national security. What may be most striking about their report is the depth of the discrepancy between the FBI's pattern of relating to, and limited ability to appreciate, Marcuse as one of the most sophisticated social theorists of the twentieth century whose writings have come to be recognized and widely discussed by scholars and students in sociology, philosophy, political science, cultural studies, and other areas as a high point of social and philosophical thought.

The remaining three chapters are "interventions" in the sense in which Adorno (1997) used the term. Steven P. Dandaneau, whose books (Dandaneau, 1996, 2001; Dandaneau & Falcone, 1998) address themes that are closely related to issues many of the authors included in this volume address, presents a reply to the introduction to vol. 25 of this series (Dahms, 2008). Vanessa Walilko, in turn, is concerned about the lack of vision in twenty-first-century feminist theory. In the interest of signaling opportunities for feminism to become more assertive in its critique of established conditions, Walilko stresses the affinities and complementarity between early, more pro-active and activist versions of feminism, and the critical theory of Herbert Marcuse. Striking a chord that is related to yet different from Bakker's rejection of postmodernist nihilism in the second part of this volume, James Block, the author of *A Nation of Agents: The American Path to a Modern Self and Society* (Block, 2002), is concerned with the need to work with the support of the lessons learned from postmodernist critics of modern theoretical practices, to appropriate their message of caution against the abuses of theory, to recover a kind of theory that is conducive to an actualizing, non-destructive kind of individually grounded, social agency.

NOTES

1. As Adorno put it distinctly in a discussion with Ernst Bloch: "Utopia is essentially in the determined negation … of that which merely is, and by concretizing itself as something false, it always points at the same time to what should be" (Bloch, 1988, p. 12).

REFERENCES

Adorno, T. W. (1997). *Critical models: Interventions and catchwords.* New York: Columbia University Press.

Antonio, R. J. (2000). Karl Marx. In: G. Ritzer (Ed.), *Blackwell companion to major social theorists* (pp. 105–143). Malden, MA: Blackwell.

Antonio, R. J. (2002). *Marx and modernity*. Malden, MA: Blackwell.
Antonio, R. J. (2005a). For social theory: Alvin Gouldner's last project and beyond. In: J. Lehmann (Ed.), *Social theory as politics in knowledge: Current perspectives in social theory* (Vol. 23, pp. 71–129). UK: JAI Press.
Antonio, R. J. (2005b). Max Weber in the post-World War Two US and after. *Ethics and Politics*, *7*(2), 1–94http://www2.units.it/~etica/
Antonio, R. J., & Glassman, R. (Eds). (1985). *A Weber-Marx dialogue*. Lawrence: University Press of Kansas.
Barry, J. (1999). *Environment and social theory*. London: Routledge.
Bloch, E. (1988). *The utopian function of art and literature: Selected essays*. J. Zipes & F. Mecklenburg (Trans.). Cambridge, MA: MIT Press.
Block, J. E. (2002). *A nation of agents: The American path to a modern self and society*. Cambridge, MA: Belknap Press of HUP.
Dahms, H. F. (2008). How social science is impossible without critical theory: The immersion of mainstream approaches in time and space. In: H. F. Dahms (Ed.), *No social science without critical theory: Current perspectives in social theory* (Vol. 25, pp. 3–61). UK: Emerald.
Dahms, H. F. (2009). Modernity. In: R. Munck & H. Fagan (Eds), *Globalisation and security: An encyclopedia* (Vol. 2, pp. 303–320). Portsmouth, NH: Praeger.
Dandaneau, S. P. (1996). *A town abandoned: Flint, Michigan, confronts deindustrialization*. Albany: State University Press of New York.
Dandaneau, S. P. (2001). *Taking it big: Developing sociological consciousness in postmodern times*. Thousand Oaks, CA: Pine Forge Press.
Dandaneau, S. P., & Falcone, M. (1998). *A wrong life: Studies in lifeworld-grounded critical theory*. Stamford, CT: JAI Press.
Habermas, J. ([1968]1972). *Knowledge and human interests* (J. Shapiro, Trans.). Boston: Beacon Press.
Hazelrigg, L. E. (1989a). A wilderness of mirrors: On practices of theory in a gray age. *Social science and the challenge of relativism* (Vol. 1). Tallahassee: Florida State University Press.
Hazelrigg, L. E. (1989b). Claims of knowledge: On the labor of found worlds. *Social science and the challenge of relativism* (Vol. 2). Tallahassee: Florida State University Press.
Hazelrigg, L. E. (1995). Cultures of nature: an essay on the production of nature. *Social science and the challenge of relativism* (Vol. 3). Gainesville: University Press of Florida.
Kellner, D. (1998). *Technology, war and fascism: Collected papers of Herbert Marcuse* (Vol. 1). New York: Routledge.
Kellner, D. (2001). *Towards a critical theory of society. Collected papers of Herbert Marcuse* (Vol. 2). New York: Routledge.
Kellner, D. (2004). *The New Left and the 1960s: Collected papers of Herbert Marcuse* (Vol. 3). New York: Routledge.
Kellner, D. (2006). *Art and liberation: Collected papers of Herbert Marcuse* (Vol. 4). New York: Routledge.
Kellner, D., Lewis, T. E., & Pierce, C. (2008). *On Marcuse: Critique, liberation and reschooling in the radical pedagogy of Herbert Marcuse*. Rotterdam: Sense.
Langman, L., & Kalekin-Fishman, D. (2006). *The evolution of alienation: Trauma, promise, and the millennium*. Lanham: Rowman and Littlefield.

Lemert, C. (2007). *Thinking the unthinkable: The riddles of classical social theories*. Boulder, CO: Paradigm Publishers.

Luke, T. W. (1997). *Ecocritique: Contesting the politics of nature, economy and culture*. Minneapolis: University of Minnesota Press.

Marcuse, H. (1968). *Negations: Essays in critical theory*. Boston: Beacon Press.

Zimmerman, M. E. (1994). *Contesting earth's future: Radical ecology and postmodernity*. Berkeley: University of California Press.

Harry F. Dahms
Editor

PART I
NATURE

CLIMATE CHANGE, THE RESOURCE CRUNCH, AND THE GLOBAL GROWTH IMPERATIVE☆

Robert J. Antonio

"until the last ton of fossil fuel has burnt to ashes."

– Max Weber

This quote precedes Weber's famous passage that culminates his Protestant ethic thesis; he said that "material goods" had attained such extraordinary power over peoples' lives that what Puritan Richard Baxter said should be worn like a light coat and easily cast aside had become a "steel-hard casing."[1] Weber implied that modern capitalism's compulsive strivers and insatiable appetites will continue unabated until the energy resources, needed for its mechanized productive forces, are exhausted. Weber expressed this view after witnessing Americans' tireless enterprise and accumulation on his turn of the 20th century visit to the United States, where industrial capitalism was only a few decades old. In a talk at the1904, Saint Louis World's Fair, he warned about "the boiling heat of modern capitalist culture" and consequent "heedless consumption of natural resources, for which there are no substitutes." He argued that capitalism's "dissolving

☆This essay is in memory of my first teacher, Pasquale Caracciolo, and my undergraduate teacher and advisor, William F. Cottrell, who did their best to instruct this slow learner about the entwinement of culture and nature.

Nature, Knowledge and Negation
Current Perspectives in Social Theory, Volume 26, 3–73

ISSN: 0278-1204/doi:10.1108/S0278-1204(2009)0000026004

effect" would someday collide with the "natural limits" of its resource base, but that might not be until fossil fuels are used up (Weber, 1958 [1904], pp. 364, 366, 369).[2] Although Weber was no ecologist, he grasped the tension between capitalist growth and the environment. Before the U.S. Trip, Weber already was familiar with Karl Marx's earlier portrayal of capitalism's growth imperative, which, arguably, also implied a tension between capitalism and nature (e.g., O'Connor, 1998; Foster, 1999, 2002).

Capitalism's compulsive drive to grow persists, but, impacting much larger spaces more intensively than ever before, its tensions with nature have increased massively. Environmental sociologists, following in Marx's and Weber's tracks, have held that contemporary capitalism's "treadmill of production" drives relentless efforts to maximize return on invested capital, dismiss threats of environmental degradation and resource depletion, and externalize costs of ecological damages. Treadmill theorists argue that nearly everyone, in globalized capitalist regimes, are dependent on growth whether it be through employment, financial investments, small business ownership, job security, retirement and health benefits, or life-style.[3] In the post–World War II era, ascendant consumer capitalism generated greatly increased growth and consequent resource pressure and environmental damage.[4] However, postwar modernization theorists argued that scientific and technological knowledge gives culture de facto autonomy from nature's limits; that is, progress in capacities to understand and engineer environments provide ample means to overcome resource constraints and ecological crises. Postwar growth and development policies presumed that "science," mediated by markets, will solve environmental and resource problems.

When former Vice President Al Gore won the 2007 Nobel Peace Prize for his climate change efforts, conservative op-ed and editorial writers called him nasty names, compared him to "worst" former winners, and challenged the legitimacy of the peace prize per se. A conservative Hoover Institution, Fellow argued that Gore is a propagandist, suffering from "Narcissistic Personality Disorder," and compared him to an earlier winner who was honored for advocating prefrontal lobotomies.[5] *The Wall Street Journal* editorial on the award did not mention Gore, but listed "worthy" 2008 nominees. Repeating President Bush's rationale for rejecting the Kyoto Protocol, Gore's critics held that efforts to reduce greenhouse gases would wreck the U.S. economy. Many of them considered anthropogenic warming (driven by human activities) to be a myth conjured up by bad science, radical environmentalists, and the anti-business left.[6]

The day before Gore received his Nobel, *The Wall Street Journal* ran retiring American Enterprise Institute, President Christopher Demuth's (2007) op-ed, saying that the secret of AEI's and other right-wing think tanks' "stupendous successes" is that they avoid "doubling back to first principles." Treating their political presuppositions as a priori truths, they focus exclusively on battles against their opposition and on policy work that furthers their ideals. Their vitriolic attacks on Gore expressed transparently their view that acknowledging anthropogenic warming would contravene their beliefs in unregulated, economic growth as *the* standard for evaluating economies, societies, and political leaders and for addressing problems, including climate change and resource shortages. Conservative, George Will (2008) charged that "green alarmism" over climate change is a leftist ploy to increase state supervision, justify unlimited regulation, and restrict individual choice (e.g., to build a "power plant" or drive an "SUV"). Will invoked free-market, economist Friedrich Hayek's views about the left's "fatal conceit" that the state could predict the future and plan for it effectively and obliviousness to consequent erosion of freedom and economic efficiency. Putting a populist halo around unregulated growth, this view appeals to many Americans, who equate, at least tacitly, the growth imperative with their "freedom." And arguably, their job security, financial investments, consumer life-styles, and prospects for a "better life" all depend on growth. Neoliberal globalization spread this "treadmill" worldwide.

This chapter will explore how globalized capitalism's growth imperative and vastly more intensive and extensive resource consumption have generated damages and endgame risks that call for fundamental rethinking of postwar ideas about economic growth. I focus especially on climate change and related ecological problems. This essay addresses the need to reshape the postwar era's foundational presuppositions about culture's autonomy from biophysical nature. Classical theory's Darwinian inspired visions of culture embedded in nature should be recovered and reformulated in light of recent science and today's ecological risks and limits.

THE GROWTH IMPERATIVE: FROM POSTWAR MODERNIZATION TO GLOBALIZATION

During the great post–World War II economic expansion, modernization theorists held that the new American capitalism balanced mass production and mass consumption, meshed profitability with labor's interests, and ended class conflict. They thought that Keynesian policies insured a near full-employment, low-inflation, continuous growth economy. They viewed the United States as the "new lead society," eliminating industrial capitalism's backward features and progressing toward modernity's penultimate "postindustrial" stage.[7] Many Americans believed that the ideal of "consumer freedom," forged early in the century, had been widely realized and epitomized American democracy's superiority to communism.[8] However, critics held that the new capitalism did not solve all of classical capitalism's problems (e.g., poverty) and that much increased consumption generated new types of cultural and political problems. John Kenneth Galbraith argued that mainstream economists assumed that human nature dictates an unlimited "urgency of wants," naturalizing ever increasing production and consumption and precluding the distinction of goods required to meet basic needs from those that stoke wasteful, destructive appetites. In his view, mainstream economists' individualistic, acquisitive presuppositions crown consumers sovereign and obscure cultural forces, especially advertising, that generate and channel desire and elevate possessions and consumption into the prime measures of self-worth. Galbraith held that production's "paramount position" and related "imperatives of consumer demand" create dependence on economic growth and generate new imbalances and insecurities.[9] Harsher critics held that the consumer culture blinded middle-class Americans to injustice, despotic bureaucracy, and drudge work (e.g., Mills, 1961; Marcuse, 1964). But even these radical critics implied that postwar capitalism unlocked the secret of sustained economic growth.

The growth imperative suffused Paul Samuelson's best-selling introductory economics textbook. The leading economies' (the United States on top) sharply up-sloping, hundred year, per capita GNP growth lines graced the book's inside, front cover. Samuelson equated *more*, or the "quantity of goods," with "increased community welfare;" he declared that the "*privilege of being able to buy a vast array of goods at low prices cannot be overestimated*" [emphasis in the original] (Samuelson, 1973, pp. 436–437). He acknowledged mounting economic problems, but still asserted that the "modern mixed economy" and Keynesian regulatory regime worked well and would endure. Expecting more long expansions and ever increasing disposable income and consumption, Samuelson insisted that the chances for depression were "negligible," recessions would be fewer and milder, and stagnancy would mean slower "absolute growth," rather than contraction. Rejecting pessimistic projections of the then nascent environmental movement and recently published *The Limits to Growth* (1973), Samuelson held that new technologies and relatively painless tradeoffs would be enough to overcome ecological and resource problems.[10]

In a magisterial 1973 work on postwar modernization, *The Coming of Post-Industrial Society*, Daniel Bell dismissed "limits to growth" claims and the "apocalyptic hysteria of the ecology movement."[11] He held that new technologies, energy sources, and recycling methods and modest tradeoffs would solve all resource and ecological problems. Bell (1977, p. 18) declared later that: "If one thinks only in physical terms, then it is likely that one does not need to worry about ever running out of resources." He shared mainstream economists' view that "resources are properly measured in economic, not physical terms." Bell heralded, as core postindustrial trends, shifts from raw material and energy technologies to information technologies and from system "design" oriented to nature to that based on a "game between persons" (Bell, 1999 [1973], p. 117). He held that limits to growth do not inhere in human nature or the natural environment, but derive largely from social constraints (especially conflictive interests, values, fears), which can be overcome. Bell asserted that we "live more and more outside nature"; our environments are technologically and scientifically mediated and, therefore, open to alteration and improvement. Bell recognized that growth was slowing, but held that it would stay robust enough so that American economic life would remain a "non-zero-sum game" in which "everyone could end up a winner, though with differential gains" (Bell, 1999 [1973], pp. 164, 274, 456–475, 488). Parsons (1971, p. 143) held that modernization would continue for, at least, a century. Bell shared his optimism.

In 1973, Samuelson portrayed the laissez faire views of Hayek and the Austrian School and of Milton Friedman and the Chicago School to be outside "mainstream economics." He argued that state regulatory and welfare policies, which counterbalance the market's vicissitudes, worked so well that "reversion" to free-market policies was out of the question (Samuelson, 1973, pp. 734, 847–848). However, U.S. corporate circles and conservative activists already were building a network of think tanks championing free-market "solutions" to curb regulatory and welfare costs, labor union power, and taxes (Akard, 1992; Jenkins & Eckert, 2000). Stagflation, oil price hikes, other economic problems and election of President Reagan paved the way for return of the ideas that postwar modernization theorists had declared dead. Business economist and Senior Fellow at the libertarian CATO institute, Julian Simon advocated a most assertive version of living outside nature, rejecting emphatically claims about population, food, energy, and natural resource limits. He argued that evolution endows humanity with creative powers, which leave each generation with more knowledge and wealth than preceding ones and, thus, provide us ever increased control over our environments. In his view, the "need for resources" generates ever greater capacities to find or create them. He declared that: "It is beyond any doubt that natural resource scarcity – as measured by the economically meaningful indicator of cost or price – has been decreasing rather than increasing in the long run for all new raw materials, with only temporary and local exceptions. And there is no reason this trend should not continue forever. The trend toward greater availability includes the most counterintuitive of all – oil" (Simon, 1995, p. 3).[12] However, Simon argued that full realization of this possibility and of human potential and social progress depend on maintaining a free-market regime that minimizes government regulation. Simon's position might sound extreme, but it justified unregulated growth. *Living outside nature is not merely an idea, it is a way of life*. The theory lags and mirrors lived social realities!

The 1989 collapse of the Soviet empire and the U.S.-led 1991 multinational intervention in Iraq opened a new era that President George H. W. Bush famously called "the New World Order." Francis Fukuyama (1989) and, later, Thomas Friedman (2000) cheered the dawn of a new U.S.-led global capitalism, which vindicated modernization theory, sans the weighty New Deal and Social Democratic baggage that they said stalled growth. The themes of libertarian think tanks became the dominant policy regime, facilitating neoliberal "globalization" (e.g., Stiglitz, 2003). Neoliberalism's deregulated "New Economy" was the type of "free-market" regime that Simon held was best suited to harness human powers for overcoming apparent, or culturally constructed ecological limits. Neoliberals generally did not take account of globalization's ecological risks.[13] Also framing new, leaner versions of postwar liberalism, "neomodernization" theorists opposed extreme libertarian views, yet still converged with broader neoliberal currents. They recognized growing resource and environmental problems and major ecological risks, but held that the new regime's scaled-down regulatory/welfare states, lowered expectations about publicly planned change, and new communications links cultivated much more active civil societies and increased sharply modernization's "reflexivity" – enhancing public awareness of risks, widening debate over them, and multiplying powers to cope with them. Neomodernizationists saw ecological problems to be "opportunities" for change, not "crises"; they claimed that new civil society dynamics, with their active citizenry, green businesses, market incentives, and public-private partnerships publicize, politicize, and address critically ecological risk, providing unparalleled, self-corrective powers. Neomodernizationalists held that ecological risk is best managed, although with some inevitable slippage, by market mechanisms, civil society, and mild regulatory intervention, *not* by extensive government intervention or planning.[14]

Globalization's *spatial-temporal compression* reduced barriers to moving information, goods, workers, and capital, extended, intensified, and sped-up production and exchange, and increased capital accumulation. In earlier capitalism, major innovations (e.g., railroads), which stirred development in many sectors and made the world smaller and faster, ignited growth spurts that overcame stagnancy, profit squeezes, and market crises (Harvey, 1989, pp. 199–323). Thomas L. Friedman's (2000, 2005) celebratory writings about globalization's new "fast," "flat" world extolled compression. Neoliberal globalization's *technological transformation,*[15] *political transformation,*[16] and *cultural transformation*[17] increased enormously capitalism's speed and reach and ended mid-1970s to early 1980s stagnancy. Capitalism's extension into less developed, highly populous parts of the world accelerated a surge in economic growth and generated much increased global opportunities for consumption. Soaring optimism about global growth ruled at the peak of the boom (e.g., Schwartz, Leyden, & Hyatt, 1999). Even the year 2000 dot-com bust and stock market retreat and post-9/11 slowdown did not extinguish the belief in globalization's new era of geometrically accelerated growth. Recall Samuelson's point about appreciating the importance of a multitude of inexpensive consumer choices. In a popular globalization text, George Ritzer (2007a, p. 115) spoke of globalization's "*monumental abundance*" (emphasis in the original), asserting that "the promise is that this is just the beginning, and the range of that which is available, and the ease with which everything it encompasses can be obtained, will increase exponentially in the coming years." He implied that growth had no limits or would not decline. Ritzer acknowledged consumption's lure and benefits, but, in a culturally critical voice, portrayed the cornucopia of mass-produced commodities as "nothing" (i.e., lacking distinct content). However, this nothing *is* something from the perspective of the natural resources used to produce, process, pack, transport, and dispose of the goods and the wastes deposited in the process. Part of the "sense of loss" that Ritzer lamented derives from visible savaging of nature.

Neoliberals and neomodernization theorists have not come to terms with globalization's enormous environmental risks and pressures.[18] Until recently, few globalization analysts of any persuasion engaged seriously nascent ecological crises.[19] Many of globalization theories retained taints of precursor approaches, framed when hinterlands and natural resource stocks seemed unlimited. Resource consumption cannot grow exponentially ad infinitum. Before the 2008 economic crisis, U.S. hyperconsumption, fueled by cheap credit, drove global economic growth. Most globalization advocates and many critics presumed that accelerated growth would be sustained. Many thinkers and policymakers still believe that modest, well-targeted regulatory change and stabilization of the banking system will restore the pre-recession pattern of global growth. However, without fundamental changes, globalization's contradictions will likely re-emerge, with even more force, after the global recovery. Contributing massively to neoliberal globalization's expansive growth, spatial-temporal compression *also greatly accelerated the speed and volume of the throughput of natural resources and creation of wastes*. My arguments below about the relationship of global growth and climate change and other related resource and environmental problems illustrate how globalization, which made the world smaller and faster, also made it "fuller," magnifying the contradiction of unplanned, exponential growth with the biophysical world and with culture, which is embedded in the biosphere and dependent upon it.[20]

CLIMATE CHANGE, ECOLOGICAL DAMAGE, AND CATASTROPHIC RISK

In the later 20th century, ecological activism and legislation arose in response to sprawl, pollution, and ecodisasters.[21] The 2007 release of the U.N. Intergovernmental Panel on Climate Change (IPCC), *Fourth Assessment Report* (*AR4*) and the IPPC Panel's and Al Gore's joint receipt of the Nobel Peace Prize brought ecology to the global public stage.[22] Although carefully calibrating claims and uncertainties, the IPCC *Synthesis Report* (*SYR)* asserted that evidence for accelerated warming, melting, and rising seas is "unequivocal."[23] It held that anthropogenic, heat-trapping, greenhouse gases (GHG) have risen an alarming 70%, between 1970 and 2004 and that average global temperatures for 11 of the last 12 years rank among the 12 warmest since records began in1850. *SYR* estimated a 90% probability that warming is due to anthropogenic GHG increases, originating largely from burning fossil fuels. The most long-lasting GHG, carbon dioxide (CO_2) was at 379 parts per million (ppm) in 2005, exceeding the normal range for the last 650,000 years.[24] Most importantly, about half of every metric ton of CO_2 remains in the atmosphere from several hundred to several thousand years (UNHDP, 2007, p. 36). Given today's rapidly rising rate of CO_2 emissions and enduring accumulated concentrations, the panel urged very prompt mitigation to minimize the already grave risks faced by future generations.

The IPCC panel reported that warming and melting will continue to be greatest in high northern latitudes and southern seas, causing high temperatures and heavy precipitation and changing storm tracks, winds, and precipitation patterns (e.g., increasing it in northerly latitudes and reducing it in most subtropical areas). They noted that the last time polar regions were substantially warmer, 125,000 years ago, melting caused a 4–6 meter sea rise. They warned that unabated climate change, continuing on the current path, likely would result in severe floods, drought, fire, famine, disease, heat waves, high storm surges, coastal and island submersions, extinctions, and systemic ecosystem damages. The panel held that poor nations in Africa, Asia, and the Arctic, which lack adequate resources to adapt to climate change, have already experienced serious human development setbacks, and, without aid, may eventually suffer forced migrations and mass deaths (IPCC, 2007a, [summary for policymakers], pp. 1, 8 passim; on rising waters, see UNDP, 2007; Berardelli, 2007b; Rohling et al., 2008). For example, very poor and densely populated Bangladesh is 70% flood plain; rising seas and Himalayan melting may cause substantial losses of land over the next two decades (Wax, 2007b; Sengupta, 2007). It is already suffering from climate change's many-sided impacts.[25]

On *SYR*'s release, UN Secretary General Ban Ki-Moon (2007) declared that, in some regions, climate change's ecological damages are already reminiscent of bad "science fiction" scenarios, but are "even more terrifying, because they are real." He had recently witnessed the Antarctica Peninsula melt, Amazon deforestation, and Chilean ozone hole. Also commenting on the report's release, Kiel Institute for World Economy's Gernot Klepper declared that: "The world is already at or above the worst case scenarios in terms of emissions" and is on the way to exceed *AR4*'s "most pessimistic estimates," or is already "above that red line" (quoted in Rosenthal, 2007b). Some climate scientists argued that achieving consensus, among *AR4*'s many experts and government appointees, muted climate change's most extreme risks.[26] Later studies warned about approaching catastrophic "tipping points," or "points of no return."[27] Shortly after the report's release, 200 leading scientists cosigned the "Bali Climate Declaration," calling on the 180 nations, at the UN sponsored conference, to embrace the goal of limiting atmospheric concentrations of $CO_{2\text{-}e}$[28] "well below" 450 ppm. In urgent tones, they held that the global average temperature rise must be kept to no more than 2°C (Celsius), or 3.6°F (Fahrenheit), above its preindustrial level.[29] To meet this *minimum* target, they argued, GHG emissions must peak and decline in 10–15 years and be cut 50% below their 1990 level by 2050 (UNSW, 2007). However, the United States resisted efforts at the UN Climate Change Conference in Bali to set even provisional GHG emission targets or frame new policies that worked toward this goal (*New York Times*, 2007; Broder & Barringer, 2007).

$CO_{2\text{-}e}$ concentrations were already 430 ppm in 2006. Although the global recession might slow emissions, they have grown about 3 ppm yearly, and the rate has been rapidly accelerating.[30] A major report estimated that we may reach 550 ppm $CO_{2\text{-}e}$ by 2035 (HMT, 2006, p. iii). The IPPC posed six "narrative storylines," combining different scenarios of possible patterns of demographic, economic, social, technological, and environmental development (exclusive of major political climate initiatives). Even the two most optimistic scenarios would likely stabilize, by 2100, $CO_{2\text{-}e}$ at somewhere between 600 and 700 ppm, and the other scenarios would likely stabilize it between 750 and 1550 ppm, increasing risks of climate catastrophes and profound human development setbacks to 95% or above and neutralizing later generations' ability ever to stabilize below 450 ppm.[31] The panel implied that even stabilizing $CO_{2\text{-}e}$ at 600 or 700 ppm would require major changes (e.g., much reduced reliance on fossil fuel, decades of declining global population, much more efficient and less materially intensive production). *AR4* held that continuing the current emissions path until 2100 would likely exceed 1000 ppm for CO_2 alone and all other GHG targets, possibly producing a 6°C temperature rise or more and generating irreversible damages – severe food shortages, water problems, very high seas, massive biodiversity losses, forced mass migrations, and mass deaths.[32] Poor nations would likely suffer most, but global ecosystems would be pushed close to or beyond their tipping points (UNHDP, 2007, pp. 1–10, 33–38, 46–48; IPCC, 2007a, [Topic 3]; GWA, 2008).[33]

An April 2008, *Nature* essay contended that the IPCC "seriously underestimated" the mitigation challenge; their scenarios and stabilization targets assume most technological change and decarbonization are spontaneous, or are not driven by public policy mandates. The authors argued that many of the scenarios are "unrealistic," and some may be impossible. All scenarios were framed on the presupposition that "energy intensity" would decrease between 2000 and 2010 and most scenarios projected "carbon intensity" to do the same, which has not been the case (the original projections did not anticipate China's and India's meteoric growth, especially China's massive GHG emissions increases). The authors contended that the technological problem of GHG stabilization may be "much larger" than what the IPCC projected, and cast doubt on the beliefs that "market incentives" would solve climate and energy problems (Pielke, Wigley, & Green, 2008; Schiermeier, 2008b; Schiermeier & Tollefson, 2008).

In December 2007, the American Geophysical Union meetings offered symposia on a "year of worsts" for climate change (Leake, 2007). The National Snow and Ice Data Center reported that the peak summer retreat of Arctic sea ice shattered the (2005) record September low by 23%, and was 39% below the 1979–2000 average for the end of the melt season and 50% below its 1950s levels (NSIDC, 2007; Revkin, 2007).[34] The melt "far exceeded" projections and was so extensive that standard ocean vessels could have sailed the Northwest Passage for the first time. The NSIDC reported that unusual atmospheric conditions contributed to the extreme melt, but warned that Arctic sea ice may be past its tipping point and that an ice free summer is possible by 2030. The IPCC (2007a, [Topic 3], p. 4) had said that "some projections" held that late-summer Arctic sea ice may be gone by the latter 21st century. The Arctic melt generated speculation about tapping underwater mineral resources and jostling between nations over rights to them (Brett, 2007). Scientists also reported record summer melting of Greenland's ice sheet, which had been accelerating for 34 years. The 2007 rate was 60% greater than the 1998 record (Mote, 2007). Another study reported a record summer melt in western Greenland and an air temperature increase of the ice sheet of 7°F since 1991 (CIRES, 2007). Greenland's coastal glaciers have been thinning, destabilizing, and breaking apart in the sea; surface water, heated by solar energy, flows down glacial ice tunnels, lubricates the glacier's base, and speeds up its slide. Although future melting is a matter of debate and inquiry, Greenland contains one-twentieth of the globe's ice; a complete melt could raise the sea 7 meters (Thompson, 2007; Witze, 2008).[35]

In early 2008, another study reported that the West Antarctic Ice Sheet (WAIS) lost 75% more ice in 2006 than in 1996; the losses have increased 59% over the previous decade. The 200 billion metric tons lost last year was enough to raise global seas "more than half a millimeter." Lead author, Eric Rignot said that the year's loss approached the Greenland melt. He warned that the western ice sheet could change quickly. He stated that he was "astonished" by the new data and that recent IPCC projections were too cautious (Hand, 2008; Kaufman, 2008a; Rignot et al., 2008).[36] His collaborator, Jonathan Bamber (quoted in Smith, 2008a) said that "dramatic change," in Antarctica, "fits a global trend – We've seen the same thing in mountain glaciers, in Greenland, Patagonia, and…in Alaska. We are seeing the same thing everywhere we look." In late winter 2008, an ice bloc, about seven times larger than Manhattan, broke from the Wilkins Ice Shelf, which is weakening at an unexpectedly rapid pace. The most southerly ice shelf on the Antarctic Peninsula, Wilkins' deterioration suggests that global warming is worsening and raises concerns about the nearby WAIS.[37]

Glaciologists fear that the 21st century sea rise may be *far greater* than the IPCC's projected, likely upper limit. A 2008 UN-sponsored study, tracking 30 glaciers around the world, reported an accelerated, record melt and warned that the glaciers could disappear in a few decades. Glacial melting adds to flooding and rising seas, but water shortages will be a long-term problem. Nine glacial-fed, Asian rivers provide water for 2.4 billion people (Jowit, 2008; AP, 2008c). In 2008, record melting of Arctic sea ice continued; northeast and northwest passages opened.[38] Also, the 4,500-year-old Markham Ice Shelf broke away from Ellesmere Island and the region suffered other major ice losses. Scientists feared that the melt threatens the Arctic interior; the permafrost melt could eventually lead to profoundly dangerous releases of GHG (Hoag, 2008a; Ljunggren, 2008; Schuur, 2008; AP, 2008e; Revkin, 2008d). As a result of sea ice loss and warmer water, fall 2008 Arctic temperatures reached a record 9°F above normal.[39] In early 2009, International Polar Year 2007–2008 researchers confirmed substantial West Antarctic melting, warming and freshening of the southern oceans, increasing methane emissions in the northern hemisphere, and accelerating Greenland and Arctic melting (ICSU, 2009; Engeler, 2009; Allison et al., 2009). A later research report moderated earlier projections of the global sea level rise that would likely follow a sudden collapse of the WAIS; it projected that the highest rise would occur along the U.S. west and east coasts.[40] MIT scientists, employing a new model, hold that the rise of global surface temperatures have been underestimated and, without major policy changes and technical breakthroughs, that there is a 90% chance that they will rise 3.5–7.4°C by 2100 (Hance, 2009).

SYNERGISTIC IMPACTS: CLIMATE CHANGE AND OTHER GROWTH-RELATED ECOLOGICAL PROBLEMS

Climate change processes interact, and have "feedback" effects that increase or decrease warming. For example, CO_2-driven melting releases water vapor, itself a GHG that intensifies warming. Also, more pine bark beetle larvae survive western North America's warmer winters. The large tree kills from these pests and other climate-related factors are turning forests from carbon sinks to a major CO_2 source (Kurz et al., 2008; Brown, 2008;

Pennisi, 2009).[41] Interactions and aggregate impacts of feedbacks are complicated and unknown.[42] Yet they increase the uncertainty of climate conditions and risks. For example, some scientists fear that Arctic warming could reduce thaw depths of permafrost and eventually release vast amounts of methane and CO_2, stored in frozen peat bogs. Although not as long lasting as CO_2, methane is an extremely potent GHG that could set off an "atmospheric tsunami," sharply raising temperatures and pushing major parts of the global ecosystem toward their tipping points. In 2009, scientists warned that permafrost melting is increasing and a grim scenario is possible (UNHDP, 2007, p. 38; Pearce, 2007, pp. 108–117; Lyderson, 2009). Positive feedbacks have been accelerating Arctic, Greenland, and Antarctic melting. Snow and ice have a powerful albedo effect, reflecting much more solar energy back into space than open water or bare ground (negative feedbacks from new "carbon sinks" would not neutralize the problem).[43] Shrinking snow and ice cover warms. Human activities also produce feedback effects. Increased biofuel production and higher food prices spurred massive cutting and burning of tropical forests to expand agricultural land, increasing CO_2 emissions and reducing carbon sinks. Rapid population growth also drives GHG and resource problems (K. Smith, 2008).

Even if the current scientific consensus about climate change gave way to more optimistic projections, other growth-related environmental risks would still call for rethinking the growth imperative and doing far more accounting and planning of natural resource throughput and waste production. Diverse anthropogenic ecosystem impacts can be mutually reinforcing. Warming is a "*synergistic driver*" that magnifies other ecological problems (Crist, 2007, p. 40). A major study, tracking nearly 30,000 "significant changes" in physical and biological systems, found more than 90% of the changes to be in the expected direction consistent with global warming hypotheses.[44] A *Science* article reported how aggregate impacts of climate change and other anthropogenic drivers are bringing a unique rural ecosystem to near "collapse." Mongolia's winter temperatures had risen a "staggering 3.6°C on average during the past 60 years," melting mountain snow and glaciers in its northern high country. Climate change increased temperatures, and generated more severe summer droughts and stronger storms. As the region's permafrost thins and recedes, surface water is absorbed more deeply into the soil. Thawing and drying stress shallow root systems of the region's flora, shrinking forests and damaging other plant life. One of the world's 17 ancient lakes (millions of years old) and most pristine outside Antarctica, Lake Hovsgol and its ecologically rich nature reserve are threatened by climate change. As the taiga recedes, the steppes dry. Mongolia's post-Soviet economy's increased market pressures and consequent opportunities swelled greatly the numbers of herders and livestock. Lacking the traditional knowledge of their precursors, the herders over grazed the dry, thin grasslands, increasing the region's vulnerability to fires and flash floods. Topsoil erosion transformed some of the area into "semidesert," and threatened to turn the entire region into "shrubby wasteland." Increased runoff into Lake Hovsgol's tributaries could ruin the nation's best source of drinking water (Bohannon, 2008).

Warming has caused many fish species to migrate to cooler waters, which could lead to extinctions (Ansari, 2009). Scientists warn that Tuna and Cod will disappear and that marine life per se will be devastated without regulation of commercial fishing and dumping, agricultural runoff, and other human sources of pollution and acidification. From 1970 to 2005, marine species declined 14% on average (WWF, 2008, p. 8). Climate change's warming, acidifying, and rising oceans have, in some locations (e.g., parts of the Great Barrier Reef), slowed coral growth and sometimes bleached, drowned, and reduced them to "rubble banks." Coral reef damages are magnified by other anthropogenic drivers' "synergies and feedbacks" (e.g., coastal runoff and deforestation). However, CO_2 levels near or above 500 ppm will likely severely damage reefs, and possibly push them to their "tipping points." Thus, even the most optimistic IPCC scenarios will imperil the reefs and livelihoods of "tens of millions of people who depend upon them."[45] A 2008 study mapped impacts of 17 anthropogenic drivers on marine ecosystems. Damages were "much worse" than expected; over 40% of the globe's oceans have been heavily impacted and pristine polar regions are threatened, as warming and melting opens them to increased human activities. The scientists concluded that global warming related drivers have had the widest impact on eroding marine ecosystems (Halpern et al., 2008; UCSB, 2008). Another research team found areas close to the U.S. west coast, just east of the California current, devoid of oxygen and aquatic life (Chan et al., 2008). A later study demonstrated that the "dead zones" were widespread across the globe and their numbers have roughly doubled every decade since the 1960s. The scientists held that the primary drivers were agricultural runoff (i.e., nitrogen and potassium) and warming.[46]

Atlanta almost ran out of water in late 2007. Some communities allowed lawn watering and filling swimming pools until Lakes Lanier and Altoona were nearly drained. Governor Sonny Perdue asked the Army Corps of Engineers to allow Georgia more water from the lakes, which it shares with Florida and Alabama. The three states have been battling over water rights for 18 years. After being turned down, Governor Perdue conducted a public prayer service for rain. He also charged that the decision was "an ill-advised choice in favor of mussel and sturgeon species over Georgia citizens…" (Perdue, 2007). All three states wanted the water for multiple purposes, but Florida had invoked the Endangered Species Act for legal leverage as well as environmental ends. Georgia wanted the law suspended. A federal appellate court ruled in favor of Florida and Alabama on other grounds, nullifying a 2003 Georgia-Army Corps of Engineers' pact that would have increased the state's share of Lake Lanier water. Fast-growing metro Atlanta would have increased its take by 65% (Abdulla, 2007; Bluestone, 2007; Goodman, 2008a; Shelton, 2008a). After the decision, Georgia lawmakers argued that the state's northern border was improperly surveyed in 1818 and that it should run a mile farther north across the middle of Tennessee's Nickajack Reservoir. A Georgia legislator proposed that residents of lands claimed "illegally" by Tennessee be given tax amnesty for unpaid back taxes (Emery, 2008; B. Smith, 2008). In the midst of the battle, the Georgia House of Representatives passed a bill to limit local communities' powers to restrict water consumption (later passed into law)! The landscaping industry claimed to have lost three billion dollars from local water usage regulation. Georgia and Tennessee seemed destined for a federal battle over water rights. In-state fighting over water grid-locked the Georgia Legislature, preventing creation of a prudent State Water Plan (Shelton, 2008b; Georgia General Assembly, 2008; Cook, 2008; Chapman, 2008).[47] A water law expert said that metro Atlanta building permits should be issued only to applicants who insure an independent, long-term water source. He added: "But pigs will fly before Atlanta does that" (quoted in Shelton, 2008a). In early 2009, court battles over metro Atlanta's water share from Lake Lanier and state plans on how to provide sufficient water for continued metro growth remained a contested, uncertain terrain. Drought conditions continued, and Lake Lanier was still low.[48]

The ongoing water battles in the southeastern United States pale by side of likely problems in the western states dependant on Colorado River water. *AR4* held that warming driven melting will lead to more winter floods, reduced summer flow, and increased "competition for over-allocated water resources" in the region.[49] In 2007, Lake Mead was down about 50%, and may eventually drop below Las Vegas's intake pipes (new ones were being drilled near the lake's bottom). Colorado River fed Lake Powell was down about the same as Mead. Seven western basin states, Mexico, various Indian nations, cities, farmers, and other entities have claims on the reduced flow.[50] Much of the region was expected to keep growing rapidly. Las Vegas developers said they will build desalinization plants in California or pump water from afar. "Water flows uphill toward the money" they say. However, the region's water managers warned that the area may be on the brink of a great water crisis that will stir internecine legal battles over water allotments and even forced migrations (Gertner, 2007). Researchers argue that warming is a primary driver of western drying, melting, and declining river flow and that Lake Mead has a 50% chance of going dry by 2021, even if demand for water does *not* increase (Barnett et al., 2008; Barnett & Pierce, 2008; Barringer, 2008; Kaufman, 2008b). Another study, held that feedback, from climate change driven, poleward widening of the tropical belt, will cause additional drying in the western United States (Seidel et al., 2008, p. 24). Other researchers held that atmospheric warming is generating more extreme precipitation in other regions and that climate change models had underestimated these events (Allan & Soden, 2008). The design and operation of water management systems assume "stationarity" – that natural systems fluctuate within predictable limits. However, climate change, powerful feedback loops, and rapid growth, which showed no sign of abating, have led scientists to declare stationarity "dead" (Milly et al., 2008).

Appearing during peak reportage about *AR4*, the Atlanta drought, and other late 2007 ecological bad news, the story, "Huge Water Park Planned for Arizona Desert," received national attention. Mesa "Waveyard" developers planned to create a coastal environment, which would consume about 60–100 million gallons of water a year, in an area that gets about 8 inches of annual precipitation, suffers from a 10 year drought, and draws water from the Colorado River and an underground aquifer, also dependent on waning mountain runoff. The Tucson to Phoenix strip had been growing explosively, and is expected to get dryer from climate change. However, Waveyard promises to provide a playground for tourists and residents and abundant economic benefits (i.e., 7,500 jobs, billion dollar revenue, and sales tax for a city without property tax).The developer said he wanted to recreate and share with local children: "the kind of lush environment he remembers from growing up in Virginia Beach…and surfing in Morocco, Indonesia, Hawaii, and Brazil" (quoted in Kahn, 2007a). In the Phoenix metro area, like metro Atlanta, growth is not pegged rigorously to long-term, sustainable planning about water resources, which have become more problematic after stationarity.[51] Waveyard is a practical manifestation of "living outside nature," employing capital and technology to create a simulated "natural" environment, heedless of the sustainability of resource throughput. It is a microcosm of the tradeoffs and risks, inhering in the deeply contradictory relationship between the growth imperative and biophysical nature.[52]

Governor Perdue's assertion about favoring mussels and sturgeon over Georgians echos common complaints about the Endangered Species Act and biodiversity preservation efforts (i.e., costly, irrational blockages to growth). However, biodiversity has feedback effects that regulate the atmosphere and waters and that provide resources for medicines, food, drink, industrial goods, and leisure/tourism. It is also a wellspring of spiritual and aesthetic value. Ecologists argue that biodepletion diminishes cultural treasures as well as natural resources. Ecosystem damages, they say, are so severe and widespread that they may be initiating a "global extinction crisis" or the "sixth great extinction" of the last 520 million years (McKibben, 2006 [1989]; Crist, 2007; Wilson, 2007, pp. 73–81, 91–99, passim; Eldrich, 2001). The International Union for Conservation of Nature and Natural Resources' "Red List" reported that 38% of all species it assessed "are threatened with extinction" (e.g., 22% of mammals, 14% of birds, 31% of amphibians, and 27% of warm water reef-building corals are threatened or extinct). Our rapidly expanding "Ecological Footprint," propelled by globalization's growth machine, squeezes out other species and imperils us. Habitat destruction and degradation, invasive species, pollution, human over-population, and overharvesting are the chief anthropogenic drivers of the biodiversity crisis.[53] A comprehensive 2008 study found that 52% of mammal species for which trend data exists are declining.[54] Another report held that the globe's wildlife has declined by a third since 1973 (WWF, 2008, p. 1). Various studies document that climate change disrupts biological systems (it is a major driver of habitat destruction, and changes breeding, migration, flowering, and distribution patterns) and threatens some species' survival.[55] A 2.5°C temperature rise could extirpate 30% of all species, and a 3.5°C rise may eliminate 70% of them. Rising temperatures already approximate these levels in some polar and mountain regions (Casey, 2007). As with water planning, instabilities generated by climate change and other pressures antiquate existent conservation plans (Eilperin, 2009).

Increasing with affluence, per capita global meat consumption has doubled since 1961, and has been projected to double again by 2050, shifting a substantial part of cereal production from human consumption to cattle feed. About 30% of the globe's ice free land has been devoted to meat production directly or indirectly, generating about 20% of GHG emissions (Bittman, 2008).[56] Also, heavily subsidized biofuel production diverts agriculture away from production for human consumption, raises food prices, and motivates more intense tropical forests cutting. In 2007, corn ethanol production helped drive up Mexican tortilla prices 400%, contributing to higher priced bread, soy, and related plant-based foods (staples of poorer peoples). A research study held that corn-based ethanol might nearly double GHG emissions in 30 years, rather than reduce them the projected 20% (even if switchgrass is substituted for corn, emissions will still likely rise 50%). Another study held that land use shifts resulting from substituting ethanol for fossil fuel will release 17–420 times more CO_2 than the GHG emissions reduction. Discharges from biodiesel plants pollute waters (Fargione et al., 2008; Rosenthal, 2008; Searchinger et al., 2008; Kleiner, 2008; Goodman, 2008b; Martin, 2008; Bradsher, 2008a; UN-Energy, 2008; Charles, 2009). In early 2008, about 30% of developing countries were food-insecure; droughts, floods, storms, and climate change driven warming already had been reducing yields in many poor nations and in some major food exporting nations. Global food supplies dwindled to dangerously low levels in 2007. In April 2008, sharply rising prices led to food riots in many poor nations, and squeezed low income people in rich nations. The cereal bill in the poorest countries rose sharply. In 2008, the UN estimated that 37 nations needed food aid and that a global food crisis was at hand (FAO, 2008, pp. 2, 11; Schiermeier, 2008a; Rosenthal, 2007c; Blas & Wiggens, 2008; Dunphy, 2008; Lacey, 2008; Fuller, 2008b; Vestal, 2008; Bradsher & Martin, 2008; Streitfeld, 2008). Future climate change impacts are projected to make matters much worse.[57]

The Millenium Ecosystem Assessment (WRI, 2005, pp. 1–2, 5–7) reported that 60% of the ecosystem services it examined are "being degraded or used unsustainably" and that the worst damages occurred in the preceding 50 years.[58] Ecosystem damages have been largely driven by economic growth, which also produced major human development gains. Thus, proposals to reduce the damages, which demand tradeoffs that threaten growth, are usually resisted. A 2008 report estimated that we are exceeding the planet's capacity to regenerate its resources by about 30%. Humanity's ecological footprint has more than doubled since 1961.[59] The United States has the largest total ecological footprint and largest per capita one of rich nations. On the "Environmental Performance Index," the United States (the richest nation) ranked 39th of 140 nations, lowest of the highly industrialized nations and lower than many developing nations (YCELP, 2008). Texas (23 million people) has a larger total CO_2 footprint than all Sub-Saharan Africa (720 million people). The United States has emitted, by far, the most CO_2 since the industrial revolution's onset. In 2008, China took the lead in CO_2 emissions, but it has been the "world's factory," producing enormous amounts of industrial exports, especially for American consumers (UNHDP, 2007, pp. 40–43, 114–115; WWF, 2008, pp. 14, 34, 36; Longfellow, 2007, p. 3).[60]

The U.S. per capita carbon footprint is almost five times that of China.[61] In 2007, the UN held that rich nations need to reduce GHG emissions 30% by 2020 and, at least, 80% by 2050 to escape disaster (UNHDP, 2007, pp. 114–115). The United States was highly resistant to setting short-term or intermediate goals for emissions reduction. In May 2008, G-8 ministers failed to agree on the UN 2020 target, which Europe supported and the United States deemed out of reach. They still claimed to have a "strong political will" to reduce their nations' emissions by 50% by 2050 (Coleman, 2008). A few months before, a study, employing new models, held that carbon emissions must be reduced to *near zero for all nations*, by mid-century, to avert catastrophe (Matthews & Caldeira, 2008; Eilperin, 2008a). In early 2009, scientists were hopeful about President Obama's support of science and intention to reduce U.S. GHG emissions. However, they also held that climate change was preceding faster than the models had predicted, that the 2007 IPCC report was too optimistic, and that GHG impacts on "surface temperature, rainfall, and sea level are largely irreversible" (Block, 2009; Lyderson, 2009; NOAA, 2009). Some critics held that atmospheric CO_2 must be reduced to 350ppm, but its concentration continued to accelerate and was nearly 385 ppm by the end of 2008 (Kleiner, 2009; Wynn, 2009).

Recall U.N. Secretary General Ban Ki-Moon's point about climate change imitating science fiction. The 1972 movie, *Silent Running's* futuristic 2008 scenario portrayed a heroic, slightly crazed ecologist and his humanized robots tending the last remains of earth's plants, carried by geodesic domes, attached to space freighters. The totally engineered Earth was devoid of flora and fauna. Scrapping plans for refoliation, the global leadership ordered the domes destroyed so that the freighters could return to commercial usage. Then pitted against postwar "living outside nature," this dystopian fantasy now expresses taints of a possibly dawning future. Eileen Crist (2003, p. 64) has held that 50–60% biodepletion might not be "pragmatically catastrophic" for humans, but would hasten the "planet's large-scale transformation into a human satellite of technological, managed, and constructed landscapes." At the opening of the Svalbard Global Seed Vault, Norwegian Prime Minister Jens Stoltenberg said that this "Noah's Ark" will secure biological diversity for future generations. Dug deep into an Arctic mountain, the "doomsday" vault is supposed to provide a 1,000 years of protection for frozen seeds. Global Diversity Crop Trust head, Cary Fowler said that a "mass extinction" of crop diversity is already underway and that remaining genetic diversity must be preserved to adapt to climate change's shifting conditions (Fouché, 2008; AP, 2008a). Fearing that the Bali targets are beyond reach, scientists now entertain "geoengineering" schemes to alter the earth's climate. This technological fix requires enormous investment, and its impacts are highly uncertain. Such tampering with the biosphere's exceptionally complex interdependence opens the way for unanticipated, dangerous feedbacks (Victor, Morgan, Apt, Steinbruner, & Ricke, 2009; Victor, 2008; Steffen, 2007; Kintisch, 2007, 2008; Dean, 2008). President Obama's science advisor said that geoengineering technologies must be studied and refined should other strategies to stem catastrophic warming fail (Revkin, 2009b).

GLOBALIZING CONSUMPTION: HYPERGROWTH AND PLANETARY OVERSHOOT

The Consumer's Guide to Effective Environmental Choices detailed the major role that U.S.-style consumption plays in the worst ecological problems (i.e., air pollution, global warming, habitat alteration, and water pollution). The authors held that ecologically destructive activities are "pretty fundamental to the American middle-class way of life" (Brower & Leon, 1999, p. 83). Their list of "most harmful activities" included the ways we transport ourselves, eat, use water, get rid of sewage, and build, equip, heat, cool, and light our homes. They held that automobiles, a prime postwar symbol of American consumer freedom, are a most ecologically destructive facet of our consumption. Motor vehicle operation, construction, and disposal entail a huge throughput of resources and enormous production of wastes.[62] Early 21st century Americans lived in more sprawling metro areas and drove heavier, more gas-guzzling vehicles longer distances to work and other activities than ever before. Suburban homes became much larger, and more ecologically costly. Sprawl makes it hard to car pool or create mass transit. Increased inequality, high upper strata incomes, an antitax and antiregulatory climate, and cultural rule of unregulated consumer choice made environmental mandates, demanding "real" costs and sacrifices, seem whimsical. Reducing GHG emissions and Ecological Footprints require a massive overhaul of American "middle-class" lifestyle. Yet the prospect of its global diffusion, which marketers have strived to realize, punctuates dramatically the depth of the contradiction between growth and nature.[63]

Like the postwar United States, owning and using a car is a badge of middle-class status in China. Shanghai was projected to add 2,600 miles of new highway in the first decade of the century (Powell, 2008). China has built highways in many regions, and even helps finance connections to markets in neighboring countries, spurring regional growth of truck and auto traffic (Fuller, 2008a). Highways, drive-ins, car washes, parking lots, billboards, road-side venders, and other facets of car culture dot China's landscape. Drive-through McDonald's has been there since 2007. Touting long holiday drives, a Chinese Website engineer and auto club member stated: "I really like what the car brings to my life-convenience, freedom, flexibility, a quick rhythm. I can't imagine life without it" (Fan, 2008). Increasing auto production by 30% in 2007, China passed Germany as the globe's third leading automaker and was expected to soon overtake Japan. First-time car buyers have spurred China's demand and vastly expanded its auto sales (AP, 2007; Gow, 2007; Bradsher, 2008b, Roberts, 2008). Increased auto and truck traffic contributed substantially to Chinese cities' choking air pollution and congestion. Already one of the world's most smoggy, gridlocked cities, in 2008, Beijing was adding 1,200 new vehicles a day. Auto-related problems threatened the Beijing Olympics, requiring reduced traffic flow (Yardley, 2007e, 2008a; Bradsher, 2007b, 2008b; Jacobs, 2008; Macur, 2008). India is also building a vibrant car culture; its Tata Motors is producing the Nano – a $2,000 "Peoples Car." Moreover, Tata bought Land Rover and Jaguar. In China and India, big cars are a popular option for those who can afford them (Kanter, 2009; Welch & Lakshman, 2008; Timmons, 2008; Giridharadas, 2008; Sengupta, 2008). Personal transportation is the largest source of oil consumption, and most rapidly growing source of CO_2 emissions.[64] In early 2009, China exceeded (recession slowed) U.S. auto sales (Kurtenbach, 2009). However, China's incentive plan to develop electric cars and major subway construction efforts indicated some awareness of the mounting ecological and social costs of a powerful car culture (Bradsher, 2009a, 2009b).

The combined population of China and India was about 2,500 billion people in 2009, and they have been projected to grow to nearly 3 billion by 2045 (U.S. Census Bureau, 2009; UNPD, 2000). Both nations had meteoric early 21st century economic growth rates (China 10%–12% and India 8%–9%). In 2007, consumption topped investment as China's "most powerful engine of growth." Its consumer market was projected to become the globe's second or third largest by 2015. Indian consumption was also expected to rise sharply (Jiang & Shen, 2007; McKinsey Global Institute, 2006, 2007; Reuters, 2007; WT, 2007b; Hamm, 2008). The globe's largest retailers, Wal-Mart and France's Carrefour, have growing operations in China. Wal-Mart also has expanded into India. Best Buy, the top U.S. consumer electronics retailer, saw China to be its number one market for expansion. China has become the third export area for U.S. goods. On a mall-building spree, China has the globe's two biggest malls and two others that rank near the top. The largest, Dongguan's South China Mall has about three times more space than the Mall of America. Described as "a jumble of Disneyland and Las Vegas" and advertised as a "Supermega Theme Shopping Park" and "Life-Style City," the South China Mall's dramatic built environments are spread thorough seven simulated "water cities," linked by canal.[65] The world's second largest mall, in Beijing, had more than a 1,000 shops. China's malls cater to its rapidly growing, younger professional middle-class and wealthy strata. Average Chinese people can afford few expensive commodities, but they enjoy mall browsing and socializing. However, rising incomes and growing ranks of shoppers fuel a rapidly expanding consumer culture.[66] Through mid-fall 2008, even slowing economic growth did not stem China's growing "conspicuous consumerism" (Cha, 2008).

A New York architect, working in China, said that Americans had become "backward-looking," whereas the Chinese "want to make everything look new." "This is their moment in time," he said. In China, urban growth has occurred with exceptional speed, on an unparalleled scale (i.e., construction of entire new down-towns, subways, highways, airports, parks, water systems) (Ouroussoff, 2008). Shanghai's high rise skyline and shiny postmodern financial district, Pudong expresses dramatically forces transforming China. On busy weekends, about 260,000 people have visited its 13 story "Super Brand Mall." Suburban homes and consumption have been transforming Chinese life. "Our local shopping Mall – which looks like it could be White Plains N.Y., or the San Fernando Valley of Los Angeles – has had a KFC and a Pizza Hut up and running for the past year" said an American journalist living in a rapidly expanding Shanghai suburb. His five-bedroom townhouse had more than double the space of downtown apartments. The subdivision resembled U.S. counterparts down to the local Starbucks, shiny recreation center, rich developers, and low-wage migrant workers. The journalist said that nearly all residents in his subdivision own cars, drive short distances rather than walk, and often pay huge fees for special license plates to commute downtown. Thirty-five other subdivisions have been finished or soon will be in this new satellite area; it is expected to grow to about a half million people. In the next decade, five million more Shanghai residents have been projected to move into similar suburban complexes. "Suburban flight" goes on at different levels of intensity in many locations in China (Powell, 2008; J. Garnaut, 2008; http://www.sohochina.com).[67] The massive suburban and metropolitan building boom spiked global demand for construction materials, and stressed local resources (e.g., water, energy). One estimate holds that China's urban population will nearly double to a billion people by 2030 (McKinsey Global Institute, 2008).

In China, modern conveniences (e.g., toilets, showers, pools, golf courses, parks, lawns, public gardens, intensive agriculture) have spiked water consumption. The South China Mall's "water cities" parallel Mesa, Arizona's Waveland project. Beijing's two main reservoirs are so polluted that the city ceased using them for residential water over a decade ago, and the third has been down nearly 70% in 10 years. To supply the Olympics' demand, China expropriated water from already hard-pressed small farmers (AP, 2008b). China's growth is not pegged to sustainable water resources. Much increased water consumption threatens the North China Plain's aquifer, which supplies 200 million people and produces half of the nation's wheat. The region's swamps, streams, and wetlands have been severely compromised by rapid growth, imprudent water policy, and climate change. However, even traditional practices can contribute to water problems. India's "hundreds of millions of cookstoves" produce an enormous amount of black carbon, or soot, which, carried by winds and deposited on the Tibetan Plateau, accelerate melting of glaciers that feed Asia's rivers.[68] The damages have been cultural and aesthetic as well as material; Tibetan Buddhism's sacred mountains are losing their glaciers and China's "mother," Yellow River and its ecosystems are threatened. India's sacred Ganges and the glaciers that feed it are also under threat (Larson, 2008; Sheehan, 2008; Gifford, 2007; Yardley, 2007a; Wax, 2007a; Pepper, 2007). China is building the Three Gorges Dam (world's largest) to increase water availability and renewable energy, but this effort and related water projects have displaced hundreds of thousands of people, imposed major ecological damages, and failed to address the root of water problem there. Many dams were imperiled by the May 12, 2008, earthquake (Yardley, 2007a, 2007b, 2007c, 2007d; J. Kahn, 2007b).[69]

Power generation is the number one source (40%) of global atmospheric CO_2 emissions. Leading power-sector CO_2 emitters, the United States, China, and India have enormous coal deposits, and are expected to use them as demand for power generation grows (UNHDP, 2007, pp. 133–134, 152). In 2008, China was bringing online two new coal burning plants weekly (Zeng, 2008, p. 730). It surpassed U.S. GHG emissions by 8% in 2006 (NEAA, 2007; GCP, 2008). China gets about 81% of its electricity from coal powered plants. Its efforts to expand middle-class consumption and improve lives of hundreds of millions of peasants and urban poor require much more power generation. How can China, India, and other developing nations provide their rural and often impoverished masses electricity and still reduce carbon emissions? (Victor, 2009; Bradsher & Barboza, 2006; Kahn & Yardley, 2007; Kahn & Landler, 2007; Yardley, 2008b; Bradsher, 2008a; R. Garnaut, 2008, pp. 15–17). Since 2000, anthropogenic CO_2 emissions have been increasing four times faster than in the 1990s. In 2007, more than 50% of global emissions came from developing nations (GCP, 2008; Revkin, 2008e). An important study estimated that China's annual CO_2 emission increases, for 2004–2010, will average at least 11% instead of the 2.5%–4.8% IPCC projection. Its 2007 CO_2 emissions accounted for about two thirds of the global 3.1% increase. China's increases likely will be greater than all Kyoto Protocol reductions (Auffhammer & Carson, 2008; Jagoda, 2008; NEAA, 2008; GCP, 2008). If China grows at 7% through 2030, with the same carbon intensity, its emissions will match that of the entire world in 2008 (Zeng et al., 2008, p. 731; Victor, 2009). China's huge scale of development spikes demand for other resources that increase GHG concentrations. Some thinkers argue that sharply increasing 2007 and 2008 oil prices were driven by Chinese demand and, thus, will eventually go higher (Mouawad, 2007; Bowley & Jolly, 2008).[70] China is projected to match the United States' total Ecological Footprint by 2015 (Dietz, Rosa, & York, 2007, p. 16). China has active renewable energy, reforestation, and other ecological programs, but it also aims to alleviate mass poverty, build a middle-class, increase the standard of living of all its citizenry, and reduce dependence on exports. Many developing nations face similar cross-pressures (Cha, 2007; French, 2007, 2008; Zeng et al., 2008; Normile, 2008; Moore, 2008; Graham-Harrison, 2009).

Much increased throughput of finite resources and production of wastes has occurred in many nations, who joined the globalization system in the later 20th century. The immense scale and speed of China's growth makes it a focal point of global growth and of ecological concern (e.g., Victor, 2009; Clark, 2009). The global economic crisis, starting in fall 2008, hurt China's massive export-oriented manufacturing sector. Other nations, in the region, experienced similar manufacturing downturns (Foley & Rosenbaum, 2008; Wassener, 2009; Barboza, 2009; Foreman, 2009). In early 2009, Chinese officials estimated that as many as 26 million migrant workers were jobless (Chang, 2009). Already slowing for six quarters, China's growth rate declined from 9% to 6.8% in the forth quarter of 2008. Fearing social unrest, China instituted a $585 billion stimulus package, and allocated funds to develop universal health care (Plafker, 2009). Some scientists held that the growth rate of Chinese power generation was dropping, and projected that the nation might emit far less CO_2 than anticipated in the latter part of the decade (Revkin, 2009a). The U.S. 2008 carbon emissions had the sharpest fall since the end of the early 1980s recession (Mufson, 2009). Depending on the recession's depth and length, slowed growth could reduce substantially global GHG emissions and the rate of increase of atmospheric CO_2.

However, other analysts suggest a contrary scenario (Wynn, 2009; Mrasek, 2009). Hopes for substantial reductions are a matter of conjecture. Strategies to restore growth might stall efforts to reduce GHG emissions and cope with climate change and other ecological problems.

Believers in the growth imperative have argued that developing nations will follow an "environmental Kuznets curve"; increased growth and incomes will make possible new public and private choices and initiatives and technological innovations that increase the efficiency of resource usage, develop substitutes for depleted resources, and diminish ecological degradation.[71] By contrast, critics hold that economic growth, increased affluence, and population growth have made matters worse (e.g., Dietz et al., 2007). The richest nation, the United States has released massive amounts of GHG, has the largest Ecological Footprint, has been the main motor of increased global resource consumption, and has resisted regulatory change to diminish these problems. The great scale and speed of anthropogenic drivers in a globalized world, in the absence of sweeping policy changes, likely will produce irreversible, profound ecological damages. Most rich nations have not made serious enough efforts to curtail GHG emissions, and accelerated development and growth are still in the early stages in much of the world. CO_2 concentrations are already high, rapidly increasing, and long-lasting. Poor nations already have suffered from climate change and other major ecological problems.[72] Resource wars, forced migrations, and mass deaths are projected if the growth-related complex of problems runs on indefinitely.[73] Some thinkers argue that many ancient wars and civilizational declines and collapses had major environmental drivers, including climate change. National Security agencies see climate change as a "threat multiplier" in unstable parts of the world and as a likely future, severe security threat almost everywhere (Diamond, 2005; Busby, 2007; CNA, 2007; NIC, 2008; D. Zhang et al., 2008a; Brown & Crawford, 2009).[74] The Arctic sea ice summer melts already have stirred fears about international conflict over mineral rights (Meilke, 2008). Waiting for the Kuznets curve and a technological fix risks catastrophe!

In 2008, UN Secretary-General Ban Ki-moon estimated that climate mitigation will cost \$15 to \$20 trillion over the next 20–25 years (Heilprin, 2008).[75] Economist William Nordhaus (2008; see Dyson, 2008) estimated that the cost of doing nothing could amount to \$23 trillion by the end the century. Another projection held that it will take a \$45 trillion investment in energy projects over several decades to cut global GHG emissions by 50% by 2050 (Murray, 2008). The Center for Research on the Epidemiology of Disasters reports "four times more weather related disasters in the last

30 years than in the previous 75 years" and that the United States has suffered more of them than any other nation. Other reports point to ongoing major American climate change problems (e.g., fire, storm, drought), which already substantially impact vulnerable regions and subgroups and will grow worse even if GHG emissions are reduced (Blow, 2008; Eilperin, 2008b; Revkin, 2008c; CCSP, 2008; NSTC, 2008; Vitello, 2007; Pachauri, 2008). Estimating mitigation costs is a guessing game, but coping with it will be very costly and require massive life-style changes. Climate change and other ecological issues have been covered prominently and well in important segments of the U.S. print media (e.g., *New York Times* and *The Washington Post*). However, after extensive coverage of the 2007 IPCC report, a 2008 Pew Center survey, found that fewer Americans believed that there is solid evidence for global warming and anthropogenic drivers than in 2006.[76] In March 2009, Gallup Polls found that more Americans believed that the climate change threat was exaggerated than in their 1998 survey and that a majority of respondents said, for the first time in the 25 years of surveying the question, that economic growth should be given priority even if the environment suffers.[77] Little can be done about the looming global ecological crises unless the leading consumer nation changes its ways. Other surveys reported that support for climate change initiatives have dropped in other parts of the globe as well (O'Neil, 2008; Pew Center, 2009).

If successful, President Obama's promised efforts to elevate the role of science and address climate change and other ecological issues might break the gridlock. In early 2009, however, restoration of economic growth was the overriding public concern, and the President's climate change proposal garnered the least support of any major initiative in his overall agenda.[78] Oil companies were not ready to embrace his green agenda (Mouawad, 2009). Ironically, Obama's effort to reverse Bush administration environmental policies and his more progressive ecological views led increasing numbers of Americans to believe that "environmental" "quality" was actually improving.[79] The global economic crisis still offers an opportunity for and may necessitate rethinking U.S. development strategies and lifestyles (e.g., Florida, 2009). Public resistance to change may weaken if the economic crisis persists, and, especially, if there is no way to return to the economic status quo ante and if these matters are complicated by intense ecological problems or a regional American catastrophe. In a *New York Times* op-ed, Thomas L. Friedman (2009) asked:

> What if the crisis of 2008 represents something much more fundamental than a deep recession? What if it's telling us that the whole growth model we created over the

> last 50 years is simply unsustainable economically and ecologically and that 2008 was when we hit the wall – when Mother Nature and the market both said: "No more."

He implied that climate change and other ecological problems call for fundamental rethinking of socio-economic policy. A few years ago, it would have been hard to imagine the leading advocate of neoliberal globalization's "golden straightjacket" ever pondering such serious doubts about the economic growth regime that he once portrayed as an inevitable, nearly universal, benign force.

CRITICAL SOCIAL THOUGHT'S ECOLOGICAL TASK: RE-EMBEDDING THE SOCIAL IN NATURE

Concluding the classic "Objectivity" essay, Weber (1949 [1904], p. 112) predicted that specialized social scientists would one day employ accepted concepts and methods to address established problems; they would "consider the analysis of data as an end in itself," and put aside reflection about the "value ideas" that make their objects of study "problematic" or "worth knowing." But he also implied that social ruptures would regenerate critical reflection about the taken-for-granted presuppositions and forge new problems, concepts, and methods. "The light of great cultural problems moves on," Weber said, and then science views events from the "heights of thought" and prepares to change its ways. Natural science studies about accelerating ecological problems and catastrophic threats already offer ample guidance for many needed policy changes and for new directions in social science and policy research. Social science research may also contribute to the immediate policy efforts.[80] However, the longer-term, reconstructive project of creating a more ecologically friendly and humane alternative to the growth imperative and neoliberal capitalism calls for the type of foundational reflection that Weber implied earlier.

Environmental sociologists, Riley E. Dunlap and William R. Catton, Jr. contended that "*exemptionalist*" presuppositions, which exclude humanity from natural constraints, pervade social theory, social science, and public policy.[81] They also held that the "strong constructivism," which stresses almost exclusively cultural definition and ignores nature, converges with and amplifies exemptionalism.[82] Exemptionalists posit almost unlimited powers to "construct" nature so that human population size, consumption levels, resource depletion rates, and waste production are limited only by scientific and technological capacities. Exemptionalist presuppositions anchor the growth imperative and many other facets of capitalist political economy, life-style, habits, built environments, consumer culture, and development schemes. "Living outside nature" is a prime manifestation of exemptionalism. Post–World War II era beliefs in growth implied that evolutionary biology, which unifies the life sciences, and naturalistic limits have little or no relevance for society and culture. Dunlap and Catton called for a "new ecological paradigm" that stresses our dependence on the biophysical world *and* recognizes human cultural capacities. They framed a critical foundational project that should not only be pursued in environmental sociology, but also in sociology and social theory per se, and especially in modernization and globalization theories and other approaches, which impact public policy debates. The severity of the ecological crisis and likely huge scope of reconstruction necessitates that this become a public project beyond academe.[83]

Dunlap and Catton criticized classical social theories, which they alleged shaped today's exemptionalist currents. They held that Durkheim's antireductionist perspective (i.e., social facts must be explained by other social facts) has been the primary root of recent sociology's exclusion of biophysical factors. However, they did not offer much argument or evidence that Durkheim's methodological imperative has been as widely diffused or as influential as they claim. Moreover, anthropogenic drivers and their ecological impacts *are* Durkheimian social facts! Only a misleadingly narrow interpretation of social facts could exclude biophysical conditions per se.[84] Dunlap and Catton also asserted that Weber, Mead, and other classical theorists, who stressed cultural construction's importance, provided additional roots of constructivist exemptionalism (Dunlap & Catton, 1979, p. 253; Catton 1994, p. 88, 2002, pp. 90–92, 108–109, note 4, 2008; Dunlap, 2002a, 2008b, pp. 16–17). However, Weber grasped culture's dependence on nature, warned against "one-sided spiritualistic analysis," and employed a Darwinian-like concept of "selection." And Mead anchored his idea of social action in Darwinian theory (see, e.g., Weber, 2002 [1920], p. 159, passim; Mead, 1962 [1934], passim). Dunlap and Catton spoke too broadly and dismissively about classical theory's responsibility for recent sociology's limited ecological purview. Their well-founded claim that the exemptionalist vision of the culture-nature relationship has been "deeply influenced by the Western Culture in which it evolved" calls for wider, more careful inquiries about classical theory and other earlier influences. But Catton said that debating Durkheim's and classical theory's contribution to current ecological blind spots is a "sterile" exercise and that this type of backward glance will reveal only distractions, which shed little "light on the human prospect."

By no means do I suggest that classical theory or its interpreters should be spared of criticism or should be the prime source of ideas for a new ecological paradigm (Catton, 2008, pp. 475–476; e.g., Dunlap, 2002a, pp. 17–18).[85] However, an effective eco-friendly, alternative ecological paradigm cannot be created de novo. Dunlap's and Catton's worthy project requires critical reflection about how previous ideas, often taken for granted and embedded in habits and institutions, may limit or aid reconstruction and how other forgotten or marginalized facets of our cultural past may be reclaimed critically and redeployed.

George Herbert Mead's and John Dewey's pragmatist social theory and social psychology provides resources that could assist in constructing a new ecological paradigm.[86] They argued emphatically that the culture–nature dualism has pervaded western philosophy and culture, from their ancient Greek, Socratic roots to the present and that it has been the source of important intellectual and policy failures.[87] If Dewey and Mead were alive, today, they would very likely embrace Dunlap's and Catton's critique of exemptionalism. Mead and Dewey located society and culture within biophysical nature. Their Darwinian-inspired "organism and environment" model treats linguistically mediated, social life as an emergent, distinctly human layer of naturalistic interdependence.[88] They contended that humanity's exceptional cultural powers make possible unique, reflexive, flexible adaptative capacities, highly complex systems of cooperation, and an enormous array of instrumentalities (material and symbolic) that provide unique means to mediate and rearrange environments. In their view, these species-capacities have biological and other physical substrates, which must be taken into account, neutralized, or mobilized to execute human ends. They held that planning and coping with even mundane personal problems or adjusting to unintended consequences of resultant actions requires assessing and dealing with naturalistic limits. Although recognizing biological bases in the "central nervous system," Dewey and Mead saw language and culture to be collective responses to recurrent features of our environments.

Strong constructivists sometimes borrow from or converge with Mead and Dewey in partial, distorted ways. However, Mead and Dewey never could have supported narrow, totalistic, constructivist culturalism or the related idea of "living outside nature," which treats the biophysical world as if it were an infinitely malleable extension of human ends. By contrast, they saw symbolic and material cultural practices to be part of nature, interacting constantly with biophysical forces and conditions – shaping them, being shaped by them, clashing with their resistant facets, and producing consequences that can foster survival and enrich life *or* result in costly, painful, or even catastrophic consequences. Dewey (1989 [1934], p. 34) said:

> Nature is the mother and the habitat of man, even if sometimes a stepmother and unfriendly home. The fact that civilization endures and culture continues – and sometimes advances – is evidence that human hopes and purposes find a basis and support in nature. As the developing growth of an individual from embryo to maturity is the result of interaction of organism with surroundings, so culture is the product not of efforts of men put forth in a void or just upon themselves, but of prolonged and cumulative interaction with environment.

Dewey deployed the organism-environment model in his critique of free-market individualism, which he held does not take account sufficiently of interdependence and the need for cooperative action. He said that the United States' abundant natural resources were being plundered imprudently, and decried American capitalism's "waste" and possible "exhaustion" of natural resources (Dewey, 1988b [1927], 1988c [1929]; Dewey & Tufts, 1985 [1932], pp. 405–406).[89]

Durkheim stressed the interdependence of the various natural realms, but did not develop a framework that could take account of the complex exchanges and feedback between them. His idea of culture's relative autonomy stressed primarily connections of symbolic culture to social structure (i.e., patterns of association), or the substrate that he held exerted the most direct and substantial casual force on symbolic practices and collective representations. He understood that articulating complex relations between symbolic culture and social structure becomes far more difficult in modern societies in which voluntary association and social differentiation increase geometrically, individuate people in much more diverse ways, and increase culture's autonomy. Durkheim lacked the social psychology to elaborate well modern culture's intricate connections to social structure and other substrata.[90] By contrast, Dewey's and Mead's social psychology not only provided ideas on how to bridge better the gap between social facts and the individual level, but they framed them in a manner that recognized directly that symbolic processes, which mediate psychological states and social action, are embedded in nature. They held that "meaning" arises from our interaction with a natural world that includes biophysical objects and other types of sentient beings as well as humans. Mead and Dewey saw symbolically mediated practices of thinking, feeling, and acting to take place in an always co-present embodied world that impacts, limits, and even directs action. They did not see nature as a mere tablet for cultural inscription by a self-enclosed "subject"; they denoted its obdurate qualities, which constrain, resist, and imprint.

Mead's and Dewey's general social theory and social psychology suggest fluid borders between natural realms, and deconstruct Western Culture's imperious, disembodied, dualistic concepts of mind and soul and consequent "epistemological" severance of thought from action, theory from practice, and self from body and the biophysical world. A 2007 *Science* article contended that human and natural systems are complexly coupled and that their intricate patterns and processes are elided when they are studied separately. It called for interdisciplinary inquiries, which probe connections between realms, usually treated as separate realities (Liu et al., 2007). Mead's and Dewey's emphasis on fluid borders and multi-directional interchanges and feedbacks between culture and biophysical nature favors interdisciplinary work and discourages reification of the culture-nature analytic divide and repetition of the later 20th century "science wars."[91] Mead and Dewey built a foundation for more refined specialized optics to analyze anthropogenic drivers and complex feedback loops. In their view, cultural construction is ubiquitous, but they stressed that the process, to be *intelligent*, must address how biophysical and social structural conditions block or facilitate implementation of cultural ends. They framed a comprehensive starting point for *postexemptionalist theory and practice* that is worthy of re-engaging critically.

Critical reflection about the foundations of a new ecological paradigm should scrutinize theories that have posed pertinent, unresolved questions that have been suppressed or distorted by faith in the growth imperative. Reconsidering Mead and Dewey on the culture–nature dualism provides a start. However, efforts to develop a new postexemptionalist paradigm should cast the nets much more widely to regain and redirect our cultural bearings in the face of the still emergent crisis. For example, Herman Daly has argued that coping with climate change and other ecological issues requires re-engaging still unsolved problems posed by Malthus (over-population), Marx (unequal distribution), and Keynes (involuntary unemployment), who have been dismissed prematurely by neoliberal theorists and policymakers holding that market-generated affluence and technical/scientific progress can solve all problems (Daly, 1996, p. 119).[92] If the latest scientific projections about climate and other ecological risks are true, the collision of capitalism and nature, of which Weber warned, will incur ever higher costs and, ultimately, shipwreck the growth imperative and global society. To have veracity, the detraditionalized, constructivist vision of culture and nature must be readjusted to treat humanity and culture as part of nature, dependent upon it, and constrained by it. We need to *frame theories and practices that stress culture's embeddedness in nature and that, consequently, manifest actively our responsibilities to distant peoples, future generations, and other life forms with which we share the planet.* Our futures are entwined. This Copernican turn calls for profound rethinking, redefining, reshaping, and redirecting of growth, *not* abandoning it.

ACKNOWLEDGMENTS

Thanks to Tony Arnold, Scott McNall, Joane Nagel, George Ritzer, and Wayne White for their critical readings and helpful suggestions. Also, thanks to Harry Dahms for his invaluable intellectual and instrumental help at several stages of the paper; he has my gratitude for his generosity and patience.

NOTES

1. Weber held that Americans pursued gain so single-mindedly that it was losing ethical meaning and taking on the character of sport (Weber, 2002 [1920], pp. 123–125, 245–246, n. 129).

2. Musing about a train ride across Oklahoma, then "Indian Territory," Weber described capitalism overtaking remnants of virgin forest, indigenous culture, and frontier life; he declared "too bad; in a year this place will look like Oklahoma [City], that is, like any other American City. With almost lightening speed everything that stands in the way of capitalistic culture is being crushed" (quoted in Marianne Weber, 1975 [1926], pp. 291–294; see Scaff, 2005, pp. 64–66 on this passage's context). Scaff's forthcoming book about the Webers' U.S. trip (*Weber in America*) will say more Max Weber's ideas about capitalism and the environment. In an email (August 17, 2007), Scaff wrote that: "There is actually an 'ecological' subtext in a good deal of what Weber says ('sustainability' and the like). Radkau [author of a Max Weber biography; 2009 {2005}] has noticed that too, and it came up in our discussion in Washington in March. Interesting, isn't it? A Weber for the 21st century and the age of global warming." Radkau's (2008 [2002]) global environmental history also has a strong Weberian thread. On Weberian themes in environmental thought, see for example, Murphy (2002), Buttel, Dickins, Dunlap, and Gijswijt (2002), Buttel (2002), and Martinez-Alier (1987, pp. 183–192).

3. Most economists equate economic growth with increasing amounts of goods and services, which they hold elevate the standard of life and overall well-being. They employ Gross Domestic Product (GDP) to measure economic growth (i.e., the percentage increase of market value per capita of goods and services). The following quote approximates my usage: "Economic growth is simply an increase in the production and consumption of goods and services. It entails increasing population and/or per capita consumption, where *consumption* refers to consumption of materials and energy by firms, households, and governments" (Czech, 2006, p. 1563).

On growth and nature, see Daly (2005) and Speth (2008). On the treadmill of production, see Schnaiberg (1980), Gould, Pellow, and Schnaiberg (2004), Buttel (2004), and Foster (2002, pp. 44–51). Ecologist Garrett Hardin's (1968) classic "tragedy of the commons" argument illuminates the treadmill's workings, especially unregulated capitalism's tendency to deplete or degrade public resources.

4. Ecological sociologist, Eileen Christ (2007, pp. 34–35) employs the apt term, "industrial-consumer civilization." Former state socialist regimes advocated a state-driven growth imperative and savaged their environments, but failed to deliver the goods and, thus, suffered profound legitimacy problems, hastening their collapse. After communism, the contradiction of growth and nature calls out for a profound rethinking of industrial-consumer civilization.

5. See Miller (2007). The same free-market journal ran the satirical "news story" – "Peace Prize Committee Disbands!" The charter termination and rescindment of Gore's award was said to be a response "from every civilized person on earth with any semblance of intelligence...." It called him an "itinerant comedian and performance artist" and an "egocentric Luddite who specializes in creating hysteria and false science." The piece "quoted" the King of Norway, who portrayed the Nobel Committee as "ridiculous, vapid, self-centered, and dim-witted"; their "bizarre, counterproductive, and down-right dangerous" choices made Norwegians seem "silly, provincial, and uniformed." "The Gore prize was what did it," they said. See W. S. Smith (2007).

6. See WSJ (2007). AEI reprinted op-eds, charging that Gore was a "panic-monger" and a "high water mark of climate hysteria" (Frum, 2007; Hayward, 2007). A *CATO* op-ed said that Gore champions "climate lunacy" (Michaels, 2007). *The Washington Times* (WT, 2007a) ran a cartoon of Gore as an overinflated hot air balloon, pumped up by the prize; its op-ed referred to his high utility bills and wasteful globetrotting and compared him to other "bad" Nobel winners – Yasser Arafat and Jimmy Carter. *National Review*, Managing Editor, Jay Nordlinger (2007) charged that the Nobel Peace Prize committee "is a standard left-wing pressure group" and compared Gore to past winners Yasser Arafat and North Vietnam's Le Duc Tho. Liberal economist, Paul Krugman (2007) called this heated response from the right "Gore Derangement Syndrome." In 2008, *Wall Street Journal* op-ed writers continued to attack Gore and charge that global warming was a product of "mass neurosis" and "yellow science" (Kerian, 2008; Stephens, 2008a, 2008b).

7. Modernization theorists held that the new order was reducing economic inequality, opening opportunities for minorities, stabilizing population growth, and making work more satisfying. See, for example, Davis (1945, 1963), Kuznets (1955), Rostow (1959), Parsons (1965, 1971), and Bell (1999 [1973]). For an overview of postwar American economic growth policies, see Collins (2000).

8. President Nixon's and Premier Nikita Khrushchev's 1959 "kitchen debate" at Moscow's American National Exhibit, which showcased U.S. household appliances, was a landmark battle over U.S. consumer freedom and cultural superiority. See Foner (1998, pp. 147–151, 262–273).

9. Galbraith explained that economists' concept of "diminishing marginal utility" holds that desire for a particular good declines with consumption and satiation, but that satisfaction of one desire does not diminish the overall "urgency of wants." He held that economists believe: "It is human nature to want *more*..." [emphasis added] (Galbraith, 1958, pp. 143; passim 121–169).

10. Samuelson (1973, pp. 195–197, 216–217, 266–267, 817–820); Meadows, Meadows, Randers, and Behrans (1972) and, for a thirty year update, see Meadows, Randers, and Meadows (2004). Rachel Carson's *Silent Spring* (1962) and Paul R. Ehrlich's *Population Bomb* (1968) earlier stirred dialogue about ecological limits. On energy, see Cottrell's (1955) pathbreaking book.

11. Bell saw claims that nature was uncontrollable to manifest "primal impulses," unleashed by "utopian" movements. He viewed the irrational fears stirred up by critical ecologists to be more dangerous than corporate capitalism's excesses. See Bell (1999 [1973], pp. 463–467, 487–489).

12. He extended his critiques of Ehrlich's *Population Bomb* (1968) and the Carter Administration's *Global 2000 Report* (CEQ, 1980) into neoclassical broadsides against environmental limits and regulation. See Simon (1996a [1981], 1996b) and Simon and Kahn (1984).

13. Neoliberals sometimes warned of globalization's risks (e.g., terrorist exploitation of new technologies, porous borders, cultural shocks). For example, Thomas L. Friedman (2000, p. 103) and John Micklethwait and Adrian Wooldridge (2000, p. 343) quoted Marx's famous passages about capitalism's power to unravel tradition ("all that is solid melts into air…."). Friedman (2000, pp. 401–405) warned presciently about Osama bin Laden and the terrorist threat, holding that globalization's cultural dislocations generate resentment, resistance, and violence. He called for state intervention to deal with this grave threat. However, neoliberals tended to ignore or diminish ecological risks, which they usually held could be solved by markets, civil society, and mild regulatory legislation (e.g., Friedman, 2000, pp. 276–291). However, even progressive liberal, Benjamin Barber (1996) did not take into account globalization's ecological risks. See Antonio (2007).

14. However, neomodernizationists have had varied views. Many of them altered their ideas in response to worsening scientific reports about the environment. The 2008–2009 economic crisis will likely lead to more changes. See Beck ([1986] 1992, 1997, 1998), Beck, Giddens, and Lash (1994), Ferry (1995), Giddens (1990, 1999, 2000), Gore (1992), Hajer (1995), *Urban Studies* (1997, Vol. 34 [5 and 6]), Buell (2001), Mol (2001, 2006), Flitner and Heins (2002), and Inglehart and Welzel (2005).

15. New forms of transport (e.g., jumbo jets, supertankers, containerization), information-communication technologies (e.g., miniturization, digitalization, personal computers, Internet, cable TV, satellite communications), production methods (e.g., programmable automation), and organizational innovations (e.g., network firms and flexible specialization) helped give rise to globalization's smaller, faster world. Electronic financial trading and other forms of E-commerce decentralize, extend, and accelerate exchange, reduce transaction costs, and increase growth. New technologies increased vastly the scope, volume, speed, and ease of intercourse, helped organize, police, and legitimize globalization, and, overall, facilitated global interdependence (Harvey, 1989; Castells, 1996; Rifkin, 2000; T. L. Friedman, 2000, 2005; Dicken, 2007).

16. Free-market and free-trade policies reduced political barriers to capitalism. "Restructuring" increased commodification, privatization, deregulation, financialization, and securitization, which shifted decision-making power and resources from the public sphere to the private sector. Unions were weakened; state regulation and support for health, education, and welfare were cut; and financial/corporate elites were empowered and enriched. Government still provided infrastructure, coordination, and

protection, subsidized private projects, and bailed-out failed, risky, elite financial investments. These changes reduced "friction" or "rigidity" and increased "flexibility," allowing capital, commodities, and labor to flow faster and more freely across "borders" than ever before. Neoliberal restructuring varied widely across the globe. Some nations retained robust welfare states, and others employed state-centered strategies. Everywhere, the state remained a vital actor, coordinating and subsidizing capitalism. Free-market doctrine obscured neoliberalism's "crony capitalism" side (pandering to corporate/financial elites) and the U.S.'s role in globalizing neoliberalism thorugh American economic and military power and influence on IMF, World Bank, and G8 policy. See, for example, Phillips (2002) and Stiglitz (2002, 2003, 2006).

17. Neoliberalism is a cultural complex as well as a policy regime and ensemble of technical, political, and economic practices. Its individualist, productivist, consumerist, growth-oriented American cultural template is flexible and easily tweaked to fit diverse locations and traditions. Living, working, and consuming in the globalization system alters self-formation, but should not be equated with narrow materialist determination of cultural matters or cultural homogeneity. In precapitalist and former communist regimes, introduction of market-based institutions erodes traditional constraints and gives rise to new cultural agents, whose tastes, desires, and habits favor growth. Globalization opens local cultures to new goods and images, and fosters niche markets and hybrid products, bodies, and selves. Global social relations of consumption forge new appetites and alter old ones. Local tradition persists, but is reshaped by new fusions and reconstructed collective identities. Integrating developing nations into global capitalism makes many millions more people dependent on the treadmill and growth imperative, including multitudinous peasants and urban poor, who aspire to better lives. Although fearing and sometimes resisting threats to traditional cultures and natural environments, poor peoples often express stronger support for neoliberal globalization than Americans (Pew Center, 2007a, 2007b).

18. By contrast, critics of neoliberal globalization portray it as *extreme capitalism*, prone to breakdowns and crises of all sorts. See, for example, Dobkowski and Wallimann (2002), Davis (2006a, 2006b), Harvey (2005, 2006), Najam, Runnalls, and Halle (2007), and Reich (2007).

19. Globalization analysts often mentioned environmental and resource problems, but they usually addressed the topics briefly or ignored them. For example, Dicken's (2007, pp. 25–29, 517–518, 540–545) comprehensive globalization text had only a few brief comments about climate change and environment. Stiglitz (2006, pp. 161–186) included a chapter on warming. Benjamin M. Friedman (2005, pp. 369–395) discussed ecology and climate change, but embraced the growth imperative and environmental Kuznets curve. Thomas L. Friedman's *The World is Flat* (2005) did not address the environment at all, and his earlier *Lexus and the Olive Tree* (2000) treated the topic passingly. David Held's co-authored globalization book included a chapter on the environment, and his later co-authored work had a brief section on the topic. His co-edited collection left out ecology (Held, McGrew, Goldblatt, & Perrraton, 1999; Held & McGrew, 2007a, 2007b, pp. 64–72). Saskia Sassen (2006, 2007) and Giovanni Arrighi (2007) did not discuss ecological issues in their globalization books. Ritzer's (2007b) collection has a brief essay on environment. There have been exceptions. For example, Michael N. Dobkowski's and Isidor

Wallimann's (2002) collection stressed ecological crises. Philip McMichael's (2007), Friedman's (2008), Jeffery Sachs' (2008), and Speth's (2008) books address climate change and environment broadly and likely are the start of a major wave of works stressing globalization and ecology.

20. Ecological economist, Herman E. Daly has contended that early modern economies were "infinitesimal relative to the natural world," but that global capitalism has radically altered the scale and speed of resource depletion, waste, and ecological damage. Daly has called for a "full world economics" that analyzes GDP in relation to the "throughput" of resources and production of wastes. He has criticized neoclassical economists for treating the economic processes as a self-enclosed circulation between firms and households and the ecosystem as if it were a subsystem (i.e., "extractive sector") of the economy, rather than the converse, and assuming that scarce items always could be attainted through trade or replaced by substitute products and technical innovation. See Daly (1995, 1996, 1999, 2005, pp. 48–60); Costanza and Daly (1992).

21. Increased energy prices, oil geopolitics, and extreme weather events (e.g., Hurricane Katrina, Europe's 2003 heat wave [27,000 deaths], the Atlanta area's 2007 drought) have raised ecological fears even more. Comparing survey data for 2002 and 2007, a Pew Foundation (2007a, p. 2) study found increasing numbers of respondents, in 30 of 35 nations, naming environmental problems as the "top global threat" (70% or more for China, South Korea, Japan). In a another Pew Center (2007b, p. 19) survey, more than 50% of respondents in all 47 nations and 70% in 29 nations agreed that the environment must be protected, even if mitigation slows growth and costs jobs.

22. The IPCC was formed, in 1988, jointly by the World Meteorological Organization and the United Nations Environment Programme.*AR4's* three working groups focused on: The Physical Science Basis; Impacts, Adaptation, and Vulnerability; and Mitigation and Climate Change. The summary *Synthesis Report* (*SYR*) was released on November 17, 2007. Previous assessment reports were completed in 1990 [supplemented 1992]; 1995; 2001. See (http://www.ipcc.ch/).

23. IPCC (2007a, p. 1) (summary for policymakers). The panel estimated uncertainty based on levels of scientific evidence and agreement, which they used in "*confidence*" judgments about their models, analyses, or claims (i.e., from "*very high confidence*" [at least 90%] to "*very low confidence*" [$<$10%]) and of the "*likelihood*" of future outcomes (i.e., from "*virtually certain*" [$>$99%] to "*exceptionally unlikely*" [$<$1%]). IPCC (2007b [uncertainty guidance notes for lead authors], pp. 3–4; [technical summary], pp. 22–23). The panel held that it is "*extremely unlikely*" that the last 50 years of unusual climate change can be explained without "external forcing" and "*very likely* that it is not due to known natural causes alone" (IPCC, 2007a, [Topic 2], p. 6).

24. IPCC, 2007a (Summary for Policymakers), pp. 1, 4. Besides CO_2, other primary atmospheric GHGs are water vapor (most common), nitrous oxide, methane, and ozone. These GHGs have natural and anthropogenic sources. There are also other entirely anthropogenic GHGs (e.g., halocoarbons) (IPCC, 2007a, [Appendices], p. 8). Also, scientists have discovered new GHGs such as nitrogen trifluoride (an extremely potent, long-lasting gas), whose extent and threat has to monitored and assessed (Hoag, 2008b). The 2007 CO_2 concentration was at 383 ppm

("37% above the concentration at the start of the industrial revolution, or about 280 ppm)" (GCP, 2008).

25. See GHF (2009) for a comprehensive report on climate change impacts on poor peoples. Also see, Inman (2009) and Vedantam (2009).

26. Hagg (2008). *AR4's* participants came from more than 130 countries, included more than 450 primary authors and 800 other contributors, and were reviewed by more than 2500 other people.

27. See, for example, L. Smith (2007), Bone (2007), and Leake (2007). "Tipping point" refers to a juncture when environmental conditions are so drastically altered that an ecosystem is vulnerable to sudden, irreversible, fundamental structural changes. See IPCC (2007b, pp. 775–777).

28. In 2007, $CO_{2\text{-}e}$ was already at 430ppm, was increasing at about 3 ppm per year and projected to accelerate (Dunlop, 2007). The $CO_{2\text{-}e}$ concentration measure takes account of other GHG gases in a single metric; that is, "the concentration of CO_2 that would cause the same amount of radiative forcing as a given mixture of CO_2 and other forcing components" *IPCC* (2007a, [Topic 2], p. 1) and Betts (2008).

29. One degree Celsius equals 1.8°F. The panel held that a 2°C rise might be enough to melt Greenland's ice sheet after 2100 and raise seas 7 meters (IPCC, 2007a [summary for policymakers], p. 13).

30. Average global temperatures have increased about 0.7.6°C from 1850–2005, but they are rising faster now (IPCC, 2007b [summary for policymakers], p. 5; see also GCP, 2008). The IPCC estimated that CO_2 concentrations alone were increasing about 1.9 per year. CO_2 levels reached 381.2 ppm in 2006, increasing 2.0 ppm above 2005. CO_2 concentration rates, for the previous decade, have risen about 30% faster than the previous 40 years; they increased only 20 ppm for 8,000 years before industrialism. At the end of 2008, atmospheric CO_2 was at about 385 ppm (WMO, 2007; UNHDP, 2007, pp. 26, 33; Wynn, 2009).

31. The six "families" of 40 scenarios were formulated prior to *AR4* (IPCC, 2000). The various scenarios were constructed to provide baselines to help estimate costs of mitigation strategies and frame policies to reach desired stabilization targets for GHG atmospheric emissions. See IPCC (2007a [Topic 3: 3 {Table 3.1, note c}]); UNHDP (2007, pp. 35–36, 46–47, 63).

32. For example, the IPCC warned, with "medium confidence" that 20–30% of species assessed would "likely" be threatened with extinction with a 1.5–2.5°C increase in global average temperature (relative to 1980–1999) and 40–70% above 3.5°C (IPCC, 2007a, [Topic 3], p. 15).

33. On climate policy and global inequality, see Roberts and Parks (2007).

34. Satellite measurements of Arctic Ice were initiated in 1979; the 1950s comparison was on based on aerial photos. On the drivers of the 2007 record Arctic ice melt, see J. Zhang et al. (2008b).

35. Melting in Greenland and Antarctica have made them "adventure destinations" for tourists. However, warming threatens Greenland's Inuit people's way of life and its characteristic flora and fauna. Some of Greenland's plant, animal, and insect species start their spring cycles 30 days earlier than a decade ago. The Obama administration has called for limits on Antarctic travel to protect its fragile ecology. See Boyd (2007), Spiegel (2007), Lyall (2007), McConnico (2007), Todras-Whitehill (2007), Revkin (2008a), and Lee (2009).

36. *AR4*'s models projected that Antarctica would stay too cold for major melting in this century and its ice sheet would likely grow. Yet *AR4* stated that the models did not take into account fully enough "ice dynamical processes," revealed in recent work on Greenland and the Antarctic and that melting could be greater in both locations (IPCC, 2007a, [Topic 3], p. 6, [Topic 5], p. 3). Monaghan et al. (2008) held that warming in Antarctica has been less than estimated by the IPCC. However, Steig et al. (2009) demonstrate significant warming in West Antarctica and continent-wide average temperature increases.

37. A later NSIDC (2009b) news release reported that a key ice bridge supporting the Wilkins Ice Shelf had collapsed. See NSIDC (2008a), Spotts (2008), AP (2008d), Kaufman (2008a), L. Smith (2008a), Steig et al. (2009), and Glasser and Scampos (2008). Antarctica holds about 90% of the world's ice. Thus, melting there is alarming (although unlikely for a very long time, a total Antarctic melt might raise the seas 61–65 meters).

38. Although summer 2008 was cloudier and cooler, Arctic sea ice reached its second lowest point since measurements began in 1979 and a record amount of multiyear ice was lost. The extensive amount of first year ice opens the way for greater peak summer melts (NSIDC, 2008b). In February 2009, only 10% of Arctic sea ice was more than two years old (NSIDC, 2009a).

39. The *Arctic Report Card 2008* (NOAA, 2008, p. 2) also reported the impacts of warming on fisheries, marine mammals, permafrost, and the Greenland melt. For a brief summary of major 2008 work on warming and its impacts, see Mascarelli (2009).

40. The estimated global, sea level rise from a WAIS collapse is about 3.3 meters (rather than the earlier projected 5–6 meters); it projects a 25% greater rise along the U.S. east and west coasts. Although evidence suggests that the WAIS may be unstable and that its melt has accelerated, scientists see its collapse as a long-term, low probability event (Bamber et al., 2009; Ivins, 2009).

41. Fire is the source of about 50% of the GHG emissions coming from deforestation, and it accounts for about 20% of all anthropocentric GHG emissions (Bowman, 2009).

42. For example, water vapor may contribute to cloud formation, which can have a cooling effect. But cloud formation's feedback impacts depend on the nature of the clouds, and clouds that cool in daytime might trap heat and warm at night (Pearce, 2007, pp. 140–150). Similar human activities may have divergent impacts in different locations. For example, expanding agricultural land and cutting forest tends to have warming effects in tropics and likely cools in middle latitudes, where there is winter snow cover (Rosenzweig et al., 2008, p. 356).

43. New Arctic vegetation would not make up for what is lost from the reduced albedo effect. Carbon sinks include plants and organisms (e.g., plankton) that use CO_2 in photosynthesis, removing it from the atmosphere, storing it in biomass, and releasing oxygen. Natural land and ocean carbon sinks contribute substantially to reducing atmospheric CO_2, but their efficiency in this removal process has been declining for a half century (GCP, 2008).

44. The scientists mapped changes in surface temperatures across the globe (1970–2004), and then analyzed whether changes of biological and physical systems, in their database, were consistent with a warming climate. See Rosenzweig et al. (2008) and Zwiers and Hegerl (2008).

45. Coral reefs protect coasts, support fisheries and tourism, and are sources of construction materials and biomedical compounds. Coral reef decline would likely cause billions of dollars of losses from tourism alone. See Hoegh-Guldberg et al. (2007) and Flannery (2005, pp. 104–113).

46. The scientists explained that the Mississippi River's coastal waters were the world's largest dead zone and that 405 similar zones could be found in other parts of the globe (Diaz & Rosenberg, 2008). Other scientists have predicted a 50% increase in dead zones by the end of the century (Armstrong, 2008). On the release of two earlier studies, National Public Radio hosted four scientists who discussed serious threats to marine life. They reported a Texas-sized body of floating, plastic trash in the Pacific Ocean. They argued that globally coordinated mitigation is needed to save the oceans. The segment was broadcast on February 16, 2008, "Talk of the Nation: Science Friday" (http://www.npr.org/templates/story/story.php?storyId = 19085884). See Hohn (2008) on the plastic trash floating the global seas and littering seashores.

47. If Georgia won the battle, it would have to pump Tennessee River water a 120 miles to metro Atlanta; the project possibly would cost billions of dollars.

48. Average rainfall in North Georgia has been declining since 1979. In the severe 2007 drought, average rainfall was down more than 36% and down nearly 18% in 2008 (Shelton, 2009a). See Shelton (2009b, 2009c, 2009d, 2009e).

49. See IPCC (2007, [Topic Three], p. 12). About 85% of the western U.S.' water comes from mountain sources threatened by warming. For example, Glacier National Park's glaciers are retreating rapidly; only 25 of its former 150 glaciers remain. Warming is likely the main driver of the snow and ice loss, but other anthropogenic feedbacks also contribute to melting (e.g., soot increases glacial and snow absorption of solar energy more than 40%) (Berardelli, 2007a).

50. California, Nevada, Arizona, New Mexico, Colorado, Utah, and Wyoming are the most impacted states. Most of the 101 U.S. cities that receive the least annual precipitation are in this region (http://www.city-data.com/top2/c463.html). See McKinley (2009). However, many other states face future water shortages (e.g., Santos, 2007; Skoloff, 2007).

51. The Atlanta metro area has two major water parks, and a third was planned in the midst of the 2008 water crisis (Redmond, 2008).

52. In early 2009, the Waveland founders had not yet secured the financing to begin the construction, which had gained the support of city officials (Nelson, 2009).

53. The 2008 IUCN, Red List includes 44,838 plant and animal species or about 2.5% of all described species; 16,928 species are threatened. The Red List distinguishes between three levels of threat: critically endangered, endangered, and vulnerable. See IUCN (2008, [State of the World's Species], 2007), Benjamin (2007), and Wilson (2007, pp. 73–81). The biodiversity crisis manifests complex anthropogenic feedbacks. For example, growing global log and charcoal demand spurs intense rainforest cutting (Laurence, 2008; Smith, 2008b; Tollefson, 2008; Wang et al., 2008). Increased biofuel production, high cereal prices, and increased demand for meat also contribute to the problem. Poor settler farmers resist efforts to protect tropical forests (e.g., Partlow, 2009). Tropical ecosystems have the most diverse species so that accelerated cutting in them is a key driver of the biodiversity crisis (Laurence, 2006). Cutting destroys a major carbon sink. Also, evidence of reduced tree growth in the Amazon, may be driven by climate change feedback, which may be

reducing tropical forests' effectiveness as a carbon sink. Warming driven drought threatens to reverse the role of these carbon sinks (Reed, 2007; Hopkin, 2007; Phillips et al., 2009).

54. This study reported that 25% of mammals (i.e., data sufficient) are "threatened" with extinction and that another 5.9% are "near threatened" (Schipper et al., 2008, pp. 227–228).

55. One study reports that "90% of the ~28,800 documented changes [between 1970–2004] in plants and animals" in its database are consistent with a warming climate (Rosenzweig et al., 2008, p 354). Another, report contends that 60% of 1598 species studied (over 20–150 years) experienced changes in distributions or phenology (CCSP, 2008, pp. 9, 150–180 passim).

56. Most corn and soy is used to feed livestock. Meat production contributes heavily to water quality problems (Blinch & Love, 2008; Steinfeld et al., 2006).

57. Battisti and Naylor (2009) documented that severe heat waves have significantly damaged agricultural production and that sharp temperature rises, projected for this century, will greatly increase food insecurity. Limited adaptive capacities and other drivers (e.g., disease) intensify the problem of malnutrition in poor countries (Lobell et al., 2008; Brown & Funk, 2008).

58. Ecosystem services refer to nature's provisioning, regulating, supporting, and preserving functions and to its cultural benefits (e.g., clean drinking water, crop pollination, outdoor recreation). Many ecosystem services are being consumed or damaged at rates undercutting their sustainability. Rapid GDP growth, insulation of rich nations from the worst problems, relative invisibility of certain damages, and far away impacts make coping with eroding ecosystem services more intractable (WRI, 2005, pp. 2, 7, 10–11). See Najam et al. (2007).

59. "The Ecological Footprint measures humanity's demand on the biosphere in terms of the area of biologically productive land and sea required to provide the resources we use and absorb our waste." It is measured in hectares (each hectare equals 2.471 acres) per person (WWF, 2008, pp. 1–3, 14; see Wackernagel & Rees, 1996).

60. About 7.5% of U.S. consumer goods spending has been on Chinese imports (about 80% of toys; 85% of footwear; 40% of clothes sold in the United States) (Barboza, 2008; Clark, 2009).

61. China's CO_2 emissions were 14% higher than the United States, but U.S. per capita emissions were nearly four times higher than China (NEAA, 2008). Also, China's energy intensity and carbon intensity have improved (GCP, 2008). By 2009, China outpaced the United States in the development of cleaner coal-fired technology, and also improved substantially its wind energy technologies and capacity. Yet China still has many inefficient coal-fired plants, produces much industrial soot, and its fossil fuel use is still rapidly on the rise. China argues that it is unfair to expect them to sacrifice growth to reduce carbon emissions, when developed countries, especially the United States is unwilling (Bradsher, 2009c; Krugman, 2009; Ramanathan & Carmichael, 2008).

62. Brower and Leon (1999, p. 55) estimated transportation's share of total consumer impact on environmental problems: about 15% of land use problems; 3% of water use problems; 23% toxic of water pollution; 7% of common water pollution; 52% of toxic air pollution; 28% of common air pollution; 32% of GHG. In 2007, transport was estimated to be responsible for about 18% of global CO_2 emissions (Sachs, 2008, p. 96). Also see EIA (2008). On consumption and development, see Arrow et al. (2004).

63. The contradiction of growth and nature is visible in the juxtaposed downward sloping "Living Planet Index" and upward sloping "Ecological Footprint" index (WWF, 2008, p. 2).

64. Chinese and Indian demand for oil is expected to double in 20 years, and the two countries will likely import as much oil as the United States and Japan by 2030. In 2004, the automotive sector of wealthy OECD nations produced about 30% of GHG emissions (UNHDP, 2007, pp. 137–138; Mouawad, 2007). Developing nations' growing car culture is changing this scenario. Transportation energy consumption of China and other developing Asian nations is projected to increase sharply from 2005–2030. Non-OECD Asian nations' use of transportation liquids is expected to increase from 12.6% to 34.5% over this period. Global non-OECD, transportation energy use has been projected to rise 3% yearly, during the same period (EIA, 2008, pp. 6–10).

65. From 2003 to mid-2008, China had built about 500 malls. However, the huge South China Mall has had much trouble filling its commercial space; it manifests planners' and investors' aspirations for China's growing consumer culture, rather than its current scope (Donohue, 2008).

66. Younger professionals bought cell phones, DVD players, and other electronic goods, clothes, household furnishings, fake antiques and, sometimes, Gucci, Louis Vuitton, and other upscale goods (Barboza, 2005, 2008; TA, 2005; Powell, 2008; SD, 2008; Xinhua, 2008a, 2008b). See http://www.southchinamall.com.cn; http://www.pdong.gov.cn: http://www.easternct.edu/depts/amerst/mallsworld.html

67. In early 2008, Shanghai suburbanites protested in front of city hall and then, to avoid the police, mixed with shopper crowds in nearby Pudong. They were protesting the planned extension of the Maglev transrapid train across their neighborhoods; they feared it would lower their property values and degrade their life-style enclave (http://shanghaiist.com/2008/01/14/maglev_protest_1.php).

68. Black carbon, deriving from incomplete burning of fossil fuel, biofuel, and other organic matter, has been estimated to be the source of about 18% of global warming and about half of Arctic warming and melting. It has yet to be included in governmental climate change efforts (Rosenthal, 2009b; Ramanathan & Carmichael, 2008)

69. Chinese officials warned that 69 dams were threatened by 2008 earthquake. Dams, reservoirs, and irrigation systems suffered "extensive" damage (Hooker, 2008; Wong & Swart, 2008).

70. Also, China uses and produces about 45% of global cement; chemical processes to produce it release about 5% of global CO_2 emissions (Rosenthal, 2007a). Moreover, China imports nearly half of all traded global timber; much of it is cut illegally in the tropics (Laurence, 2008). Finally, China's growing meat consumption has helped increase cereal prices and GHG emissions (Blinch & Love, 2008).

71. Simon Kuznets (1955) speculated that development would reverse the growing inequality of industrializing nations. For an example of the environmental version of Kuznets' thesis, see, B. M. Friedman (2005, pp. 369–395).

72. Ecomigrations have begun; as many as 17 million people have already been displaced by Bangladeshi ecological disasters alone, and many millions more people, suffering from various environmental catastrophes, have been forced to move in other poor nations (Vedantam, 2009). For more on inequality and climate change, see Roberts and Parks (2007).

73. Estimates of forced migrations, by mid-century, range from the tens of millions to a billion or more people. See *Forced Migration Review's* (October 2008

[Issue 31]) special issue on "climate change and displacement." Also, see Bogardi and Warner (2009).

74. Ecological problems even extend into space; NASA has estimated that 19,000 objects (more than four inches) are orbiting the earth (Oleskyn, 2009).

75. The IPCC (2007a [Summary for Policymakers], pp. 22–23) estimated earlier that the macroeconomic costs, in 2050, for stabilizing $CO_{2\text{-}e}$ 445 ppm and 700 ppm would be in a range from "1% gain and 5.5% decrease" of GDP.

76. In April 2008, 71% of Americans surveyed agreed that there is strong evidence for global warming, and 47% of the believers said the warming is due to human activity (down respectively from 79% and 50% in July 2006). More strikingly, 13% fewer Republicans surveyed agreed that there is strong evidence for warming in January 2007 (Pew Center, 2008b). In October 2007, 66% of the people surveyed said that the environment should be protected even if it reduces growth (Pew Center, 2007b, p. 19). Other surveys indicated growing awareness of environmental problems, but when growth slowed and recession threatened, only 12% of Republicans and 47% of Democrats saw climate change to be a "top priority" (Pew Center, 2008a). In another 2008 poll, only 1% of the respondents considered the environment to be an important priority, and climate change did not register all (WP, 2008). Also, it is unclear if Americans, who believe in anthropogenic warming, would accept the substantial costs and life-style changes that genuine mitigation would require.

77. In 1985, 61% of the Gallup respondents said that the environment ought to have priority over economic growth, while, in 2009, only 42% took that position (Saad, 2009). In 1998, 61% of the Gallup respondents said that news about the seriousness of the global warming threat was either correct or underestimated, and 31% thought it was exaggerated. In 2009, the figures on this issue were respectively 57% and 41%. Moreover, the survey demonstrated that concern about climate change had receded in other important respects. For example, fewer respondents believed that the process has started or would impact during their lifetimes (Saad, 2009). These shifts might reflect increased disinformation efforts as well as the weak economy. On right-wing disinformation: for example, see McCright and Dunlap (2000), Demeritt (2006), Rogers (2007), Gunter (2008), Revkin (2008b), and UCS (2007). Polls indicate a sharp Democratic-Republican divide on the topic (Dunlap, 2008a; Newport, 2009; Silver, 2009a).

78. Senator Jim Webb (D-Va.) identified Obama's climate change proposal to be the facet of his agenda most likely to fail. Webb added that it would fail because of inadequate support, not because it was infeasible. Smith and Raju (2009); also see Branigin, Eilperin, and Mufson (2009), Overbye (2009), *New York Times* (2009), Landler (2009), Rosenthal (2009a), Silver (2009a), and Stein (2009).

79. In an April 2009 Gallup Poll, 26% more Americans (mostly Democrats and Independents) than in a 2008 poll said that environmental quality was improving (Jones, 2009; Silver, 2009b).

80. The suffusion of published work on climate change and other ecological problems in natural science journals such as *Science* and *Nature* should be instructive to cultural critics, who sometimes have held that such science is inherently cacophonous and slow to respond to human problems. However, I will leave it to natural scientists and philosophers of science to determine whether this new work

has led to critical rethinking of the foundations of their practices. See, for example, Nagel, Dietz, and Broadbent (2009) on sociological perspectives on climate change.

81. See the special section on the important contributions of Dunlap and Catton in *Organization & Environment*, 2008, 21:4 (December), pp. 446–487.

82. Dunlap's and Catton's critique of constructivism has pertinence for types of environmental sociology and ecological modernization theories, which have drawn heavily from late 20th century cultural sociology and cultural theory, stressing strong or totalistic versions of cultural autonomy. Many cultural theorists attributed linguistic practices enormous power, and decoupled them from the biophysical world and social structure. For example, Jeffrey C. Alexander's (2003, pp. 13, passim) "strong program in cultural sociology" stresses "radical uncoupling of cognitive content from natural determination" and "a sharp analytical uncoupling of culture from social structure, which is what we mean by cultural autonomy." This double analytical decoupling has utility in cultural analyses, which for some specified purpose, limit focus strictly to meanings or connections between representations. However, universalizing this approach elides focus on the issue of symbolic culture's pragmatic relationship with obdurate features of the world, which is of fundamental importance for public issues like the global recession or ecological crisis. This solipsistic "culturalism" plagued many "postmodernist" approaches. See Antonio (1998), and see Crist (2004) for incisive criticism of constructivism in environmental and ecological thought.

83. Environmental sociologists' ideas about exemptionalism and other important facets of the ecological crisis (e.g., "treadmill of production") have not been diffused to other sociological subareas or social theorists (Lewis & Humphrey, 2005). On exemptionalism and the new ecological paradigm see, for example, Dunlap and Catton (1979, 1994), Catton (1994, 2002, 2008), Buttel (1987, 2004), Murphy (1995), Daly (1996), Dunlap (2002a, 2002b, 2008b), and Wilson (2005, 2007, 2008).

84. Durkheim (1953 [1898], pp. 23, 1–34 passim) acknowledged that culture's sui generis reality has "relative autonomy" from its "substrata," implying dependence on biophysical as well as social conditions. He referred to biophysical factors in his work (e.g., suicide rates and population dynamics are social facts and biological events), and sometimes held that they determined social relationships (e.g., his problematic claims about biological differences shaping the gendered division of labor and other cultural differences between men and women). Even his analyses of clan cultures' totems imply interaction between cultures and their environments. Durkheim (1964 [1893], p. 195) said that the "determining causes of social evolution" are "in the environment that surrounds" us, but, qualifying the point, he held that the "physical environment remains relatively constant" and, thus, cannot explain rapid successive social and cultural changes. His assertion excludes nonhuman, multiphase catastrophic or irruptive biophysical causes, but today's rapidly increasing GHG levels and deepening ecological crises have anthropogenic drivers and are social facts. See Durkheim (1966 [1897], pp. 215–216, 269–276, passim; 1964 [1893], pp. 16–24, 200–225); Durkheim and Mauss (1963 [1903]), and Rosa and Richter (2008).

85. Dunlap's and Catton's critical points have some credence. For example, Durkheim's claim that social differentiation frees us "from the yoke of the organism" or "body" does diminish nature's constraints (Durkheim, 1964 [1893],

pp. 345–346, 329–350, passim). Moreover, too often mid- and late-20th century theorists overextended Weber's arguments about "*verstehen*" and the causal force of Protestant culture, and understated the structural and material threads of his work. "Symbolic interactionists" often have ignored Mead's references to the "central nervous system" and external "environment."

86. Dunlap's co-edited collection, *Sociological Theory and the Environment* (2002) does not mention Mead and Dewey (Dunlap, Buttel, Dickins, & Gijswijt, 2002; Buttel et al., 2002; Catton, 2002; Dunlap, 2002b).

87. Reiterating the desire to avert dualism near the end of his life, Dewey said that he wished that he could retitle one of his most comprehensive works on the topic *Culture and Nature* (Dewey, 1988a [1925], appendices 1 and 2, pp. 361, 329–392 passim, 1988d [1929]).

88. Mead and Dewey were influenced deeply by William James's engagement of Darwin. Mead (1964b [1929–1930], pp. 283–284) stated: "But for James the act is a living physiological affair, and must be placed in the struggle for existence, which Darwinian evolution had set up as the background of life. Knowledge is an expression of the intelligence by which animals meet the problems with which life surrounds them." Mead and Dewey did not embrace the idea of progressive evolution. Their organism-environmental model stressed inquiry into particular conditions of specific situations, and did not posit a direction to history (although they stressed that useful cultural instrumentalities, over the long term, tend to accumulate and diffuse) (see, e.g., Mead, 1964a [1927], 1962 [1934]; Dewey, 1983a [1904], 1983b [1909], 1985 [1916]).

89. See Rorty (1995) on Darwinian and Dewey-Mead pragmatism; Armitage (2003) and Williams (2003); also see McDonald (2004) on Dewey and environmentalism, and Brulle (2000) for a related Habermasian view.

90. Durkheim (1964 [1893], p. 286) said that "socio-psychology" is needed to bridge the gap between "psycho-physiological causes" and "psychological facts," deriving "from social sources."

91. These wars pitted postmodernist cultural theorists, stressing strong constructivist positions, against scientific "realists" (Antonio, 1998). The conflicts and polarized positions hardened unfortunate stereotypes of natural science and social science, which undercut the type of collaborative work across disciplines, which the ecological crisis demands.

92. On Marx and Weber and ecology, see, for example, O'Connor (1998), Foster (1999, 2002), Dickens (2002b), Clark and York (2005), Murphy (1994, 2002), and York and Mancus (2009).

REFERENCES

Abdulla, H. (2007). Southeast drought fuels battle over water rights. *McClatchy Newspapers*, October 19. Available at http://www.mcclatchydc.com/homepage/story/20708.html. Retrieved on October 20, 2009.

Akard, P. J. (1992). Corporate mobilization and political power: The transformation of U.S. economic policy in the 1970s. *American Sociological Review*, *57*, 597–615.

Alexander, J. C. (2003). *The meanings of social life: A cultural sociology*. Oxford and New York: Oxford University Press.

Allan, R. P., & Soden, B. J. (2008). Atomospheric warming and the amplification of precipitation extremes. *Science*, *321*(September 12), 1481–1484.

Allison, I., Béland, M. et al. (2009). *The state of polar research: A statement from the International Council for Science/World Meteorological Organization Joint Committee for the International Polar Year 2007–2008*. Available at http://www.wmo.ch/pages/mediacentre/press_releases/documents/IPY_StateofPolarResearch_EN_web.pdf. Retrieved on May 25, 2009.

Ansari, A. (2009). Fish migrating to cooler waters, study says. *CNN*, February 12. Available at http://www.cnn.com/2009/TECH/science/02/12/fish.migration.study/index.html. Retrieved on February 12, 2009.

Antonio, R. J. (1998). Mapping postmodern social theory. In: A. Sica (Ed.), *What is social theory? The philosophical debates* (pp. 22–75). Malden, MA, and Oxford, UK: Blackwell Publishers.

Antonio, R. J. (2007). The cultural construction of neoliberal globalization. In: G. Ritzer (Ed.), *The Blackwell companion to globalization* (pp. 67–83). Malden, MA, and Oxford UK: Blackwell Publishing.

AP. (2007). McDonald's serves China's car culture. *MSNBC*, January 19. Available at http://www.msnbc.msn.com/id/16709901/. Retrieved on January 22, 2009.

AP. (2008a). "Doomsday" seed vault opens in Arctic. *MSNBC*, February 27. Available at http://www.msnbc.msn.com/id/23352014/. Retrieved on August 22, 2009.

AP. (2008b). Water is Beijing's dirty little secret. *MNSBC*, March 7. Available at http://www.msnbc.msn.com/id/23373004/. Retrieved on August 22, 2009.

AP. (2008c). Glaciers shrinking at record rates, UN agency reports. *New York Times*, March 16. Available at http://www.nytimes.com/2008/03/16/health/16iht-glacier.1.11132352.html. Retrieved on March 16, 2008.

AP. (2008d). Runaway ice chunk in Antarctica worries scientists. *The New York Times*, March 26. Available at http://www.nytimes.com/2008/03/26/health/26iht-ice.1.11432642.html. Retrieved on March 27, 2008.

AP. (2008e). 4500-year-old ice shelf breaks away. *CNN*, September 3. Available at http://climate-change.suite101.com/article.cfm/markham_ice_shelf_breaks_away. Retrieved on September 4, 2009.

Armitage, K. C. (2003). The continuity of nature and experience: John Dewey's pragmatic environmentalism. *Capitalism Nature Socialism*, *14*(3), 49–72.

Armstrong, A. (2008). Doubling dead zones. *Nature Reports*, *2*(September 4). Available at http://www.nature.com/climate/2008/0810/full/climate.2008.92.html. Retrieved on September 5, 2008.

Arrighi, G. (2007). *Adam Smith in Beijing: Lineages of the twenty-first fentury*. New York and London: Verso.

Arrow, K., et al. (2004). Are we consuming too much? *The Journal of Economic Perspectives*, *18*(3), 147–172.

Auffhammer, M., & Carson, R. T. (2008). Forecasting the path of China's CO_2 emissions using province-level information. *Journal of Environmental Economics and Management*, *55*(3), 229–247.

Bamber, J. L., et al. (2009). Reassessment of the potential sea-level rise from a collapse of the West Antarctic Ice Sheet. *Science*, *324*(May 15), 901–903.

Barber, B. R. (1996). *Jihad vs. McWorld: How globalism and tribalism are reshaping the world.* New York: Ballantine Books.

Barboza, D. (2005). China, new land of shoppers, builds malls on gigantic scale. *The New York Times*, May 25. Available at http://www.nytimes.com/2005/05/25/business/worldbusiness/25mall.html. Retrieved on August 20, 2008.

Barboza, D. (2008). China's inflation hits American price tags. *The New York Times*, February 1. Available at http://www.nytimes.com/2008/02/01/business/worldbusiness/01inflate.html. Retrieved on February 1, 2008.

Barboza, D. (2009). A textile capital of China is hobbled by a downturn gone global. *The New York Times*, February 28. Available at http://www.nytimes.com/2009/02/28/business/worldbusiness/28textile.html. Retrieved on March 1, 2009.

Barnett, T. P. et al. (2008). Human-induced changes in the hydrology of the western United States. *Science Express*, January 31. Available at http://www.sciencemag.org/cgi/content/full/319/5866/1080. Retrieved on February 2, 2008.

Barnett, T. P., & Pierce, D. W. (2008). When will Lake Mead go dry? *Water Resources Research*, *44*(March 29), W03201.

Barringer, F. (2008). Lake Mead could be within a few years of going dry, study finds. *The New York Times*, February 13. Available at http://www.nytimes.com/2008/02/13/us/13mead.html. Retrieved on February 13, 2008.

Battisti, D. S., & Naylor, R. L. (2009). Historical warnings of future food insecurity with unprecedented seasonal heat. *Science*, *323*(January 9), 240–244.

Beck, U. (1992 [1986]). *Risk society: Towards a new modernity* (M. Ritter, Trans.). London & Thousand Oaks, CA: SAGE Publications.

Beck, U. (1997). *The reinvention of politics: Rethinking modernity in the global social order.* (M. Ritter, Trans.). Oxford, UK, and Cambridge MA: Polity.

Beck, U. (1998). Politics of the risk society. In: J. Franklin (Ed.), *The Politics of the risk society* (pp. 7–22). Oxford, UK, and Cambridge MA: Polity.

Beck, U., Giddens, A., & Lash, S. (1994). *Reflexive modernization: Politics, tradition and aesthetics in the modern social order*. Stanford, CA: Stanford University Press.

Bell, D. (1977). Are there social limits to growth? In: K. D. Wilson (Ed.), *Prospects for growth: Changing expectations for the future* (pp. 13–26). New York: Praeger Publishers.

Bell, D. (1999 [1973]). *The coming of post-industrial society: A venture in social forecasting*. New York: Basic Books.

Benjamin, A. (2007). Threatened species Red List shows escalating "global extinction crisis." *Guardian Unlimited*, September 12. Available at http://www.guardian.co.uk/environment/2007/sep/12/internationalnews.greenpolitics. September 13, 2007.

Berardelli, P. (2007a). Trouble in them thar hills. *ScienceNOW Daily News*, December 12. Available at http://www.sciencenow.sciencemag.org/cgi/content/full/2007/1212/3. Retrieved on February 1, 2008.

Berardelli, P. (2007b). Lessons from the interglacial past. *ScienceNOW Daily News*, December 18. Available at http://sciencenow.sciencemag.org/cgi/content/full/2007/1218/2. Retrieved on December 19, 2007.

Betts, R. (2008). Comparing apples with oranges? *Nature Reports*, *2*(January), 7–8. Available at http://www.nature.com/climate/2008/0801/full/climate.2007.74.html. Retrieved on January 10, 2008.

Bittman, M. (2008). Rethinking the meat-guzzler. *The New York Times*, January 27. Available at http://www.nytimes.com/2008/01/27/weekinreview/27bittman.html. Retrieved on January 27, 2008.

Blas, J., & Wiggens, J. (2008). Expensive tastes: Rising costs force food prices up the political agenda. *Financial Times*, March 17. Available at http://www.ft.com/cms/s/0/f733766a-f449-11dc-aaad-0000779fd2ac.html?nclick_check = 1. Retrieved on March 18, 2008.

Blinch, R., & Love, B. (2008). World food prices soar as Asia consumes more. *New York Times*, March 31. Available at http://www.nytimes.com/2008/03/31/business/worldbusiness/31iht-food.1.11548804.html. Retrieved on March 31, 2009.

Block, B. (2009). Climate change outpaces predictions. *Worldwatch Institute*, January 14. Available at http://www.worldwatch.org/node/5990. Retrieved on January 15, 2009.

Blow, C. M. (2008). Farewell, fair weather. *The New York Times*, May 31. Available at http://www.nytimes.com/2008/05/31/opinion/31blow.html. Retrieved on June 2, 2008.

Bluestone, G. (2007). Reservoirs almost down to the dregs. *Lawrence Journal World*, November 17. Available at http://www2.ljworld.com/news/2007/nov/17/reservoirs_almost_down_dregs/. Retrieved on November 18, 2007.

Bogardi, J., & Warner, K. (2009). Here comes the flood. *Nature Reports*, *3*(January), 9–11.

Bohannon, J. (2008). The big thaw reaches Mongolia's pristine north. *Science*, *319*(February 1), 567–568.

Bone, J. (2007). A global warning for UN chief from the ice floes at the foot of the world. *The Times*, November 12. Available at http://www.timesonline.co.uk/tol/news/world/article2852908.ece. Retrieved on Novermber 12, 2007.

Bowley, G., & Jolly, D. (2008). Economic toll mounts from high oil prices. *The New York Times*, May 23. Available at http://www.nytimes.com/2008/05/23/business/worldbusiness/23oilweb.html?hp#. Retrieved on May 23, 2008.

Bowman, D. M. J. S. (2009). Fire in the earth system. *Science*, *324*(April 24), 481–484.

Boyd, R. S. (2007). Scientists: Greenland ice melting at a record rate. *Lawrence Journal World*, December 11. Available at http://www2.ljworld.com/news/2007/dec/11/scientists_greenland_ice_melting_record_rate/. Retrieved on December 11, 2007.

Bradsher, K. (2007a). China's green energy gap. *The New York Times*, October 24. Available at http://www.nytimes.com/2007/10/24/business/worldbusiness/24power.html. Retrieved on October 24, 2007.

Bradsher, K. (2007b). Trucks power China's economy, at a suffocating cost. *The New York Times*, December 8. Available at http://www.nytimes.com/interactive/2007/12/08/world/asia/choking_on_growth_7.html. Retrieved on December 9, 2008.

Bradsher, K. (2008a). An oil quandary: Costly fuel makes costly calories. *The New York Times*, January 19. Available at http://www.nytimes.com/2008/01/19/business/worldbusiness/19palmoil.html. Retrieved on January 19, 2008.

Bradsher, K. (2008b). With first car, a new life in China. *The New York Times*, April 28. Available at http://www.nytimes.com/2008/04/24/business/worldbusiness/24firstcar.html. Retrieved on April 28, 2008.

Bradsher, K. (2009a). Clash of subways and car culture in Chinese cities. *The New York Times*, March 27. Available at http://www.nytimes.com/2009/03/27/business/worldbusiness/27transit.html. Retrieved on March 27, 2009.

Bradsher, K. (2009b). China details incentive plan for electric cars. *The New York Times*, April 10. Available at http://www.nytimes.com/2009/04/11/business/energy-environment/11electric.html. Retrieved on April 11, 2009.

Bradsher, K. (2009c). China outpaces U.S. in cleaner coal-fired plants. *The New York Times*, May 11. Available at http://www.nytimes.com/2009/05/11/world/asia/11coal.html. Retrieved on May 10, 2009.

Bradsher, K., & Barboza, D. (2006). Pollution from Chinese coal casts a global shadow. *The New York Times*, February 23. Available at http://www.nytimes.com/2006/06/11/business/worldbusiness/11chinacoal.html. Retrieved on August 22, 2009.

Bradsher, K., & Martin, A. (2008). World's poor pay price as crop research cut. *The New York Times*, May 18. Available at http://www.nytimes.com/2008/05/18/business/worldbusiness/18focus.html. Retrieved on May 25, 2009.

Branigin, W., Eilperin, J., & Mufson, S. (2009). Obama announces new energy, environmental policies. *The Washington Post*, January 26. Available at http://www.washingtonpost.com/wp-dyn/content/article/2009/01/26/AR2009012601157_pf.html. Retrieved on Jan 26, 2009.

Brett, P. (2007). Global warming opens Arctic seabed to search for oil and gas. *New York Times*, October 30. Available at http://www.nytimes.com/bin/printfriendly.php?id = 8110116. Retrieved on November 10, 2007.

Broder, J. M., & Barringer, F. (2007). E.P.A. says states can't set emission rules. *The New York Times*, December 20. Available at http://www.nytimes.com/2007/12/20/washington/20epa.html. Retrieved on December 20, 2007.

Brower, M., & Leon, W. (1999). *The consumer's guide to effective environmental choices: Practical advice from the Union Of Concerned Scientists*. New York: Three Rivers Press.

Brown, M. E., & Funk, C. C. (2008). Food security under climate change. *Science*, *319*(February 1), 580–581.

Brown, O., & Crawford, A. (2009). *Climate change and security in Africa: A study for the Nordic-African foreign ministers meeting*, March. Available at http://www.iisd. Retrieved on April 1, 2009.

Brown, S. (2008). Beetle tree kill releases more carbon than fires. *NatureNews*, *452*(April 23). Available at http://www.nature.com/news/2008/080423/full/news.2008.771.html. Retrieved on April 23, 2009.

Brulle, R. J. (2000). *Agency, democracy, and nature: The U.S. environmental movement from a critical theory perspective*. Cambridge, MA: The MIT Press.

Buell, F. (2001). Globalization without environmental crisis: The divorce of two discourses in U.S. Culture. *Symploke*, *9*(1–2), 45–73.

Busby, J. W. (2007). Climate change and national security: An agenda for action. *Council on Foreign Relations*, November 30. Available at http://www.cfr.org/publication/14862). Retrieved on August 22, 2009.

Buttel, F. H. (1987). New directions in environmental sociology. *Annual Review of Sociology*, *13*, 465–488.

Buttel, F. H. (2002). Environmental sociology and the classical sociological tradition: Some observations on current controversies. In: R. E. Dunlap, F. H. Buttel, P. Dickins & A. Gijswijt (Eds), *Sociological theory and the environment: Classical foundations and contemporary insights* (pp. 35–50). Lanham, MD: Rowman & Littlefield Publishers.

Buttel, F. H. (2004). The treadmill of production: An appreciation, assessment, and agenda for research. *Organization & Environment*, *17*(3), 323–336.

Buttel, F. H., Dickins, P., Dunlap, R. E., & Gijswijt, A. (2002). Sociological theory and theory environment: An overview and introduction. In: R. E. Dunlap, F. H. Buttel, P. Dickins & A. Gijswijt (Eds), *Sociological theory and the environment: Classical foundations and contemporary insights* (pp. 3–32). Lanham, MD: Rowman & Littlefield Publishers.

Carson, R. (1962). *Silent spring*. Boston: Houghton Mifflin.

Casey, M. (2007). Global warming already taking a toll on world's plants and animals. *Lawrence Journal World*, December 6. Available at http://www2.ljworld.com/news/2007/dec/06/global_warming_already_taking_toll_worlds_plants_a/. Retrieved on December 7, 2007.

Castells, M. (1996). *The information age: Economy, society and culture. The rise of network society* (Vol. 1). Oxford, U.K. and Cambridge MA: Blackwell Publishers.

Catton, W. R., Jr. (1994). Foundations of human ecology. *Sociological Perspectives*, *37*(1), 75–95.

Catton, W. R., Jr. (2002). Has the Durkheim legacy misled sociology. In: R. E. Dunlap, F. H. Buttel, P. Dickins & A. Gijswijt (Eds), *Sociological theory and the environment: Classical foundations and contemporary insights* (pp. 90–115). Lanham, MD: Rowman & Littlefield Publishers.

Catton, W. R., Jr. (2008). A retrospective view of my development as an environmental sociologist. *Organization & Environment*, *21*(4), 471–477.

CCSP. (2008). *The effects of climate change on agriculture, land resources, water resources, and biodiversity in the United States*. U.S. Climate Change Program and the Subcommittee on Global Change Research (May). Available at http://www.climatescience.gov/Library/sap/sap4-3/final-report/sap4.3-final-all.pdf. Retrieved on June 1, 2008.

Cha, A. E. (2007). In China, a green awakening. *The Washington Post*, October 6. Available at http://www.washingtonpost.com/wp-dyn/content/article/2007/10/05/AR2007100502472.html. Retrieved on October 7, 2007.

Cha, A. E. (2008). Chinese consumers eager to excel at the American pastime. *The Washington Post*, November 15. Available at http://www.washingtonpost.com/wp-dyn/content/article/2008/11/14/AR2008111403648.html. Retrieved on November 15, 2008.

Chan, F., et al. (2008). Emergence of anoxia in the California current large marine ecosystem. *Science*, *319*(February 15), 920.

Chang, A. (2009). China: Up to 26 million migrants now jobless. *Deseret News*, January 4. Available at http://www.deseretnews.com/article/print/705282131/China-Up-to-26-million-migrants-now-jobless.html. Retrieved on January 17, 2009.

Chapman, D. (2008). Feds may be key to tapping Tennessee water. *The Atlanta Journal-Constitution*, March 27. Available at http://www6.lexisnexis.com/publisher/EndUser?Action = UserDisplayFullDocument&orgId = 574&topicId = 100020422&docId = l:765886525 &isSearch = true. Retrieved on April 24, 2008.

Charles, D. (2009). Corn-based ethanol flunks key test. *Science*, *324*(May 1), 587.

CIRES. (2007). Greenland melt accelerating, according to CU-Boulder study. *News & Events*, December 10. Available at http://www.colorado.edu/news/releases/2007/481.html. Retrieved on December 12, 2007.

Clark, B., & York, R. (2005). Carbon metabolism: Global capitalism, climate change, and the biospheric rift. *Theory and Society*, *34*, 391–428.

Clark, D. (2009). China's increasing carbon emissions blamed on manufacturing for West. *Guardian*, February 23. Available at http://www.guardian.co.uk/environment/2009/feb/23/china-carbon-emissions. Retrieved on February 24, 2009.

CNA. (2007). National security and the threat of climate change. *The CNA Corporation*, November 30. Available at http://securityandclimate.cna.org/report/National%20Security%20and%20the%20Threat%20of%20Climate%20Change.pdf. Retrieved on January 20, 2008.

Coleman, J. (2008). G-8 fails to agree on 2020 emissions goals. *The Washington Post*, May 27. Available at http://www.washingtonpost.com/wp-dyn/content/article/2008/05/26/AR2008052601927.html. Retrieved on August 22, 2008.

Collins, R. M. (2000). *More: The politics of economic growth in postwar America*. Oxford, GB, and New York: Oxford University Press.

Cook, J. (2008). Ga. officials take wrong tack on water. *The Atlanta Journal-Constitution*, April 22. Available at http://www.ajc.com/ajccars/content/printedition/2008/04/22/coosaed0422.html?cxntlid = inform_artr. Retrieved on April 23, 2008.

Costanza, R., & Daly, H. E. (1992). Natural capital and sustainable development. *Conservation Biology*, *6*(1), 37–46.

Cottrell, F. W. (1955). *Energy and society*. New York: McGraw Hill.

Crist, E. (2003). Limits-to-growth and the biodiversity crisis. *Wild Earth*, *13*(1), 62–65.

Crist, E. (2004). Against the social construction of nature and wilderness. *Environmental Ethics*, *26*(1), 5–24.

Crist, E. (2007). Beyond the climate crisis; A critique of climate change discourse. *Telos*, *141*, 29–55.

Czech, B. (2006). If Rome is burning, why are we fiddling? *Conservation Biology*, *20*(6), 1563–1565.

Daly, H. E. (1995). The irrationality of homo economicus (interview). *Developing Ideas Digest*, February 8. Available at http://www.iisd.org/didest/special/daly.html. Retrieved on October 20, 2008.

Daly, H. E. (1996). *Beyond growth: The economics of sustainable development*. Boston: Beacon Press.

Daly, H. E. (1999). Uneconomic growth: In theory, in fact, in history, and in relation to globalization. *Clemens Lecture Series 11* (Saint Johns University), October 25. Available at http://www.publicpolicy.umd.edu/faculty/daly/Dalypapers.html. Retrieved on October 20, 2008.

Daly, H. E. (2005). Economics in a full world. *Scientific American*, *293*(3), 100–107.

Davis, K. (1945). The world demographic transition. *Annals of the American Academy of Political and Social Science*, *237*(January), 1–11.

Davis, K. (1963). The Theory of change and response in modern demographic history. *Population Index*, *29*(4), 345–366.

Davis, M. (2006a). *Late Victorian holocausts: El Ñino famines and the making of the Third World*. London and New York: Verso.

Davis, M. (2006b). *Planet of slums*. London and New York: Verso.

Dean, C. (2008). Handle with care. *The New York Times*, August 18. Available at http://www.nytimes.com/2008/08/12/science/12ethics.html. Retrieved on August 23.

Demeritt, D. (2006). Science studies, climate change and prospects for constructivist critique. *Economy and Society*, *35*(3), 453–479.

Demuth, C. (2007) Think-tank confidential. *The Wall Street Journal*, October 11. Available at http://www.opinionjournal.com/extra/?id = 110010718. October 12, 2009.

Dewey, J. (1983a [1904]). The philosophical work of Herbert Spencer. In: J. A. Boydston (Ed.), *John Dewey: The middle works, 1899–1924* (Vol. 3, pp. 193–209). Carbondale and Edwardsville, IL: Southern Illinois University Press.

Dewey, J. (1983b [1909]). The influence of Darwinism on philosophy. In: J. A. Boydston (Ed.), *John Dewey: The middle works, 1899–1924* (Vol. 4, pp. 3–14). Carbondale and Edwardsville, IL: Southern Illinois University Press.

Dewey, J. (1985 [1916]). Progress. In: J. A. Boydston (Ed.), *John Dewey: The middle works, 1899–1924* (Vol. 10, pp. 234–243). Carbondale and Edwardsville, IL: Southern Illinois Press.
Dewey, J. (1988a [1925]). Experience and nature. In: J. A. Boydston (Ed.), *1925–1927 of John Dewey: The later works 1925–1953* (Vol. 1). Carbondale and Edwardsville, IL: Southern Illinois University Press.
Dewey, J. (1988b [1927]). The public and its problems. In: J. A. Boydston (Ed.), *1925–1927 of John Dewey: The later works 1925–1953* (Vol. 2, pp. 235–372). Carbondale and Edwardsville, I.: Southern Illinois University Press.
Dewey, J. (1988c [1929]). Individualism, old and new. In: J. A. Boydston (Ed.), *1929–1930 of John Dewey: The later works 1925–1953* (Vol. 5, pp. 41–123). Carbondale and Edwardsville, IL: Southern Illinois University Press.
Dewey, J. (1988d [1929]). The quest for certainty. In: J. A. Boydston (Ed.), *John Dewey: The later works, 1925–1953* (Vol. 4). Carbondale and Edwardsville, IL: Southern Illinois University Press.
Dewey, J. (1989 [1934]). Art as experience. In: J. A. Boydston (Ed.), *John Dewey: The later works, 1925–1953* (Vol. 10). Carbondale and Edwardsville, IL: Southern Illinois University Press.
Dewey, J., & Tufts, J. H. (1985 [1932]). Ethics. In: J. A. Boydston (Ed.), *John Dewey: The later works, 1925–1953* (Vol. 7). Carbondale and Edwardsville, IL: Southern Illinois Press.
Diamond, J. (2005). *Collapse: How societies choose to fail or succeed.* New York and London: Penguin Books.
Diaz, R. J., & Rosenberg, R. (2008). Spreading dead zones and consequences for marine ecosystems. *Science, 321*(August15), 926–929.
Dicken, P. (2007). *Global shift: Mapping the changing contours of the world economy* (5th ed). New York and London: The Guilford Press.
Dickens, P. (2002b). A green Marxism? Labor processes, alienation, and the division of labor. In: R. E. Dunlap, F. H. Buttel, P. Dickins & A. Gijswijt (Eds), *Sociological theory and the environment: Classical foundations and contemporary insights* (pp. 51–72). Lanham, MD: Rowman & Littlefield Publishers.
Dietz, T., Rosa, E. A., & York, R. (2007). Driving the human ecological footprint. *Frontiers in Ecology and the Environment, 5*(1), 13–18.
Dobkowski, M. N., & Wallimann, I. (2002). *On the edge of scarcity: Environment, resources, population, sustainability, and conflict*. Syracuse, NY: Syracuse University Press.
Donohue, M. (2008). Mall of misfortune. *The National*, June 12. Available at http://www.thenational.ae/article/20080612/REVIEW/206990272/1042. Retrieved on June 13, 2008.
Dunlap, R. E. (2002a). Environmental sociology: A personal perspective on its first quarter century. *Organization & Environment, 15*(1), 10–29.
Dunlap, R. E. (2002b). Paradigms, theories, and environmental sociology. In: R. E. Dunlap, F. H. Buttel, P. Dickins & A. Gijswijt (Eds), *Sociological theory and the environment: Classical foundations and contemporary insights* (pp. 329–347). Lanham, MD: Rowman & Littlefield Publishers.
Dunlap, R. E. (2008a). Climate-change views: Republican-democratic gaps expand. *Gallup*, May 29. Available at http://www.gallup.com/poll/107569/climatechange-views-republicandemocratic-gaps-expand.aspx. Retrieved on June 1, 2009.
Dunlap, R. E. (2008b). Promoting a paradigm change: Reflections on early contributions to environmental sociology. *Organization & Environment, 21*(4), 478–487.

Dunlap, R. E., Buttel, F. H., Dickins, P., & Gijswijt, A. (Eds). (2002). *Sociological theory and the environment: Classical foundations and contemporary insights)*. Lanham, MD: Rowman & Littlefield Publishers.

Dunlap, R. E., & Catton, W. R., Jr. (1979). Environmental sociology. *Annual Review of Sociology*, *5*, 243–273.

Dunlap, R. E., & Catton, W. R., Jr. (1994). Struggling with human exemptionalism: The rise, decline and revitalization of environmental sociology. *The American Sociologist*, *25*(1), 5–30.

Dunlop, I. T. (2007). APEC-Don't blow it: Good planets are hard to find. *The Futures Foundation*, September 8. Available at http://www.futuresfoundation.org.au/content/view/466/1/. Retrieved on September 20, 2007.

Dunphy, H. (2008). World Bank chief calls for immediate action on deepening global food Crisis. *The Washington Post*, April 14. Available at http://www.washingtonpost.com/wp-dyn/content/article/2008/04/13/AR2008041301972.html. Retrieved on April 14, 2008.

Durkheim, E. (1966 [1897]). *Suicide* (J. A. Spaulding & G. Simpson Trans.). New York: The Free Press.

Durkheim, E. (1953 [1898]). Individual and collective representations. In: *Sociology and philosophy* (G. Simpson, Trans., pp. 1–34). Glencoe, IL: The Free Press.

Durkheim, E. (1964 [1893]). *The division of labor in society* (G. Simpson, Trans.). New York: The Free Press.

Durkheim, E., & Mauss, M. (1963 [1903]). *Primitive classification* (R. Needham, Trans.). Chicago: The University of Chicago Press.

Dyson, F. (2008). The question of global warming. *The New York Review of Books*, June 12. Available at http://www.nybooks.com/articles/21494. Retrieved on June 20, 2008.

Ehrlich, P. R. (1968). *The population bomb*. New York: Ballantine Books.

EIA. (2008). International energy outlook 20008. *Energy Information Administration*, September. Available at http://www.eia.doe.gov/oiaf/ieo/pdf/0484(2008).pdf. Retrieved on September 10, 2008.

Eilperin, J. (2008a). Carbon output must near zero to avert danger, new studies say. *The Washington Post*, March 10. Available at http://www.washingtonpost.com/wp-dyn/content/article/2008/03/09/AR2008030901867.html. Retrieved on March 11, 2008.

Eilperin, J. (2008b). Report details effects of climate change across U.S. *The Washington Post*, May 28. Available at http://www.washingtonpost.com/wp-dyn/content/story/2008/05/27/ST2008052702928.html. Retrieved on May 28, 2008.

Eilperin, J. (2009). Warming trends alter conservation. *The Washington Post*, January 25. Available at http://www.washingtonpost.com/wp-dyn/content/story/2009/01/25/ST2009012500606.html. Retrieved on January 26, 2009.

Eldrich, N. (2001). The sixth extinction. *American Institute of Biological Sciences*. Available at http://www.actionbioscience.org/newfrontiers/eldredge2.html. Retrieved on January 20, 2008.

Emery, T. (2008). Georgians want access to Tenn. water. *The Tennessean*, February 8. Available at http://www.tennessean.com/apps/pbcs/articla?Aid = AID/200802/NEWS02210/8021804/&Temp.... Retrieved on February 8, 2008.

Engeler, E. (2009). Glaciers melting faster than thought. *Lawrence Journal World*, February 26. Available at http://www2.ljworld.com/news/2009/feb/26/glaciers-melting-faster-thought/. Retrieved on February 26, 2008.

Fan, M. (2008). Creating a car culture in China. *The Washington Post*, January 21. Available at http://www.washingtonpost.com/wp-dyn/content/article/2008/01/20/AR2008012002388.html. Retrieved on January 22, 2008.

FAO. (2008). Crop prospects and food situation (No. 2: April). *Food and Agriculture Organization of the United Nations*. Available at http://www.fao.org/docrep/010/ai465e/ai465e08.htm. Retrieved on May 1, 2008.

Fargione, J. et al. (2008) Land clearing and the biofuel debt. *Science Express*, February 7. Available at http://www.sciencemag.org/cgi/content/abstract/1152747. Retrieved on February 8, 2008.

Ferry, L. (1995). *The new ecological order*. Chicago & London: University of Chicago Press.

Flannery, T. (2005). *The weather makers: How man is changing the climate and what it means for life on earth*. New York: Atlantic Monthly Press.

Flitner, M., & Heins, V. (2002). Modernity and life politics: Conceptualizing the biodiversity crisis. *Political Geography*, *21*(2), 319–340.

Florida, R. (2009). How the crash will reshape America. *The Atlantic*, March. Available at http://www.theatlantic.com/doc/200903/meltdown-geography. Retrieved on April 6, 2009.

Foley, J., & Rosenbaum, A. (2008). China walks a fine line. *The New York Times*, December 22. Available at http://www.nytimes.com/2008/12/23/business/23views.html. Retrieved on December 22, 2008.

Foner, E. (1998). *The story of American freedom*. New York and London: W.W. Norton & Company.

Foreman, W. (2009). *Chinese officials grapple with slowdown in south* (February 27. Available at http://newshopper.sulekha.com/topic//economic-jitters/news/chinese-officials-grapple-with-slowdown-in-south.htm. Retrieved on February 27, 2009). Associated Press.

Foster, J. B. (1999). Marx's theory of metabolic rift: Classical foundations for environmental sociology. *American Journal of Sociology*, *105*(2), 366–405.

Foster, J. B. (2002). *Ecology against capitalism*. New York: Monthly Review Press.

Fouché, G. (2008). Svalbard's giant cold store. *Guardian*, February 26. Available at http://www.guardian.co.uk/environment/2008/feb/26/food.conservation?gusrc = rss&feed = science. Retrieved on August 22, 2009.

French, H. W. (2007). Far from Beijing's reach, officials bend energy rules. *The New York Times*, November 24. Available at http://www.nytimes.com/2007/11/24/world/asia/24evaders.html. Retrieved on November 24, 2007.

French, H. W. (2008). Dichotomies endure, but the pressure grows. *International Herald Tribune*, January 18. Available at http://www.nytimes.com/2008/01/18/world/asia/18iht-letter.1.9319660.html. Retrieved on January 19, 2008.

Friedman, B. M. (2005). *The moral consequences of economic growth*. New York: Alfred A. Knopf.

Friedman, T. L. (2000). *The Lexus and the olive tree*. New York: Anchor Books.

Friedman, T. L. (2005). *The world is flat: A brief history of the twenty-first century*. New York: Farrar, Strauss, and Giroux.

Friedman, T. L. (2008). *Hot, flat, and crowded*. New York: Farrar, Strauss, and Giroux.

Friedman, T. L. (2009). The inflection is near. *The New York Times*, March 7. Available at http://www.nytimes.com/2009/03/08/opinion/08friedman.html. Retrieved on March 8, 2009.

Frum, D. (2007). Honouring a panic-monger. *American Enterprise Institute for Policy Research*, October 15. Available at http://www.aei.org/article/26955. Retrieved on October 15, 2007.

Fukuyama, F. (1989). The end of history? *National Interest*, *16*(Summer), 3–18.

Fuller, T. (2008a). A highway that binds China and its neighbors. *The New York Times*, March 30. Available at http://www.nytimes.com/2008/03/30/world/asia/30iht-road.1.11530886.html. Retrieved on March 30, 2008.

Fuller, T. (2008b). Asian food crisis has political and civil implications. *The New York Times*, April 18. Available at http://www.nytimes.com/2008/04/18/world/asia/18iht-food.1.12130435.html. Retrieved on April 19, 2008.

Galbraith, J. K. (1958). *The affluent society*. Boston: Houghton Mifflin Company.

Garnaut, J. (2008). Developer works on future for Beijing's high-end housing market. *The New York Times*, April 24. Available at http://www.nytimes.com/2008/04/24/realestate/24iht-resoho.1.12305449.html. Retrieved on April 24, 2008.

Garnaut, R. (2008). *Garnaut climate change review: Interim report to the commonwealth, state and territory governments of Australia* (Executive summary, February). Available at http://www.garnautreport.org.au/CA25734E0016A131/WebObj/GarnautClimateChangeReviewInterimReport-Feb08/$File/Garnaut%20Climate%20Change%20Review%20Interim%20Report%20-%20Feb%2008.pdf. Retrieved on March 5, 2008.

GCP. (2008). Carbon budget 2007: An annual update of the global carbon budget and trends. *The Global Carbon Project*, September 26. Available at http://www.globalcarbonproject.org/carbonbudget/07/index.htm. Retrieved on October 1, 2008.

Georgia General Assembly. (2008). House Bill 1281 (As passed by the House and Senate). May 24. Available at http://www.legis.ga.gov/2007_88/sum/b281.tm. Retrieved on June 10, 2008.

Gertner, J. (2007). The future is drying up. *The New York Times*, October 21. Available at http://query.nytimes.com/gst/fullpage.html?res = 9C0CEFDA103CF932A15753C1A9619C8B63&sec = &spon = &pagewanted = 1. Retrieved on November 8, 2007.

GHF. (2009). *Climate change: The anatomy of a silent crisis (human impact report)*. Global Humanitarian Forum. Available at http://ghfgeneva.org/Portals/0/pdfs/human_impact_report.pdf. May 29. Retrieved on June 10, 20009.

Giddens, A. (1990). *The consequences of modernity*. Stanford, CA: Stanford University Press.

Giddens, A. (1999). *The third way: The renewal of social democracy*. Cambridge, UK, and Malden, MA: Polity Press.

Giddens, A. (2000). *Runaway world: How globalization is reshaping our lives*. New York: Routledge.

Gifford, R. (2007). China's mother river under threat. *NPR*, December 10. Available at http://www.npr.org/templates/story/story.php?storyId = 16918503. Retrieved on December 11, 2007.

Giridharadas, A. (2008). Four wheels for the masses: The $2,500 car. *The New York Times*, January 8. Available at http://www.nytimes.com/2008/01/08/business/worldbusiness/08indiacar.html. Retrieved on January 15, 2008.

Glasser, N. F., & Scampos, T. A. (2008). A structural glaciological analysis of the 2002 Larson B Ice Shelf collapse. *Journal of Glaciology*, *34*(184), 3–16.

Goodman, B. (2008a). Georgia loses federal case in a dispute about water. *The New York Times*, February 6. Available at http://www.nytimes.com/2008/02/06/us/06water.html. Retrieved on February 7, 2008.

Goodman, B. (2008b). Pollution is called a byproduct of a clean fuel. *The New York Times*, March 11. Available at http://www.nytimes.com/2008/03/11/us/11biofuel.html. Retrieved on March 12, 2008.

Gore, A. (1992). *Earth in the balance: Ecology and the human spirit*. Boston: Houghton Mifflin, & Company.

Gould, K. A., Pellow, D. N., & Schnaiberg, A. (2004). Interrogating the treadmill of production: Everything you wanted to know about the treadmill but were afraid to ask. *Organization & Environment*, *17*(3), 296–316.

Gow, D. (2007). China overtakes Germany as world's No 3 carmaker. *Guardian Unlimited*, July 20. Available at http://www.guardian.co.uk/business/2007/jul/20/china.germany. Retrieved on July 21, 2007.

Graham-Harrison, E. (2009). China's energy saying a silver lining to economic gloom. January 22. Available at http://www.reuters.com/article/latestCrisis/idUSPEK33394. Retrieved on January 23.

Gunter, L. (2008). Forget global warming: Welcome to the new ice age. *National Post*, February 25. Available at http://www.nationalpost.com/opinion/story.html?id = 332289. Retrieved on March 10, 2008.

GWA. (2008). Image: IPCC scenarios for 2100.png. *Global Warming Art*. Available at http://www.globalwarmingart.com/wiki/Image:IPCC_Scenarios_for_2100_png. Retrieved on January, 20 2009.

Hagg, A. L. (2008). What's next for the IPCC? *Nature Reports*, *2*(January), 4–6.

Hajer, M. A. (1995). *The politics of environmental discourse: Ecological modernization and the policy process*. New York: Oxford University Press.

Halpern, B. S., et al. (2008). A global map of human impact on marine ecosystems. *Science*, *319*(February 15), 948–952.

Hamm, S. (2008). Young and impatient in India. *BusinessWeek*, January 17. Available at http://www.businessweek.com/magazine/content/08_04/b4068045084263.htm. Retrieved on January 18, 2008.

Hance, J. (2009). Global warming estimates double in severity according to new MIT modeling. *Mongabay.com*, May 20. Available at http://news.mongabay.com/2009/0520-hance_doublewarming.html. Retrieved on May 20, 2009.

Hand, E. (2008). Antarctic ice loss speeding up. *Nature News*, January 13. Available at http://www.nature.com/news/2008/080113/full/news.2008.438.html. Retrieved on January 15, 2008

Hardin, G. (1968). The tragedy of the commons. *Science*, *162*(December), 1243–1248.

Harvey, D. (1989). *The condition of postmodernity: An enquiry into the origins of cultural change*. Oxford, UK, and Cambridge, MA: Blackwell Publishers.

Harvey, D. (2005). *A brief history of neoliberalism*. Oxford and New York: Oxford University Press.

Harvey, D. (2006). *Spaces of global capitalism*. London: Verso.

Hayward, S. F. (2007). Environmental Gore. *American Enterprise Institute for Policy Research*, October 15. Available at http://jobs.aei.org/publications/pubID.26954,filter.all/pub_detail.asp. Retrieved on October 15, 2007.

Heilprin, J. (2008). *U.N.: Climate change may cost $20 trillion* (January 31. Available at http://www.commondreams.org/archive/2008/01/31/6751. Retrieved on February 1, 2008). Associated Press.

Held, D., & McGrew, A. (2007a). *Globalization/antiglobalization: Beyond the great divide* (2nd ed.). Cambridge, UK, and Malden MA: Polity Press.

Held, D., & McGrew, A. (Eds). (2007b). *Globalization theory: Approaches and controversies*. Cambridge, UK, and Malden MA: Polity Press.

Held, D., McGrew, A., Goldblatt, D., & Perrraton, J. (1999). *Global transformations: Politics, economics, and culture*. Stanford, CA: Stanford University Press.

HMT. (2006). Stern review: The economics of climate change (executive summary). *HM Treasury*, October 30. Available at http://www.hm-treasury.gov.uk/d/Executive_Summary.pdf. Retrieved on August 22, 2009.

Hoag, H. (2008a). Rapid ice retreat threatens Arctic interior. *Nature Reports*, *2*(June 26), 83–84.

Hoag, H. (2008b). The missing greenhouse gas. *Nature Reports*, *2*(July 10), 99–100.

Hoegh-Guldberg, O., et al. (2007). Coral reefs under rapid climate change and ocean acidification. *Science*, *318*(December 14), 1737–1742.

Hohn, D. (2008). Sea of trash. *The New York Times*, June 22. Available at http://www.nytimes.com/2008/06/22/magazine/22Plastics-t.html. Retrieved on June 22, 2008.

Hooker, J. (2008). Toll rises in China quake. *The New York Times*, May 26. Available at http://www.nytimes.com/2008/05/26/world/asia/26quake.html. Retrieved on May 26, 2008.

Hopkin, M. (2007). Rising temperatures will stunt rainforest growth. *Nature News*, August 10. Available at http://www.nature.com/news/2007/070810/full/news070806-13.html. Retrieved on August 10, 2008.

ICSU. (2009). Ice declining faster than expected in both Arctic and Antarctic Glaciers. *ScienceDaily*, February 26. Available at http://www.sciencedaily.com/releases/2009/02/090225073215.htm. Retrieved on February 27, 2009.

Inglehart, R., & Welzel, C. (2005). *Modernization, cultural change, and democracy: The human development sequence*. Cambridge, UK, and New York: Cambridge University Press.

Inman, M. (2009). Where warming hits hard. *Nature Reports*, *3*(January 15). Available at http://www.nature.com/climate/2009/0902/full/climate.2009.3.html. Retrieved on January 16, 2009.

IPCC. (2000). *Emissions scenarios* (Summary for policymakers) (Special Report of Working Group III). Intergovernmental Panel on Climate Change. Available at http://www.ipcc.ch/pdf/special-reports/spm/sres-en.pdf. Retrieved on August 10, 2008.

IPCC. (2007a). *Fourth assessment report: Climate change 2007: Synthesis report*. Intergovernmental Panel on Climate Change. Available at http://www.ipcc.ch/publications_and_data/publications_ipcc_fourth_assessment_report_synthesis_report.htm. Retrieved on August 10, 2008.

IPCC. (2007b). *Fourth assessment report: Climate change 2007: Working group I report: The physical science basis*. Intergovernmental Panel on Climate Change. Available at http://www.ipcc.ch/publications_and_data/publications_ipcc_fourth_assessment_report_wg1_report_the_physical_science_basis.htm. Retrieved on August 10, 2008.

IUCN. (2007). Extinction crisis escalates: Red List shows apes, corals, vultures, dolphins all in danger. *International Union for Conservation of Nature and Natural Resources*, September 12. Available at http://www.iucn.org/es/medios/comunicados/?81/Extinction-crisis-escalates-Red-List-shows-apes-corals-vultures-dolphins-all-in-danger. Retrieved on September 13, 2007.

IUCN. (2008). The IUNC Red List threatened species. *International Union for Conservation of Nature and Natural Resources*. Available at http://www.iucnredlist.org/. Retrieved on November 1, 2008.

Ivins, E. R. (2009). Ice sheet stability and sea level. *Science*, *324*(May 15), 888–889.

Jacobs, A. (2008). Beijing stops construction for Olympics. *The New York Times*, April 28. Available at http://www.nytimes.com/2008/04/15/world/asia/15china.html. Retrieved on April 28, 2008.

Jagoda, B. (2008). New UC analysis shows alarming increase in expected growth of China's carbon dioxide emissions. March 10. Available at http://ucsdnews.ucsd.edu/newsrel/international/03-08ChinasCarbonDioxideEmissions.asp. Retrieved on March 12, 2008.

Jenkins, J. C., & Eckert, C. M. (2000). The right turn in economic policy: Business elites and the new conservative economics. *Sociological Forum*, *13*(2), 307–338.

Jiang, J., & Shen, S. (2007). Sales soar for Wal-Mart and Carrefour in China. *International Herald Tribune*, March 27. Available at http://www.iht.com/bin/print.php?id = 5043495. Retrieved on March 28, 2007.

Jones, J. M. (2009). In U.S., outlook for environmental quality improving. *Gallup Poll*, April 21. Available at http://www.gallup.com/poll/117769/outlook-environmental-quality-improving.aspx. Retrieved on April 22, 2008.

Jowit, J. (2008). Melting glaciers start countdown to climate chaos. *The Observer*, March 16. Available at http://www.guardian.co.uk/environment/2008/mar/16/glaciers.climate-change. Retrieved on March 17, 2008.

Kahn, C. (2007a). *Huge water park planned for Arizona desert* (November 20. Available at http://www.waveyard.com/ap_111907.pdf. Retrieved on November 21). Associated Press.

Kahn, J. (2007b). In China, a lakes champion imperils himself. *The New York Times*, October 14. Available at http://www.nytimes.com/2007/10/14/world/asia/14china.html. Retrieved on November 6, 2007.

Kahn, J., & Landler, M. (2007). China grabs west's smoke-spewing factories. *The New York Times*, December 21. Available at http://www.nytimes.com/2007/12/21/world/asia/21transfer.html. Retrieved on December 21, 2007.

Kahn, J., & Yardley, J. (2007). As China roars, pollution reaches deadly extremes. *The New York Times*, August 26. Available at http://www.nytimes.com/2007/08/26/world/asia/26china.html. Retrieved on August 27, 2007.

Kanter, J. (2009). World's cheapest car: Boom or bust. *The New York Times*, March 23. Available at http://greeninc.blogs.nytimes.com/2009/03/23/worlds-cheapest-car-boon-or-bane/. March 23, 2009.

Kaufman, M. (2008a). Escalating ice loss found in Antarctica. *The Washington Post*, January 14. Available at http://www.washingtonpost.com/wp-dyn/content/article/2008/01/13/AR2008011302753.html. Retrieved on January 14, 2008.

Kaufman, M. (2008b). Decline in snowpack is blamed on warming. *The Washington Post*, February 1. http://www.washingtonpost.com/wp-dyn/content/article/2008/01/31/AR2008013101868.html. Retrieved on February 1, 2008.

Kerian, J. (2008). Yellow science. *The Wall Street Journal*, June 25. Available at http://online.wsj.com/article/SB121433436381900681.html. Retrieved on June 25, 2008.

Ki-Moon, B. (2007). Secretary General's address to the IPCC upon release of the fourth assessment synthesis report (November 17) United Nations Environmental Programme. Available at http://www.unep.org/Documents.Multilingual/Default.asp?DocumentID = 522&ArticleID = 5708&l = en. Retrieved on December 10, 2007.

Kintisch, E. (2007). Giving climate change a kick. *ScienceNOW Daily News*, November 9. Available at http://sciencenow.sciencemag.org/cgi/content/full/2007/1109/1. Retrieved on November 10, 2007.

Kintisch, E. (2008). Giving Earth an umbrella. *ScienceNOW Daily News*, February 22. Available at http://sciencenow.sciencemag.org/cgi/content/full/2008/222/1. Retrieved on February 23, 2008.

Kleiner, K. (2008). The backlash against biofuels. *Nature Reports*, *2*(January), 9–11.

Kleiner, K. (2009). Peak energy: promise or peril? *Nature Reports Climate Change*, February 19. Available at http://www.nature.com/climate/2009/0903/full/climate.2009.19.html. Retrieved on February 20, 2009.

Krugman, P. (2007). Gore derangement syndrome. *The New York Times*, October 15. Available at http://www.nytimes.com/2007/10/15/opinion/15krugman.html. Retrieved on October 16, 2007.

Krugman, P. (2009). Empire of carbon. *The New York Times*, May 15. Available at http://www.nytimes.com/2009/05/15/opinion/15krugman.html. Retrieved on May 15, 2009.

Kurtenbach, E. (2009). Reports: China auto sales surpass US for 3rd month. *Seattle Times*, April 8. Available at http://seattletimes.nwsource.com/html/businesstechnology/2009005161_apaschinaautosales.html. Retrieved on August 20, 2009.

Kurz, W. A., et al. (2008). Mountain pine beetle and forest carbon feedback to climate change. *Nature*, *452*(April 24), 987–990.

Kuznets, S. (1955). Economic growth and income inequality. *The American Economic Review*, *45*(1), 1–28.

Lacey, M. (2008). Across globe, empty bellies bring rising anger. *The New York Times*, April 18. Available at http://www.nytimes.com/2008/04/18/world/americas/18food.html. Retrieved on April 18, 2008.

Landler, M. (2009). Clinton paints China policy with new green hue. *The New York Times*, February 22. Available at http://query.nytimes.com/gst/fullpage.html?res = 980DE6DC113AF931A15751C0A96F9C8B63&scp = 4&sq = In%20China,%20Clinton%20focuses%20on%20climate&st = cse. Retrieved on February 24, 2009.

Larson, C. (2008). On Chinese water project, a struggle over sound science. *Yale Environment, 360*. Available at http://e360.yale.edu/content/feature.msp?id = 2103. Retrieved on January 20, 2009.

Laurence, W. F. (2006). Have we overstated the tropical biodiversity crisis? *ScienceDirect*, September 29. Available at http://striweb.si.edu/tropical_extinction/Laurance2007biodiversitycrisis.pdf. Retrieved on December 20, 2008.

Laurence, W. F. (2008). The need to cut China's illegal timber imports? *Science*, *319*(February 29), 1184.

Leake, J. (2007). 2007: Worst year for the environment. *The Times*, December 14. Available at http://www.timesonline.co.uk/tol/news/environment/article3050718.ece. Retrieved on December 15, 2007.

Lee, M. (2009). US seeking to limit Antarctic travel: Obama pushing to protect fragile environment. *The Boston Globe*, April 6. Available at http://www.boston.com/news/nation/washington/articles/2009/04/06/us_seeking_to_limit_antarctica_tourism/. Retrieved on April 10, 2009.

Lewis, L. T., & Humphrey, C. R. (2005). Sociology and the environment: An analysis of coverage in introductory sociology textbooks. *Teaching Sociology*, *33*(2), 154–169.

Liu, J., et al. (2007). Complexity of coupled and natural systems. *Science*, *317*(September 14), 1513–1517.

Ljunggren, D. (2008). Massive Arctic ice shelf breaks away. September 3. Available at http://www.reuters.com/article/environmentNews/idUSN0246073320080903. Retrieved on September 4, 2008.

Lobell, D. B., et al. (2008). Prioritizing climate change adaptation needs for food security in 2030. *Science, 319*(February 1), 607–610.

Longfellow, B. (2007). Weather report: Images from the climate crisis. In: L. Panitch & C. Leys (Eds), *Coming to terms with nature: Socialist register 2007* (pp. 1–15). New York: Monthly Review Press.

Lyall, S. (2007). As temperatures rise, a greening of Greenland. *New York Times*, October 28. Available at http://www.nytimes.com/2007/10/28/world/europe/28iht-greenland.5.8086142.html. Retrieved on October 29, 2007.

Lyderson, K. (2009). Scientists: Pace of climate change increases. *The Washington Post*, February 15. Available at http://www.washingtonpost.com

Macur, J. (2008). Olympians prepare for battle with Beijing's smog. *The New York Times*, January 24. Available at http://www.nytimes.com/2008/01/24/sports/24iht-24mask.9461918.html. Retrieved on January 25, 2008.

Marcuse, H. (1964). *One-dimensional man: Studies in the ideology of advanced industrial society*. Boston: Beacon Press.

Martin, A. (2008). Biofeuls getting the blame for high fuel prices. *New York Times*, April 15. Available at http://www.nytimes.com/2008/04/15/business/worldbusiness/15iht-15food.11988995.html. Retrieved on April 15, 2008.

Martinez-Alier, J. (1987). *Ecological economics: Energy, environment and society*. Oxford, UK, and New York: Basil Blackwell.

Mascarelli, A. L. (2009). What we've learned in 2008. *Nature Reports, 3*(January), 4–6.

Matthews, H. D., & Caldeira, K. (2008). Stabilizing climate requires near-zero emissions. *Geophysical Research Letters, 35*(February 27), L04705.

McConnico, J. (2007). Climate change alters way of life for Greenland's Inuit. *New York Times*, August 26. Available at http://www.nytimes.com/2007/08/26/world/europe/26iht-climate.1.7256431.html. Retrieved on August 26, 2007.

McCright, A. M., & Dunlap, R. E. (2000). Challenging global warming as a social problem: An analysis of the conservative movement's counter-claims. *Social Problems, 47*(4), 499–522.

McDonald, H. P. (2004). *John Dewey and environmental philosophy and ethics*. Albany: State University of New York Press.

McKibben, B. (2006 [1989]). *The end of nature*. New York: Random House.

McKinley, J. (2009). Drought adds to hardships in California. *The New York Times*, February 22. Available at http://www.nytimes.com/2009/02/22/us/22mendota.html. Retrieved on February 22.

McKinsey Global Institute. (2006). From made in China to sold in China: The rise of the Chinese urban consumer (executive summary) (November). Available at http://www.mckinsey.com/mgi/publications/china_consumer/index.asp. Retrieved on December 1, 2008.

McKinsey Global Institute. (2007). The bird of gold: The rise of India's consumer market (executive summary) (May). Available at http://www.mckinsey.com/mgi/publications/india_consumer_market/executive_summary.asp. Retrieved on December 1, 2008.

McKinsey Global Institute. (2008). Preparing for China's urban billion (March 23). Available at http://www.mckinsey.com/mgi/publications/china_urban_summary_of_findings.asp. Retrieved on March 24, 2008.

McMichael, P. (2007). *Development and social change: A global perspective* (4th ed.). Thousand Oaks, CA: Pine Forge Press.

Mead, G. H. (1962 [1934]). *Mind, self and society*. Chicago: University of Chicago Press.

Mead, G. H. (1964a [1927]). The objective reality of perspectives. In: A. J. Reck (Ed.), *Selected Writings* (pp. 306–319). Indianapolis: Bobbs-Merrill Company.

Mead, G. H. (1964b [1929–1930]). The philosophies of Royce, James, and Dewey in their American setting. In: A. J. Reck (Ed.), *Selected Writings* (pp. 371–391). Indianapolis: Bobbs-Merrill Company.

Meadows, D., Randers, J., & Meadows, D. (2004). *Limits to growth: The thirty year update*. White River Junction, VT: Chelsea Green Publishing Company.

Meadows, D. H., Meadows, D. L., Randers, J., & Behrans, W. W., III. (1972). *The limits to growth: A report for the Club of Rome's project on the predicament of mankind*. New York: Universe Books.

Meilke, J. (2008). Researchers chart fuel-rich areas in Arctic for first time. *Guardian*, August 6. Available at http://www.guardian.co.uk/global/2008/aug/06/russia. Retrieved on August 6, 2008.

Michaels, P. J. (2007). Gore's Nobel challenge. October 12. Available at http://www.cato.org/pub_display.php?pub_id = 8744. Retrieved on November 20, 2007.

Micklethwait, J., & Wooldridge, A. (2000). *A future perfect: The essentials of globalization*. New York: Crown Business.

Miller, H. I. Dr. (2007). Gore and peace. *TCSDAILY*, October 15. Available at http://www.tcsdaily.com/article.aspx?id = 101507D. Retrieved on November 20, 2007.

Mills, C. W. (1961). *The sociological imagination*. New York: Grove Press.

Milly, P. C. D., et al. (2008). Stationarity is dead: Whither water management? *Science*, *319*(February 1), 573–574.

Mol, A. P. J. (2001). *Globalization and environmental reform: The ecological modernization of the global economy*. Cambridge, MA: MIT Press.

Mol, A. P. J. (2006). From environmental sociologies to environmental sociology?: A comparison of U.S. and European environmental sociology. *Organization & Environment*, *19*(1), 5–27.

Monaghan, A. J., et al. (2008). Twentieth century Antarctic air temperature and snowfall simulations by IPCC climate models. *Geophysical Research Letters*, *35*(April 5), L07502.

Moore, M. (2008). China to become world's largest investor in green energy. *Telegraph*, September 17. Available at http://www.telegraph.co.uk/earth/earthnews/3351835/China-to-become-worlds-largest-investor-in-green-energy.html. Retrieved on September 18, 2008.

Mote, T. L. (2007). Greenland surface melt trends 1973–2007: Evidence of large increase in 2007. *Geophysical Research Letters*, *34*(November 30), L22507. Available at http://www.agu.org/pubs/crossref/2007/2007GL031976.shtml. Retrieved on December 15, 2007.

Mouawad, J. (2007). Rising demand for oil provokes new energy crisis. *The New York Times*, November 9. Available at http://www.nytimes.com/2007/11/09/business/worldbusiness/09oil.html. Retrieved on November 10, 2007.

Mouawad, J. (2009). Oil companies loath to follow Obama's green lead. *The New York Times*, April 8. Available at http://www.nytimes.com/2009/04/08/business/energy-environment/08greenoil.html?fta = y. Retrieved on April 9, 2009.

Mrasek, V. (2009). Grim new report: China's greenhouse gas emissions threaten to double. *Spiegel*, March 6. Available at http://www.spiegel.de/international/world/0,1518,611818,00.html. Retrieved on March 7, 2009.

Mufson, S. (2009). U.S. carbon emissions fall most since '82: Recession, oil prices in 2008 decline. *The Washington Post*, May 21. Available at http://www.washingtonpost.com/wp-dyn/content/article/2009/05/20/AR2009052003655.html. Retrieved on May 25, 2009.

Murphy, R. (1994). *Rationality and nature: A sociological inquiry into a changing relationship.* Boulder, CO: Westview.

Murphy, R. (1995). Sociology as if nature did not matter: An ecological critique. *The British Journal of Sociology*, *46*(4), 688–707.

Murphy, R. (2002). Ecological materialism and the sociology of Max Weber. In: R. E. Dunlap, F. H. Buttel, P. Dickins & A. Gijswijt (Eds), *Sociological theory and the environment: Classical foundations and contemporary insights* (pp. 73–89). Lanham, MD: Rowman & Littlefield Publishers.

Murray, J. (2008). IEA calls for 45 trillion energy "revolution." *Bussinesgreen*, June 6. Available at http://www.businessgreen.com/business-green/news/2218500/iea-calls-trillion-energy. Retrieved on August 20, 2009.

Nagel, J., Dietz, T., & Broadbent, J. (2009). *Workshop on sociological perspectives on climate change* (May 30–31, 2008). Sociology Program (Directorate of Social, Behavioral, and Economic Sciences), National Science Foundation. Available at http://ireswb.cc.ku.edu/~crgc/NSFWorkshop/Readings/NSF_WkspReport_09.pdf.

Najam, A., Runnalls, D., & Halle, M. (2007). *Environment and globalization: Five propositions* (International Institute of Sustainable Development). Available at http://www.iisd.org/pdf/2007/trade_environment_globalization.pdf. Retrieved on January 25, 2008.

NEAA. (2007). China now no. 1 in CO_2 emissions; USA in second position. *Netherlands Environmental Assessment Agency*, June 19. Available at http://www.pbl.nl/en/news/pressreleases/2007/20070619Chinanowno1inCO2emissionsUSAinsecondposition.html. June 20, 2007.

NEAA. (2008). China contributing two thirds to increase in CO_2 emissions. *Netherlands Environmental Assessment Agency*, June 13. Available at http://www.pbl.nl/en/news/pressreleases/2008/20080613ChinacontributingtwothirdstoincreaseinCO2emissions.html. Retrieved on June 14, 2008.

Nelson, G. (2009). Economy stalls Waveland project but developers still optimistic. *The Arizona Republic*, January 4. Available at http://www.azcentral.com/arizonarepublic/business/articles/2009/01/04/20090104biz-mr-waveyard0104.html. Retrieved on February 20, 2009.

New York Times. (2007). Disappointments of climate (Editorial) (December 17). Available at http://www.nytimes.com/2007/12/17/opinion/17mon1.html. Retrieved on December 17, 2007).

New York Times. (2009). New day on climate change (Editorial) (January 27). Available at http://www.nytimes.com/2009/01/27/opinion/27tue1.html. Retrieved on January 27.

Newport, F. (2009). Americans: Economy takes precedence over environment. *Gallup Poll*, March 19. Available at http://www.gallup.com/poll/116962/americans-economy-takes-precedence-environment.aspx. Retrieved on March 20, 2009.

NIC. (2008). *Global trends 2025*. National Intelligence Council (November). Available at http://www.dni.gov/nic/PDF_2025/2025_Global_Trends_Final_Report.pdf. Retrieved on December 1, 2008.

NOAA. (2008). *Arctic report card 2008*. National Oceanic and Atmospheric Association (October). Available at http://www.arctic.noaa.gov/reportcard/. November 1, 2008.

NOAA. (2009). New study shows climate change largely irreversible. *National Oceanic and Atmospheric Association*, January 26. Available at http://www.noaanews.noaa.gov/stories2009/20090126_climate.html. Retrieved on January 27, 2009.

Nordhaus, W. (2008). *A question of balance: Weighing the options on global warming policies.* New Haven and London: Yale University Press.

Nordlinger, J. (2007) Rules of the game, & c. *National Review*, October 15. Available at http://article.nationalreview.com/?q = ZDUwZjczYTA2Yjk2ODM0Nzg4MTlhYTNjNzBlZGQwNmE = . Retrieved on November 20, 2008.

Normile, D. (2008). China's living laboratory in urbanization. *Science, 319*(February 8), 740–743.

NSIDC. (2007). Arctic sea ice shatters all previous record lows. *National Snow and Ice Data Center*, October 1. Available at http://nsidc.org/news/press/2007_seaiceminimum/20071001_pressrelease.html. Retrieved on October 2, 2007.

NSIDC. (2008a). Antarctic ice shelf disintegration underscores a warming world. *National Snow and Ice Data Center*, March 25. Available at http://nsidc.org/news/press/20080325_Wilkins.html. Retrieved on April 7, 2008.

NSIDC. (2008b). Arctic sea ice down to second-lowest extent; Likely record-low volume. *National Snow and Ice Data Center*, October 2. Available at http://nsidc.org/news/press/20081002_seaice_pressrelease.html. Retrieved on October 6, 2008.

NSIDC. (2009a). Arctic sea ice younger, thinner as melt season begins. *National Snow and Ice Data Center*, April, 6. Available at http://nsidc.org/news/press/20090406_seaice.html. Retrieved on April 7, 2009

NSIDC. (2009b). Media advisory: Ice bridge supporting Wilkens Ice Sheet collapses. *National Snow and Ice Data Center*, April 8. Available at http://nsidc.org/news/press/20090408_Wilkins.html. Retrieved on April 9, 2009.

NSTC. (2008). Scientific assessment of the effects of global change on the United States: A report of the Committee on Environment and Natural Resources. National Science and Technology Council (May). Available at http://www.climatescience.gov/Library/scientific-assessment/Scientific-AssessmentFINAL.pdf. Retrieved on June 20, 2008.

Oleskyn, V. (2009). What a mess! Experts ponder space junk problem. *Lawrence Journal World*, February 20. Available at http://www2.ljworld.com/news/2009/feb/20/what-mess-experts-ponder-space-junk-problem/. Retrieved on February 20.

Ouroussoff, N. (2008). The new, new city. *The New York Times*, June 8. Available at http://www.nytimes.com/2008/06/08/magazine/08shenzhen-t.html. Retrieved on June 8, 2008.

Overbye, D. (2009). Elevating science, elevating democracy. *New York Times*, January 26. Available at http://www.nytimes.com/2009/01/27/science/27essa.html. Retrieved on January 26, 2009.

O'Connor, J. (1998). *Natural causes: Essays in ecological Marxism*. New York and London: The Guilford Press.

O'Neil, P. (2008). Efforts to support global climate-change falls: Poll. *Canada.com*, November 27. Available at http://www.canada.com/windsorstar/news/story.html?id = f0a1687c-decd-4c72-9d0e-7e6dd92d4ebe. Retrieved on December 1, 2008.

Pachauri, R. (2008). IPCC chairman address to World Economic Forum. January 23. Available at http://www.ipcc.ch/graphics/speeches/pachauri-davos-january-2008.pdf. Retrieved on February 1, 2008.

Parsons, T. (1965). Evolutionary universals in society. *American Sociological Review*, *29*(3), 339–357.

Parsons, T. (1971). *The system of modern societies*. Englewood Cliffs, NJ: Prentice-Hall.

Partlow, J. (2009). A protected forest's fast decline. *The Washington Post*, February 6. Available at http://www.washingtonpost.com/wp-dyn/content/story/2009/02/06/ST2009020600778.html. Retrieved on February 6, 2009.

Pearce, F. (2007). *The last generation: How nature will take revenge for climate change*. Toronto: Key Porter Books.

Pennisi, E. (2009). Western forests suffer death by degrees. *Science*, *323*(January 23), 447.

Pepper, D. (2007). Dead rivers and raw sewage: Choking on pollution in India. *Spiegel*, July 6. Available at http://www.spiegel.de/international/world/0,1518,493033,00.html. Retrieved on July 7, 2007.

Perdue, G. S. (2007). Governor seeks injunction to protect Georgia's water resources. *Office of The Governor*, October 17. Available at http://gov.georgia.gov/00/press/detail/0,2668,78006749_96092834_96286007,00.html. Retrieved on November 13, 2007.

Pew Center. (2007a). Global unease with major world powers: rising environmental concern in 47-nation survey. *Global Attitudes Project*, June 27. Available at http://pewglobal.org/reports/display.php?ReportID = 256. Retrieved on June 27, 2007.

Pew Center. (2007b). World publics welcome global trade – but not immigration. *Global Attitudes Project*, October 4. Available at http://pewglobal.org/reports/pdf/258.pdf. Retrieved on October 4, 2007.

Pew Center. (2008a). Election year economic ratings lowest since '92. January 24. Available at http://pewresearch.org/pubs/704/election-year-economic-ratings-lowest-since-92. Retrieved on January 24, 2008.

Pew Center. (2008b). An increase in GOP doubt in global warming deepens partisan divide. May 8. Available at http://pewresearch.org/pubs/828/global-warming. Retrieved on May 8, 2008.

Pew Center. (2009). Economy, jobs trump all other policy priorities in 2009. January 25. Available at http://people-press.org/report/485/economy-top-policy-priority. Retrieved on January 25, 2009.

Phillips, K. (2002). *Wealth and democracy*. New York: Broadway Books.

Phillips, O. L., et al. (2009). Drought sensitivity of the Amazon rainforest. *Science*, *323* (March 6), 1344–1347.

Pielke, R., Jr., Wigley, T., & Green, C. (2008). Dangerous assumptions. *Nature*, *452*(April 3), 531–532.

Plafker, T. (2009). China finds no immunity from global crisis. *The New York Times*, January 27. Available at http://www.nytimes.com/2009/01/25/business/worldbusiness/25iht-rdavchina.1-413512.html. Retrieved on January 27.

Powell, B. (2008). The short march. *Time*, February 14. Available at http://www.time.com/time/magazine/article/0,9171,1713336,00.html. Retrieved on March 1, 2008.

Radkau, J. (2008 [2002]). *Nature and power: A global history of the environment* (T. Dunlap Trans.). Cambridge, UK, and New York: German Historical Institute and Cambridge University Press.

Radkau, J. (2009 [2005]). *Max Weber* (P. Camiller, Trans.). Cambridge, UK, and Malden, MA: Polity Press.

Ramanathan, V., & Carmichael, G. (2008). Global and regional climate changes due to black carbon. *Nature Geoscience*, *1*(3), 221–227.

Redmond, J. (2008). Buford OK's water park despite drought. *The Atlanta Journal-Constitution*, May 4. Available at http://www.ajc.com/ajccars/content/metro/stories/2008/05/04/metwaterpark.html?cxntlid = inform_artr. Retrieved on August 22, 2009.

Reed, C. (2007). Sinking carbon-sink hopes? *Harvard Magazine*, *110*(1), 38–39.

Reich, R. (2007). *Supercapitalism*. New York: Alfred A. Knopf.

Reuters. (2007). Investors beware as China charts consumer boom. March 23. Available at http://www.reuters.com/article/businessNews/idUSPEK15571720070323. Retrieved on March 24, 2007.

Revkin, A. C. (2007). The Arctic's alarming sea change. *The New York Times*, October 2. Available at http://www.nytimes.com/2007/10/02/world/americas/02iht-environ.1.7713707.html. Retrieved on. December 11, 2007.

Revkin, A. C. (2008a). In Greenland, ice and instability. *The New York Times*, January 8. Available at http://www.nytimes.com/2008/01/08/science/earth/08gree.html?_r = 1. Retrieved on January 10, 2008.

Revkin, A. C. (2008b). Skeptics on climate seize on cold spell. *The New York Times*, March 2. Available at http://www.uvm.edu/cosmolab/papers/Revkin_2008_3888.pdf. Retrieved on March 2, 2008.

Revkin, A. C. (2008c). Under pressure, White House issues climate change report. *The New York Times*, May 30. Available at http://www.nytimes.com/2008/05/30/washington/30climate.html. Retrieved on June 4, 2008.

Revkin, A. C. (2008d). Warmth opens Arctic routes, scientists say. *New York Times*, September 7. Available at http://www.nytimes.com/2008/09/07/health/07iht-arctic.1.15942381.html. Retrieved on September 7, 2008.

Revkin, A. C. (2008e). CO_2 flow speeds up: Poor countries in lead. *New York Times*, September 26. Available at http://dotearth.blogs.nytimes.com/2008/09/26/co2-flow-speeds-up-poor-countries-now-lead/. Retrieved on September 26, 2008.

Revkin, A. C. (2009a). China's power surge ends (for now). *New York Times*, January 6. Available at http://dotearth.blogs.nytimes.com/2009/01/06/chinas-power-surge-ends-for-now/. Retrieved on January 6, 2009.

Revkin, A. C. (2009b). Science advisor lays out climate and energy plans. *New York Times*, April 9. Available at http://dotearth.blogs.nytimes.com/2009/04/09/science-adviser-lists-goals-on-climate-energy/. Retrieved on April 10, 2009.

Rifkin, J. (2000). *The age of access: The new culture of hypercapitalism: Where all of life is a paid-for experience*. New York: Jeremy P. Tarcher/Putnam Book.

Rignot, E., et al. (2008). Recent Antarctic ice mass loss from radar interferometry and regional climate modeling. *Nature Geoscience*, *1*(2), 106–110.

Ritzer, G. (2007a). *The globalization of nothing 2*. Thousand Oaks, CA: Pine Forge Press.

Ritzer, G. (Ed.) (2007b). *The Blackwell companion to globalization*. Malden, MA, and Oxford, UK: Blackwell Publishing.

Roberts, D. (2008). China autos: Chery keeps accelerating. *BusinessWeek*, January 4. Available at http://www.businessweek.com/globalbiz/content/jan2008/gb2008014_784117.htm. Retrieved on January 5, 2008.

Roberts, J. T., & Parks, B. C. (2007). *A climate of injustice: Global inequality, north-south politics, and climate policy*. Cambridge, MA and London: The MIT Press.

Rogers, H. (2007). Garbage capitalism's green commerce. In: L. Panitch & C. Leys (Eds), *Coming to terms with nature: Socialist register 2007* (pp. 231–253). New York: Monthly Review Press.

Rohling, E. J., et al. (2008). High rates of sea-level rise during the last interglacial period. *Nature Geoscience*, *1*(1), 38–42.

Rorty, R. (1995). Dewey between Hegel and Darwin. In: H. J. Saatkamp, Jr. (Ed.), *Rorty & pragmatism: The philosopher responds to his critics* (pp. 1–15). Nashville and London: Vanderbilt University Press.

Rosa, E. A., & Richter, L. (2008). Durkheim on the environment: Ex libris or ex cathedra? Introduction to Inaugural Lecture to a Course in social science 1887–1888. *Organization & Environment*, *21*(2), 182–187.

Rosenthal, E. (2007a). Cement industry is at center of climate change debate. *The New York Times*, October 26. Available at http://www.nytimes.com/2007/10/26/business/worldbusiness/26cement.html. Retrieved on October 26, 2007.

Rosenthal, E. (2007b). U.N. report describes risks of inaction on climate change. *The New York Times*, November 17. Available at http://www.nytimes.com/2007/11/17/science/earth/17climate.html. Retrieved on November 17, 2007.

Rosenthal, E. (2007c). Global food supply is dwindling rapidly, UN agency warns. *New York Times*, December 17. Available at http://www.nytimes.com/2007/12/17/world/europe/17iht-food.html. Retrieved on December 17, 2007.

Rosenthal, E. (2008). Studies deem biofuels a greenhouse threat. *The New York Times*, February 8. Available at http://www.nytimes.com/2008/02/08/science/earth/08wbiofuels.html. Retrieved on February 8, 2008.

Rosenthal, E. (2009a). Obama's backing raises hopes of climate pact. *The New York Times*, March 1. Available at http://www.nytimes.com/2009/03/01/science/earth/01treaty.html. Retrieved on March 10, 2009.

Rosenthal, E. (2009b). Third-world stove soot is target in climate fight. *The New York Times*, April 16. Available at http://www.nytimes.com/2009/04/16/science/earth/16degrees.html. Retrieved on April 16, 2009.

Rosenzweig, C., et al. (2008). Attributing physical and biological impacts to anthropogenic climate change. *Nature*, *453*(May 15), 353–357.

Rostow, W. W. (1959). The stages of economic growth. *The Economic History Review*, *12*(1), 1–16.

Saad, L. (2009). Increased number think warming is "exaggerated." *Gallup Poll*, March 11. Available at http://www.gallup.com/poll/116590/increased-number-think-global-warming-exaggerated.aspx. Retrieved on March 12, 2009.

Sachs, J. D. (2008). *Common wealth: Economics for a crowded planet*. New York: Penguin Press.

Samuelson, P. A. (1973). *Economics* (9th ed.). New York: McGraw-Hill.

Santos, F. (2007). Inch by inch, great lakes shrink, and cargo carriers face loss. *New York Times*, October 22. Available at http://query.nytimes.com/gst/fullpage.html?res = 9501E0DD103BF931A15753C1A9619C8B63 & sec = & spon = & pagewanted = 1. Retrieved on October 24. 2007.

Sassen, S. (2006). *Territory, authority, rights: From medieval to global assemblages*. Princeton and Oxford: Princeton University Press.

Sassen, S. (2007). *A sociology of globalization*. New York and London: W.W. Norton & Company.

Scaff, L. A. (2005). Remnants of romanticism: Max Weber in Oklahoma and Indian Territory. *Journal of Classical Sociology*, *5*(1), 53–72.

Schiermeier, Q. (2008a). A long dry summer. *Nature*, *452*(March 20), 270–273.

Schiermeier, Q. (2008b). Climate challenge underestimated? *NatureNews*, *452*(April 2). Available at http://www.nature.com/news/2008/080402/full/news.2008.728.html. Retrieved on April 3, 2008.

Schiermeier, Q., & Tollefson, J. (2008). Are the IPCC scenarios unachievable? *Nature*, *452*(April 2), 508–509.

Schipper, J., et al. (2008). The status of the world's land and marine mammals: Diversity, threat, and knowledge. *Science*, *322*(October 10), 225–230.

Schnaiberg, A. (1980). *The environment: From surplus to scarcity*. Oxford and New York: Oxford University Press.

Schuur, E. A. G., et al. (2008). Vulnerability of permafrost carbon to climate change: Implications for the global carbon cycle. *Bioscience*, *58*(8), 701–714.

Schwartz, P., Leyden, P., & Hyatt, J. (1999). *The long boom: A vision for the coming age of prosperity*. Reading, MA: Perseus Books.

Searchinger, T. et al. (2008). Use of U.S. croplands for biofuels increases greenhouse gases through emissions from land use change. *Science Express*, February 7. Available at http://www.sciencemag.org/cgi/content/full/319/5867/1238. Retrieved on February 8, 2008.

Seidel, D. J., et al. (2008). Widening of the tropical belt in a changing climate. *Nature Geoscience*, *1*(1), 21–24.

Sengupta, S. (2007). In Bangladesh, picking through the pieces. *New York Times*, November 21. Available at http://www.nytimes.com/2007/11/21/world/asia/21bangladesh.html. Retrieved on November 21, 2007.

Sengupta, S. (2008). Indians hit road amid the occasional elephant. *New York Times*, January 11. Available at http://www.nytimes.com/2008/01/11/world/asia/11indiacar.html. Retrieved on January 12, 2008.

Shanghai Daily. (2008). Best Buy targets China for growth. February 25. Available at http://china.org.cn/business/2008-02/25/content_10662485.htm. Retrieved on March 5, 2008.

Sheehan, M. F. (2008). China climate change: A sacred glacier recedes. *The Nature Conservancy*. Available at http://www.nature.org/wherewework/asiapacific/china/features/glacier.html. Retrieved on January 20, 2009.

Shelton, S. (2008a). Georgia loses major ruling on rights to Lanier water. *The Atlanta Journal-Constitution*, February 5. Available at http://www.alabamarivers.org/current-work/water-wars/georgia-loses-lanier-water-rights-case. Retrieved on February 5, 2008.

Shelton, S. (2008b). Legislature 2008: Bill loosens water restrictions. *The Atlanta Journal-Constitution*, March 5, D4.

Shelton, S. (2009a). Less rainfall may be new "normal" for north Georgia. *The Atlanta Journal-Constitution*, January 1. Available at http://www.ajc.com/ajccars/content/metro/stories/2009/01/01/metro_atlanta_rainfall.html?cxntlid = inform_artr. Retrieved on February 21, 2009.

Shelton, S. (2009b). High court won't review water pact on Lanier. *The Atlanta Journal-Constitution*, January 13. Available at http://www.ajc.com/thrashers/content/metro/stories/2009/01/13/water.html?cxntlid = inform_artr. Retrieved on February 21, 2009.

Shelton, S. (2009c). Metro Atlanta water plan draws protesters. *The Atlanta Journal-Constitution*, January 15. Available at http://www.ajc.com/eveningedge/content/printedition/2009/01/15/waterplan.html. Retrieved on February 21, 2009.

Shelton, S. (2009d). Water shortage still a worry for Georgians. *The Atlanta Journal-Constitution*, February 2. Available at http://www.ajc.com/tgif/content/metro/stories/2009/02/02/georgia_drought.html?cxntlid = inform_sr. Retrieved on February 21, 2009.

Shelton, S. (2009e). Gold Dome: Senators ask for reservoir in national forest. *The Atlanta Journal-Constitution*, February 3. Available at http://www.ajc.com. Retrieved on February 20, 2009.

Silver, N. (2009a). Health care > environment? It's the economy stupid. *FiveThirtyEight: Politics Done Right*, March 20. Available at http://www.fivethirtyeight.com/2009/03/health-care-cap-and-trade-its-economy.html. Retrieved on March 20, 2009.

Silver, N. (2009b). When hope is the enemy of change. *FiveThirtyEight: Politics Done Right*, April 22. Available at http://www.fivethirtyeight.com/2009/04/when-hope-is-enemy-of-change.html. Retrieved on April 22, 2009.

Simon, J. L. (1995). The state of humanity: Steadily improving. *CATO policy report*, September/October. Available at http://www.cato.org/pubs/policy_report/pr-so-js.html. Retrieved on January 20, 2008.

Simon, J. L. (1996a [1981]). *The ultimate resource II: People, materials, and environment*. Available at http://www.juliansimon.com/writings/Ultimate_Resource/. Retrieved on January 20, 2009.

Simon, J. L. (Ed.) (1996b). *The state of humanity*. Malden MA and Oxford UK: Blackwell Publishers.

Simon, J. L., & Kahn, H. (Eds). (1984). *The resourceful earth: A response to Global 2000*. Oxford UK and New York: Basil Blackwell.

Skoloff, B. (2007). *Much of U.S. could see a water shortage* (October 26. Available at http://www.truthout.org/article/much-us-could-see-a-water-shortage. Retrieved on November 20, 2007). Associated Press.

Smith, B. (2008). A deal for folks on the edge 'amnesty': Lawmaker wants to erase Georgia back taxes for those who call Tennessee home. *The Atlanta Journal-Constitution*, March 5, D1.

Smith, B., & Raju, M. (2009). Some Dems want break in Obama plans. *Politico*, March 18. Available at http://www.politico.com/news/stories/0309/20150.html. Retrieved on March 18, 2009.

Smith, K. (2008). The population problem. *Nature Reports*, *2*(May 15), 72–74.

Smith, L. (2007). "Humanity's very survival" is at risk. *The Times*, October 26. Available at http://www.timesonline.co.uk/tol/news/uk/science/article2739926.ece. Retrieved on December 15, 2007.

Smith, L. (2008a). Antarctic losing ice nearly twice as fast as ten years ago. *The Times*, January 14. Available at http://www.timesonline.co.uk/tol/news/environment/article3181944.ece. Retrieved on January 20, 2008.

Smith, L. (2008b). Satellite images show widespread destruction of Amazon rainforest. *The Times*, January 24. Available at http://www.timesonline.co.uk/tol/news/world/us_and_americas/article3245635.ece. Retrieved on January 24, 2008.

Smith, W. S. (2007). Peace Prize Committee disbands! *TCSDAILY*, October 19. Available at http://tcsdaily.com/article.aspx?id = 101907A. Retrieved on October 19, 2007.

Speth, J. G. (2008). *The bridge at the edge of the world: Capitalism, the environment, and crossing from crisis to sustainability*. New Haven and London: Yale University Press.

Spiegel. (2007). Birds and bees prematurely active in Greenland (June 19). Available at http://www.spiegel.de/international/world/0,1518,489508,00.html. Retrieved on June 20, 2007.

Spotts, P. N. (2008). Antarctica's Wilkens Ice Shelf eroding at an unforseen pace. *The Christian Science Monitor*, March 28. Available at http://www.csmonitor.com/2008/0328/p25s10-wogi.html. Retrieved on March 28, 2008.

Steffen, A. (2007). Why geo-engineering is a debate whose time has gone. *WorldChanging*, December 21. Available at http://www.worldchanging.com/archives/007704.html. Retrieved on August 20, 2009.

Steig, E. J., et al. (2009). Warming of the Antarctic ice-sheet surface since the 1957 International Geophysical Year. *Nature*, *457*(January 22), 459–463.

Stein, R. (2009). Obama aims to shield science from politics. *The Washington Post*, March 9. Available at http://www.washingtonpost.com/wp-dyn/content/article/2009/03/08/AR2009030801476.html. Retrieved on March 9, 2009.

Steinfeld, H. et al. (2006). *Livestock's long shadow: Environmental issues and options*. Available at http://www.fao.org/docrep/010/a0701e/a0701e00.HTM. Retrieved on January 20, 2008.

Stephens, B. (2008a). Global warming as mass neurosis. *The Wall Street Journal*, July 1. Available at http://online.wsj.com/article/SB121486841811817591.html. Retrieved on July 1, 2008.

Stephens, B. (2008b). Al Gore's doomsday clock. *The Wall Street Journal*, July 22. Available at http://online.wsj.com/article/SB121668313890771925.html. Retrieved on July 22, 2008.

Stiglitz, J. E. (2002). *Globalization and its discontents*. New York: W.W. Norton & Company.

Stiglitz, J. E. (2003). *The roaring nineties: A new history of the world's most prosperous decade*. New York and London: W.W. Norton & Company.

Stiglitz, J. E. (2006). *Making globalization work*. New York and London: W.W. Norton & Company.

Streitfeld, D. (2008). A global need for grain that farms can't fill. *New York Times*, March 9. Available at http://www.nytimes.com/2008/03/09/business/worldbusiness/09crop.html. Retrieved on March 9, 2008.

TA. (2005). The Great Mall of China. *Things Asian*, August 31. Available at http://www.thingsasian.com/stories-photos/3446. Retrieved on January 20, 2007.

Thompson, A. (2007). Summertime snowmelt. *Nature Reports*, *2*(December 12), 2–3.

Timmons, H. (2008). Ford and Tata said near a deal on luxury brands. *New York Times*, March 24. Available at http://www.nytimes.com/2008/03/24/business/worldbusiness/24iht-24cndauto.11383079.html. Retrieved on March 24, 2008.

Todras-Whitehill, E. (2007). Adventure destination/Greenland; As ice recedes, interest surges. *New York Times*, December 9. Available at http://travel.nytimes.com/2007/12/09/travel/09adventure.html. Retrieved on December 9, 2007.

Tollefson, J. (2008). Save the trees? *Nature*, *452*(March 6), 8–9.

UCS. (2007). Smoke, mirrors & hot air. *Union of Concerned Scientists*, January. Available at http://www.ucsusa.org/assets/documents/global_warming/exxon_report.pdf. Retrieved on March 10, 2007.

UCSB. (2008). Scientists reveal first-ever global map of total human effects on oceans (Press Release). University of California, Santa Barbara, February 14. Available at http://www.ia.ucsb.edu/pa/display.aspx?pkey = 1726. Retrieved on February 15, 2008.

UNDP. (2007). *Global outlook for ice and snow*. United Nations Development Programme. Available at http://arctic-council.org/filearchive/IceandSnowHandout.pdf. Retrieved on January 20, 2008.

UN-Energy. (2008). *Sustainable bioenergy: A framework for decision makers*. UN Food and Agriculture Organization. Available at ftp://ftp.fao.org/docrep/fao/010/a1094e/a1094e00.pdf. Retrieved on January 20, 2009.

UNHDP. (2007). *Human development report 2007/2008*. United Nations Development Programme. Available at http://hdr.undp.org/en/media/HDR_20072008_EN_Complete.pdf. Retrieved on January 18, 2008.

UNPD. (2000). India becomes a billionaire. United Nations Population Division. Available at http://www.un.org/esa/population/pubsarchive/india/yrtrend.htm. Retrieved on June 15, 2007.

UNSW. (2007). Bali climate declaration by scientists. *University of New South Wales Climate Change Research Centre*, December 6. Available at http://www.ccrc.unsw.edu.au/news/2007/Bali.html. Retrieved on December 7, 2007.

U.S. Census Bureau. (2009). Countries and areas ranked by population. Available at http://www.census.gov/ipc/www/idb/ranks.php. Retrieved on. May 15, 2009.

Vedantam, S. (2009). Climate fears are driving "ecomigration" across globe. *The Washington Post*, February 23. Available at http://www.washingtonpost.com/wp-dyn/content/story/2009/02/23/ST2009022301515.html. Retrieved on February 23, 2009.

Vestal, C. (2008). States push food stamps for the needy. Pew Research Center Publications, April 14. Available at http://pewresearch.org/pubs/798/states-push-food-stamps-for-the-needy. Retrieved on April 15, 2008.

Victor, D. G. (2008). On the regulation of geoengineering. *Oxford Review of Economic Policy*, *24*(2), 322–336.

Victor, D. G. (2009). China's energy challenge: Setting the scene. Stanford Program on Energy and Sustainable Development, January. Available at http://iis-db.stanford.edu/pubs/22382/_China_energy_and_environment_FACES.pdf. Retrieved on May 1, 2009.

Victor, D. G., Morgan, G. M., Apt, J., Steinbruner, J., & Ricke, K. (2009). The geoengineering option. *Foreign Affairs*, March/April. Available at http://iis-db.stanford.edu/pubs/22456/The_Geoengineering_Option.pdf. Retrieved on June 1, 2009.

Vitello, P. (2007). Home insurers canceling in east. *New York Times*, October 16. Available at http://www.nytimes.com/2007/10/16nyregion/16insurance.html. Retrieved on October 16, 2007.

Wackernagel, M., & Rees, W. (1996). *Our ecological footprint: Reducing human impact on the Earth*. Gabriola Island, British Columbia, Canada: New Society Publishers.

Wang, G., et al. (2008). Response [to Laurance]. *Science*, *319*(February 29), 1184–1185.

Wassener, B. (2009). Manufacturing reports show depth of global downturn. *The New York Times*, January 2. Available at http://www.nytimes.com/2009/01/03/business/worldbusiness/03yuan.html. Retrieved on January 20, 2009.

Wax, E. (2007a). A sacred river endangered by global warming. *The Washington Post*, June 17. Available at http://www.washingtonpost.com/wp-dyn/content/article/2007/06/16/AR2007061600461.html. Retrieved on June 17, 2007.

Wax, E. (2007b). In flood-prone Bangladesh, a future that floats. *The Washington Post*, September 27. Available at http://www.washingtonpost.com/wp-dyn/content/article/2007/09/26/AR2007092602582.html. Retrieved on September 27, 2007.

Weber, M. (1975 [1926]). *Max Weber: A biography* (H. Zohn, Trans.). New York: John Wiley & Sons.

Weber, M. (1949 [1904]). Objectivity in the social sciences. In: E. A. Shils & H. A. Finch (Ed. & Trans.), *The methodology of the social sciences* (pp. 49–112). New York: The Free Press.

Weber, M. (1958 [1904]). Capitalism and rural society in Germany. In: H. H. Gerth & C. W. Mills (Ed. & Trans.), *From Max Weber* (pp. 363–85). New York: Oxford University Press.

Weber, M. (2002 [1920]). *The Protestant ethic and the spirit of capitalism* (S. Kalberg, Trans.). Los Angeles: Roxbury Publishing Company.

Welch, D., & Lakshman, N. (2008). My other car is a Tata. *BusinessWeek*, January 3. Available at http://www.businessweek.com/magazine/content/08_02/b4066033073739.htt. Retrieved on August 22, 2009).

Will, G. (2008). An environmental license to intrude. *Lawrence Journal World*, May 23. Available at http://www2.ljworld.com/news/2008/may/23/environmental_license_intrude/. Retrieved on May 23, 2008.

Williams, J. (2003). Natural and epistemological pragmatism: Democracy and environmental problems. *Sociological Inquiry*, *73*(4), 529–544.

Wilson, E. O. (2005). Is humanity suicidal. *Cosmos Magazine*. Available at http://www.cosmosmagazine.com/node/1298. Retrieved on January 20, 2008.

Wilson, E. O. (2007). *Creation: An appeal to save life on earth*. New York and London: W.W. Norton and Company.

Witze, A. (2008). Losing Greenland. *Nature*, *452*(April 17), 798–802.

WMO. (2007). Greenhouse gas bulletin: The state of greenhouse gases in the atmosphere using global observations through 2006. *World Meteorological Organization*, November. Available at http://www.wmo.int/pages/prog/arep/gaw/ghg/documents/ghg-bulletin-3.pdf. Retrieved on January 20, 2008.

Wong, E., & Swart, J. (2008). Chinese soldiers rush to bolster weakened dams. *The New York Times*, May 15. Available at http://www.nytimes.com/2008/05/15/world/asia/15china.html. Retrieved on May 15, 2008.

WP. (2008). *Washington Post-ABC* poll. *The Washington Post*, February 3. Available at http://www.washingtonpost.com. Retrieved on February 3, 2008.

WRI. (2005). *Ecosystems and human well-being: synthesis (A Report of the Millennium Ecosystem Assessment)* World Resources Institute. Available at http://www.wri.org/publication/millennium-ecosystem-assessment-ecosystems-and-human-well-being-synthesis. Retrieved on November 20, 2007.

WSJ. (2007). Not Nobel winners: Some nominees for next year. *The Wall Street Journal*, October 14. Available at http://www.opinionjournal.com/editorial/feature.html?id = 110010732. October 20, 2007.

WT. (2007a). The Goracle. *The Washington Times*, October 15. Available at http://www.washingtontimes.com. Retrieved on October 16.

WT. (2007b) Consumer market to rival the U.S. *The Washington Times*, March 25. Available at http://www.washtimes.com.com/functions/print.php?storyID = 200703324-105827-1374r. Retrieved on March 25, 2007.

WWF. (2008) *Living planet report 2008*. World Wildlife Fund. Available at http://assets.panda.org/downloads/living_planet_report_2008.pdf. Retrieved on January 20, 2009.

Wynn, G. (2009). CO_2 rise in atmosphere accelerates in 2008, February 25. Available at http://in.reuters.com/article/oilRpt/idINLO80884620090225. Retrieved on February 26. 2009.

Xinhua. (2008a). Consumption becomes the no. 1 engine of economic growth. *Xinhua News Agency*, January 30. Available at http://www.china.org.cn/english/business/241360.htm. Retrieved on March 1, 2008.

Xinhua. (2008b). China now No. 3 export area for US. *Xinhua News Agency, March 3*. Available at http://shanghaidaily.com

Yardley, J. (2007a). Beneath booming cities, China's future is drying up. *The New York Times*. September 28. Available at http://www.nytimes.com/2007/09/28/world/asia/28water.html. Retrieved on September 28, 2007

Yardley, J. (2007b). Chinese dam projects criticized for their human costs. *The New York Times*, November 19. Available at http://www.nytimes.com/2007/11/19/world/asia/19dam.html. Retrieved on November 19, 2007.

Yardley, J. (2007c). Chinese officials defend dam. *The New York Times*, November 28. Available at http://www.nytimes.com/2007/11/28/world/asia/28iht-28gorges.8506060.html. Retrieved on November 28, 2007.

Yardley, J. (2007d) China's turtles, emblems of a crisis. *The New York Times*, December 5. Available at http://www.nytimes.com/2007/12/05/world/asia/05turtle.html. Retrieved on December 5, 2007.

Yardley, J. (2007e). Beijing's Olympic quest: Turn smoggy sky Blue. *The New York Times*, December 29. Available at http://www.nytimes.com/2007/12/29/world/asia/29china.html. Retrieved on January 6, 2008.

Yardley, J. (2008a). Smoggy Beijing to cut traffic by half for Olympics, paper says. *The New York Times*, January 24. Available at http://www.nytimes.com/2008/01/24/world/asia/24beijing.html. Retrieved on January 24, 2008.

Yardley, J. (2008b). As most of China celebrates the new year, a scramble continues in coal country. *The New York Times*, February 9. Available at http://www.nytimes.com/2008/02/09/world/asia/09iht-09china.9891162.html. Retrieved on February 9, 2008.

YCELP. (2008). Environmental performance index: Summary for policymakers. *Yale Center for Environmental Law and Policy*. Available at http://sedac.ciesin.columbia.edu/es/epi/papers/2008EPIPolicymakerSummary.pdf. Retrieved on January 20, 2009.

York, R., & Mancus, P. (2009). Critical human ecology: Historical materialism and natural laws. *Sociological Theory*, *27*(2), 122–149.

Zeng, N., et al. (2008). Climate change- the Chinese challenge. *Science*, *319*(February 8), 730–731.

Zhang, D. D., et al. (2008a). Global climate change, war, and population decline in recent human history. *Proceedings of the National Academy Sciences*, *104*(49), 19214–19219. Available at http://www.pnas.org/content/104/49/19214.full.pdf. Retrieved on December 5, 2008.

Zhang, J., et al. (2008b). What drove the dramatic retreat of arctic ice during summer 2007. *Geophysical Research Letters*, *35*(June 11), L11505.

Zwiers, F., & Hegerl, G. (2008). Attributing cause and effect. *Nature*, *453*(May 15), 296–297.

SOCIAL THEORY, CLIMATE CHANGE, AND THE HUMANITY–NATURE RELATION

Michael E. Zimmerman

In his article, Robert J. Antonio reviews the recent history of the growth imperative that has led techno-industrial societies to use such enormous amounts of CO_2-producing fossil fuels. Then, he discusses scientific studies supporting the anthropogenic global warming hypothesis (AGWH), according to which global temperature increases in the past decades of the 20th century have been "forced" (caused) by CO_2 introduced into the atmosphere primarily by burning coal, oil, and natural gas. He then discusses the extent to which politico-economic globalization is leading to "hypergrowth," which will lead humanity to exhaust natural resources and to overshoot planet Earth's carrying capacity, thereby devastating the ecosystems needed to support human life and other forms of life. Finally, Antonio maintains that a crucial step in avoiding the nightmarish consequences of techno-industrial, globalizing modernity is for critical social theorists to re-embed the social in nature. In what follows, I comment briefly on three of these themes, while electing not to discuss the issue of overshoot, which would require a separate article to address.

By way of preface, let me state that I agree with Professor Antonio's major concern, namely, that humankind must marshal intelligence, wisdom, energy, and money to identify and to address effectively the many enormous socio-ecological challenges facing us. I would also like to note that because

Nature, Knowledge and Negation
Current Perspectives in Social Theory, Volume 26, 75–92

ISSN: 0278-1204/doi:10.1108/S0278-1204(2009)0000026005

my remarks are *occasioned* by Antonio's article, not every criticism that I make or question that I raise is addressed to the specific positions that he develops.

1. THE GROWTH IMPERATIVE

Antonio cites J. K. Galbraith (among others) as having criticized a view central to mainstream economics, namely, that "human nature dictates an unlimited 'urgency of wants,' naturalizing ever increasing production and consumption and precluding the distinction of goods required to meet basic needs from those that stoke wasteful, destructive appetites." In citing Galbraith, Antonio evidently has some sympathy for the former's viewpoint, which I now examine.

First, what exactly are "basic needs"? Presumably, they include adequate food, clean water, clothing, shelter, and security. Surely, some people would like to make additions to this list, but the bare essentials mentioned here are provided for even in forward-thinking prisons. Decades ago, Abraham Maslow argued that after people meet basic physiological and survival needs, other needs emerge, including the need for love, esteem, purpose, and self-actualization. The 1960s countercultural rebellion arose in part because children of wealth were not satisfied with the economic security prized by parents who had experienced the Great Depression. The 1960s also saw the emergence of First World environmental movements, which provided a sense of purpose and esteem to millions and which required as its condition a significant level of national wealth. Maslow's point, of course, is that as wealth increases, the need to "have" things gives way to wanting to "be" more fully human.[1]

Long before the advent of modernity, of course, a great many human animals enjoyed – and still enjoys – personal and cultural adornment, celebration, and even excess, as displayed in the widespread practice of potlatch. Humans are endowed with imagination that opens up whole new realms of beauty, terror, and longing for immortality, as evidenced by the gigantic pyramids and temples, construction of which must have caused human suffering and environmental damage. People have been given to excess long before the rise of "consumer society," which was made possible by industrial production that allowed more people to mimic at a lower scale the excess that at one time only the wealthy could afford. To be sure, modern advertising plays upon human inclinations to acquire, to adorn, and to feast in various media, but advertising did *not* create such inclinations, nor will they be eliminated by new social structures organized in accordance with a principle of Green parsimony. Less is not more for people who – in their own assessment – do not have enough, nor is small more beautiful. Of course, we may *wish* that things were otherwise, that the bulk of humanity had achieved a level of social and moral development consistent with living simply and with less, so as to taken into account the interests of other people and the biosphere. Socio-cultural reality, however, does not accord with such a wish.

What examples do we have of people for whom fulfilling basic needs supposedly suffices? Hunter-gatherer tribes of yore perhaps, or their contemporary surrogates, in the form of scattered populations of indigenous peoples, whose survival is threatened by the quest for resources needed to enable production of ever more consumer goods? At one time, everyone lived in tribes. Gradually, tribes were absorbed into larger social units, sometimes by force, sometimes by choice on the part of those who concluded that the (often more powerful) strangers had something valuable to offer. In addition to violence and threats of violence, then, tribal people may have been lured from the old ways by more complex social arrangements and practices, ranging from cooking to education, not to mention advanced technologies of celebration and ritual. Of course, much was lost along the way, including languages, practices, divinities, and what may have been a profound wisdom regarding the relation among humans and other living beings. Much was lost, and is still being lost, as people leave their rural homes and head for the opportunities offered by cities, despite the risks involved in such a dramatic move. The tide of history has pulled us away from tribal and agrarian ways, about which Marx and many other moderns were not loathe to express their critical views. Agriculture made possible cities, which in turn produced priests and philosophers who eventually distinguished between the human and the merely natural, between history and nature. Humanity–nature dualism in turn helped to make possible and also to justify the practices that critics of modernity's dark side call domination of external nature as well as dissociation from internal nature, including emotions and desires.

Although there is much to criticize about modernity, we ought not to lose sight of what there is also much to celebrate and retain, even as we seek to move beyond its limitations. Likewise, while there is much to celebrate about tribal and premodern peoples, we ought not to harbor any illusions about their life conditions and chances. Retro-romanticism is endemic to much of environmentalism. If only we could live "close to the land" as did tribal people, we could overcome that baleful humanity–nature dualism.

Perhaps, but would not this overcoming likely include psychological and socio-cultural regression, such that we would neither recognize nor be capable of supporting aspects of ourselves that we now regard as crucial to our very identities? Would people reading this article prefer to live in communities that lack access to the vaccines that dramatically limit infant mortality and that involve social "arrangements" that we would find deeply problematic? One does not have to be blind to contemporary problems, including serious environmental issues, to acknowledge the achievements of modernity, including the differentiation of domains that tend to be unified in premodern societies, including religion, economics, aesthetics, politics, and morality. Contemplating contemporary Iran, we are reminded of the significance of the differentiation between church and state.

A major condition necessary for the development of modern public health care measures and medicine – which despite their various shortcomings have greatly increased human life spans and well being – were growth-oriented economies that produced the money and institutions needed for such developments. The rise of modern personal and political freedom was inextricably intertwined with the development of modern science, industry, and economic institutions. A vaccine for polio would not have been made produced by medieval Benedictine monks, by 19th-century American utopian communities, or by1970s back to nature communes, however admirable the intentions and in some cases the practices of such communities may have been. Historically, people have been attracted to life ways that provide more opportunity for leisure, personal development, and disposable income. Economic growth and modern social arrangements have made lives that are arguably better in many respects, although not without negative consequences for people and habitats that have all too often been blithely steamrolled in the name of economic progress. Of course, for those *dark* Greens whose major concern is protecting "nature" from the human cancer plaguing it, public health measures and increased crop yields are regarded as catastrophes, since they are largely to explain the human population explosion in the 20th century.

As we head toward a population of 9 billion people by mid-century, what would be the profile of societies living in accordance with "basic needs"? Clearly, the hunter-gatherer and horticultural options are off the table, though following a monumental human population collapse – triggered perhaps by nuclear war, resource exhaustion, and/or anthropogenic environmental dysfunction – residual human populations might return to a version of earlier social formations. (*Mad Max* and other highly popular dystopian films offer grim depictions of the post/techno-industrial future.)

In the context of Antonio's article, organizing no-frills societies that satisfy basic needs would seem to be imperative, given the alleged threats posed by anthropogenic global warming (AGW).

I would like to have a preview of what would likely be the draconian methods needed to "persuade" people to limit their aspirations to "basic needs." My surmise is that state-enforced penalties for exceeding one child per family in countries like China would look mild in comparison. I would also like for someone to sketch for me the intra- and inter-national discord that would ensue from efforts to deprive rich populations of their current wealth and wanna-be-rich countries of their *aspirations* for such wealth. In the next few decades, how will we re-invent human societies in ways that reduce CO_2 emissions by 50% or even 80% (as some experts call for), without in that very process bringing about an economic collapse that would be as bad (for humankind, at least) as the dire consequences projected by various eco-overshoot scenarios? In light of the mixed track record of government environmental policies – whether democratic/constitutional or authoritarian – I do not have a great deal of confidence that such policies will be sound.

Here, I invite readers to consider the political horse-trading and log-rolling that not only guts the 2009 Waxman–Markey cap-and-trade bill of any serious prospect of limiting greenhouse gas (GHG) emissions but also provides only a small fraction of the funds desperately needed for research into post-carbon energy sources. Clearly, political action is needed to address the social and environmental problems that beg for attention, but we ought not to be naïve about the limits of such action. About twenty years ago, I took part in a small conference at the YMCA of the Rockies in Estes Park, Colorado. Those present included deep ecologist George Sessions and poet Gary Snyder. When Sessions claimed that only far more vigorous government intervention could save the planet, Snyder – in his wry way – intoned that calling on the government would be like asking the foxes to guard the henhouse.

This issue involved here call to mind Marx's query in his *Theses on Feuerbach*: Who will educate the educators, those who will devise and promulgate the "sweeping policy changes" that Antonio claims will be needed if humankind is to avoid eco-catastrophe? Who is currently in charge of raising the generation of policy makers and enforcers who somehow will have been liberated from the "wasteful, destructive appetites" decried by Galbraith and so many others as not part of "human nature"? Have we learned nothing from the consequences of Soviet efforts to create a new socialist "man" devoid of the "destructive appetites" of bourgeois subjects constructed by capitalist economic and social arrangements? How will those in charge of designing the new living-in-accord-with-basic-needs "ecological man" manage to avoid resorting to a new Gulag to silence dissenters? Those most alarmed by AGW have called for dissenters to be put on Nuremberg-like trials for crimes against humanity. Have we learned nothing from the history of religions, which have constantly preached gospels of forgiveness, peace, and generosity, while faced with – and all too often contributing to – recurring tides of hatred, violence, and greed?

Presumably, Antonio envisions the enactment of policies *today* that would preclude the need to adopt "lifeboat ethics" (Garrett Hardin) *tomorrow*. How to identify the appropriate policies? Is it not possible that policies undertaken today will be misguided, leading to unanticipated and harmful future consequences? Social theorists, philosophers, policy makers, economists, and those working from other perspectives must think through these matters carefully. Moreover, various options must be clearly presented to the public for discussion, if there is any interest in forging a consensus about what is to be done. Unfortunately, those calling for a go-slow approach to addressing global warming – including members of Congress – are sometimes labeled as "traitors" to planet Earth.[2]

2. ANTHROPOGENIC GLOBAL WARMING?

Elsewhere, I have reviewed what I consider to be some of the potentially serious flaws regarding the validity of the AGWH.[3] Although not a scientist, I am quite capable of reading the claims and counter-claims made by scientists with very different points of view about AGWH. Were AGW not portrayed as the latest harbinger of environmental doom, demanding that all political attention be focused on it, scientists would engage in a far less politicized debate about alternative accounts of recent warming, and we would hear far less about an alleged "consensus" about the validity of AGWH.

Having experienced four decades of predictions of environmental catastrophe, none of which panned out as envisioned, I fear that environmentalism would suffer a terrible blow if credible scientific findings pulled the rug out from under AGWH. Such a blow would at least be counterbalanced by some abatement of alarm about climate, thereby permitting a perhaps more orderly transition to a post-fossil fuel era.

Of all people, social theorists should subject to scrutiny the origins, structure, and pronouncements of the Intergovernmental Panel on Climate Change (IPCC), a political body brought into being specifically to find evidence supporting the view that fossil fuel use is forcing global warming.

Having initially assumed that the validity of AGWH was a no-brainer, I subsequently spent a year delving into criticism of and alternatives to it. My research has led me to conclude that no one really knows enough about how climate works to make *reliable* predictions about what global temperatures will be like a century from now, and just about everyone agrees that it is impossible to make any serious predictions about the effects of climate change on a regional basis. It's not that research in support of AGWH isn't important, or that there is no cause for concern, but that the phenomenon under investigation – terrestrial climate – is too complex to be understood by the current state of science. That humans have affected climate on a regional basis is more clearly supported by the evidence than is AGWH. Uncertainty about a potentially major consequence of human activity doesn't mean we should conduct business as usual, however! There are *many good reasons other than climate change to find alternatives to fossil fuels as soon as possible.*

More discussion is needed to bring to light the cultural, historical, and psychological factors that lead people to frame climate change in terms of calamity, apocalypse, and the end of the world.[4] Some critics maintain that anxiety once associated with the prospect of nuclear war was projected onto global warming, as the Cold War ended around 1990. In the mid-1980s, Carl Sagan and others brought to public attention the probability that all-out nuclear war would entrain a "nuclear winter" that would destroy much of life on Earth, thereby putting an end to the illusion that either the United States or the USSR could "win" a nuclear war.[5] Given that the nuclear winter hypothesis did help to undermine the credibility of nuclear war, some scientists and others may be hoping to accomplish something similar with regard to AGW, by posting ever more alarming warnings about the consequences of global warming. A recent statement by atomic physicists, according to whom the consequences of global warming would be comparable to that of nuclear war, goes much too far however (St. James's Palace Memorandum, 2009). Certainly, IPCC scenarios do not project scenarios in which global warming would cause many hundreds of millions of people to die within a few weeks (even without nuclear winter!), as would have been the case (and could still be the case) with an all-out nuclear war.

Surely, the great majority of people do not share the kind of anxiety or "imagination of disaster" that many wealthy First World people have about climate change. For billions of poor people, at least, concerns about basic needs – health, food, clean water, safety, shelter – trump considerations about what the weather may be like in 2090. Such people seriously "discount" the future, with good reason. They are concerned about making it through today, while preferring to let people two or three generations from now cope with things as best they can.

Antonio cites one source according to which $45 *trillion* would have to be invested to achieve 50% global GHG reductions by 2050. If we focused less on dramatic reductions in GHG, and more investing in hoped-for breakthroughs in alternative energy sources, perhaps this extravagant sum could be reduced, even while allowing for the economic development needed to improve the lot of billions of people alive today and still to be born. As things currently stand, however, fossil and nuclear fuels will remain important ingredients in energy production for decades to come. Consider the following data, assembled by energy consultant Robert Bryce. Currently, solar and wind plants generate about 1.1% of the total electricity consumed in the United States. As is well known, the drawback of solar and wind is that they are both intermittent – the sun goes down, the wind stops blowing. The problem goes well beyond this; however, it is a matter of scale. Bryce demonstrates this by converting energy produced at solar and wind plants to oil equivalents. Solar and wind generate the equivalent of 76,000 barrels of oil per day, which scarcely puts a dent in "America's total primary energy use [of] about 47.4 million barrels of oil per day." Bryce goes on to observe that 76,000 barrels of oil per day is "approximately equal to the raw energy output of one average-sized coal mine" (Bryce, 2009).

Despite the many groups – including IPCC – voicing optimism that energy efficiency and non-fossil energy sources can achieve great reductions in fossil fuel use, some people who do the math arrive at a rather different conclusion. For instance, Department of Energy secretary Steven Chu claims that enormous, Nobel Prize-scale breakthroughs in our current understanding of physics and chemistry would be necessary for the efficiency increases needed for solar to make a big difference. Cal Tech professor Nate Lewis, interviewed by *Newsweek*'s Sharon Begley, has remarked, "It's not true that all technologies are available and we just need the political will to deploy them. My concern, and that of most scientists working on energy, is that we are not anywhere close to where we need to be" Begley (2009). We may be able to reduce CO_2 emissions by 20% by the year 2020 through various means, including efficiency, but reaching 80% emissions cuts – from much greater energy usage – by 2050 would be another matter altogether. Relying of Lewis' math, here is what Begley comes up with. In 2006, the world used 14 trillion watts – 14 terawatts – of energy. Making minimal assumptions about the rate of population and economic growth, and very optimistic assumptions about energy efficiency, Begley writes that the world would use 28 terawatts of energy in the year 2050. (28 terawatts is a great reduction from the 45 terawatts that would be needed without huge gains in energy efficiency!) Now, if the goal in all of this is to keep CO_2 levels to 450 ppm, then 26.5 of those 28 terawatts must come from sources that produce zero CO_2. What would it take for non-fossil energy sources to meet the energy demand in 2050?

> Are you a fan of nuclear? To get 10 terawatts, less than half of what we'll need in 2050, Lewis calculates, we'd have to build 10,000 nuclear reactors, or one every other day starting now. Do you like wind? If you use every single breeze that blows on land, you'll get 10 or 15 terawatts. Since it's impossible to capture all the wind, a more realistic number is 3 terawatts, or 1 million state-of-the-art turbines, and event that requires storing the energy – something we don't know how to do – for when the wind doesn't blow. Solar? To get 10 terawatts by 2050, Lewis calculates, we'd need to cover 1 million roofs with panels every day from now until then. "It would take an army," he says.[6]

In coming decades, a planet committed even to slow economic growth, 1.5% per year, would soak up *all* new energy provided by alternative energy sources (solar, wind, tidal, etc.), while continuing to demand all existing energy production from fossil, nuclear, and hydroelectric sources. Arguably, the major problem of the 21st century is *not* climate change but rather providing enough energy to meet demand, even at a slower rate of growth than currently projected.

Moving to a more sustainable and more equitable set of social, political, and economic arrangements is a worthy and vital goal, but it will not be easily or quickly achieved – if ever. In any event, some degree of economic growth will be required to move in the right direction. The growth imperative includes rising demand for not only "wasteful" and "tasteless" consumer goods but also health care, rapid transportation, digital communications, better education, and other things typically regarded as salutary. Surely the great majority of people living in slums and villages of Third World countries would be happy to have their "basic needs" met, but such a laudatory achievement would not long satiate the yearning for something more. The political demands for growth to meet basic and more-than-basic needs explains the statement made, June 30, 2009, by India's Environment Minister, Jairam Ramesh: "India will not accept any emission-reduction target – period. This is a non-negotiable stand" Pradhan (2009). Despite North American, European, or Australian Green hopes to the contrary, India and the vast majority of other wealth-aspiring countries will not abandon fossil fuels without having access to competitively priced alternatives that allow for economic growth. Leaders of such countries will not accept the "tutelage" offered by Western policy wonks and social theorists, according to whom people should be satisfied with having their basic needs met.

Nor is it likely that First World populations will voluntarily lower their GHG emissions and accept significant reductions in material living standards, so that billions of people elsewhere can achieve their own aspirations. In the United States, only about 20% of the population operates at the Green level of psychological development, which emphasizes social justice consistent with environmentally sound practices, whereas the great majority of the population operates from either premodern or modern personal and cultural perspectives that regard Green ideology with suspicion. The socio-cultural revolution that some Greens hoped would come from AGW will not come to pass any time soon. Economists point out that *price* is a concern that is shared by most people, regardless of their developmental level or political view. Hence, people will move away from fossil fuels only when the latter become sufficiently expensive to make alternative energy sources competitive. Unfortunately, this market mechanism has been distorted by various factors, including the political refusal to take into account the military costs involved in guaranteeing the United States access to oil.

What steps should be taken to deal with global warming, which seems likely to continue regardless of what drives it? Because even substantial decreases in GHG emissions will not affect global temperatures for several decades, and even then only in terms of small temperature decreases, prudence indicates that available funds should be divided between what is called *mitigation*, that is, reducing GHG emissions, on the one hand, and *adaptation*, that is, preparing for the possible consequences of climate change. Since no country can know in advance what its future climate will be, countries and whole regions should develop different scenarios for actions needed to accommodate different possible futures. I agree with those who argue that instead of putting all our eggs in the basket of mitigation, most investment should go to adaptation. Moreover, because wealthier countries would inevitably be more capable of adapting, economic growth should be encouraged particularly in currently poor countries that might be most affected by climate change. Obviously, however, such growth would require continued use of fossil fuels, until alternatives are available.

Pace Julian Simon, infinite economic growth is not plausible, but I see no way around the need for significant economic growth in the 21st century. Moreover, G8 countries cannot force poor countries – except through crippling tariffs or outright war – to dramatically curb fossil fuels and thus abandon hopes for greater wealth. At least, however, countries rich and poor should seek to grow in ways that are smarter and more efficient, that limit environmental damage, and that promote greater social justice. If economic growth were dramatically curbed in future decades – by energy shortages, not to mention by war, disease, famine, climate, and/or other as yet unknown apocalyptic factors – the consequences would be very tragic indeed. We must keep this sobering prospect in mind, without deluding ourselves about the scale of the tasks facing us.

Social scientists should also spend time investigating the infiltration – some would say the takeover – of the CO_2 reduction movement by big-money interests. One social scientist, Bjorn Lomborg, goes so far as to compare the close ties among regulatory agencies, private equity firms, and corporations involved in promoting the Waxman–Markey cap-and-trade bill as akin to what President Eisenhower once warned against, namely, the untoward influence of "the military-industrial complex." Many Greens, of course, regard with suspicion anything said by Lomborg, author of *The Skeptical Environmentalist*, a book that was unfairly attacked because it argued that there was some good news to report about social and environmental trends, although serious problems persist. Lomborg, however, is not alone in arguing that the "climate-industrial complex" has in effect co-opted the climate change debate, with the aim of representing the only possible solution as one that happens to be compatible with the interests of the organizations who stand to benefit from costly and still-speculative schemes to reduce carbon emissions.[7]

Consider *left-wing* historian of technology David Noble, who poses this rhetorical question: How did global warming, "an arcane subject only yesterday of interest merely to a handful of scientific specialists so suddenly come to dominate our discourse?"[8] (Noble, 2007). He answers that only governments, corporations, and their media allies have "the reach and resources to place so alien an idea in so many minds simultaneously so quickly." Al Gore and other AGWH proponents rightly insist that fossil fuel companies used advertising, public relations, political cronies, and front organizations in a well-orchestrated effort to lull people to sleep about AGW. Noble points out, however, that Gore et al. have used the same means, although employed "by different corporate hands," to "drum into our heads" their own "alarmist" message. The inside economic-ideological story of the AGWH propaganda campaign, however, has been little reported. Here is Noble's take on the matter.

In recent years, a number of the fossil fuel corporations that had once criticized AGWH concluded that further resistance was both futile and a public relations disaster. Some of these corporations realized that serious money could be made by helping to define economic-technical measures (such as cap and trade bills) to mitigate AGW. If Noble is right, then, during the past 15 years, two different corporate campaigns have been taking place, which reveal a (healable) split within elite circles. The fossil fuel industry–financed Global Change Coalition (GCC) organized the first campaign, a negative effort to rebut, silence, and sew seeds of doubt about AGWH, with such effect that even today AGWH skeptics are identified with this negative campaign. The positive campaign, in contrast, which arose after the Kyoto negotiations, has been financed by a competing group that sought to "hijack" AGW so as to turn it to corporate advantage. Co-opting global warming would enable leading corporations to undermine the radical anti-globalization movement of the 1990s, by emphasizing that only market-based (and thus globalization-friendly) solutions could deal with AGW. Corporate forces and their media agents turned AGW into "a totalistic preoccupation" that gave rise to environmental "hysteria," which has led many people in wealthy countries to favor corporate-friendly, bureaucratic solutions "at the expense of any serious confrontations with corporate power."

The growing anti-globalization movement, when combined with the Kyoto accords, generated enough concern in corporate circle to cause a split within GCC. New organizations emerged, including the Pew Center for Global Climate Change, funded and headed by people associated with major energy and financial institutions. The Pew Center gave rise to the Business Environmental Leadership Council (BELC), whose early members included – among others – Sunoco, Dupont, Duke Energy, BP, and Royal Dutch/Shell. BELC argued that science gave compelling reasons for supporting anti-AGW measures and that corporations that were early investors in market-based climate strategies would enjoy "sustained competitive advantage over their peers." Noble describes how other organizations, including the Partnership for Climate Action (PCA), formed with the apparently noble intent of "solving" the climate problem with market-based programs, happened to promise enormous profits, as investment bankers were quick to realize.

Noble then points out that Al Gore's family has long-standing ties to the key figures in the energy industry, particularly, the late oil magnate Armand Hammer. In 2004, Gore formed an environmental investment firm, Generation Investment Management, and began urging investors to become wealthy by doing the right thing environmentally. In *An Inconvenient Truth* (2006), Gore emphasized that capitalism could and should become an ally in the fight against warming. Noble goes on to detail more recent eco-investment strategies, some of which are included in the flawed Waxman–Markey climate bill.

In view of Gore's important contributions to environmentalism, it would be unfair to suggest that he has acted cynically by urging market forces to help deal with global warming. Noble's point, however, is not to personalize the issues but instead to disclose the enormous – and heretofore largely hidden – social forces at work. According to Noble, the corporate campaign against global warming, by striking a devastating blow against the worldwide, anti-globalization, social justice movement,

> has restored confidence in those very faiths and forces which that movement had worked so hard to expose and challenge: globe-straddling profit-maximizing corporations and their myriad agencies and agendas; the unquestioned authority of science and the corollary belief in deliverance through technology, and the beneficence of the self-regulating market with its panacea of prosperity through free trade ....

In the "apocalyptic rush to fight global warming," injustices and injuries brought about by global market forces are swept aside, as are any need "to question a deformed society or re-examine its underlying myth," to focus all attention on the "epic challenge" posed by global warming. Noble concludes that this war-footing mentality has created a "Manichean contest between mean and mindless deniers, on the one hand, and enlightened global warming advocates, on the other," leading even leading left-wing journalists to issue embarrassing pro-AGWH manifestos.

One does not have to accept all aspects of Noble's critique to recognize that AGWH's rise to prominence has been aided at least in part by major financial organizations about which the general public (not to mention most climate scientists) is largely unaware. Nor does one have to dismiss out of hand all market-oriented approaches to solving eco-problems, to take seriously the need for healthy skepticism regarding the promises made by such approaches. Market approaches can in fact ameliorate a number of environmental problems, though even here players can distort market mechanisms in ways that preclude environmentally optimal outcomes.

Because of the twin problems posed by the growing need for energy, on the one hand, and the environmentally harmful consequences of many aspects of energy-fueled economic growth, on the other, coming decades will be fraught with difficulties and dangers. All available approaches – market, policy oriented, legislative, educational, and cultural – will inevitably be brought to bear on the multiple problems that already confront us and new ones just around the corner. A contribution that academics can make is to scrutinize these efforts critically, while also offering positive alternatives to practices that seem unwieldy, ineffective, or corrupt.

3. RE-EMBEDDING THE SOCIAL IN NATURE: PROCEED WITH CARE

Having participated in the process by which philosophy was made aware of nature as an "other" in addition to being raw material for enhancing the power and security of the modern human subject, I can well remember parallel developments in social theory, which was arguably even more strongly anthropocentric than philosophy, in part because of the major influence of European social theorists who emphasized the difference between "history" and "nature." Salutary efforts by early environmentally oriented social theorists like Riley E. Dunlap and William R. Catton, Jr., were met with resistance from the sociological establishment, which in the 1970s and 1980s tended to regard eco-sociology as an ally of the reductionist program of sociobiology. It took a considerable struggle for eco-sociology to attain credibility within sociology, but there is no need to rehearse the *dénouement* of that accomplishment for readers already familiar with it. Suffice it to say that it is crucial for virtually all academic fields both to acknowledge and to have some understanding of humanity's dependence on a healthy biosphere and our interdependence with other forms of life. Much of what Antonio says in the final part of his article makes a great deal of sense to me. However, I would like to offer two brief remarks.

First, even though it was laudatory for eco-social theory to challenge both strong constructivism, which views nature as little more than a cultural "product," and "'*exemptionalist*' presuppositions, which exclude humanity from natural constraints," I believe that social theorists should be cautious about the discourse of "re-embedding" the human in nature. There *is* something exceptional about human beings, although this fact does not exempt humankind from the consequences of behavior that ignores the interdependence of terrestrial life. History reveals a number of episodes in which humans have altered their regional climates in ways that had very serious social consequences. We are all familiar with the industrial pollution and habitat destruction that have spread across the planet because of a growing human population, which is increasingly committed to and thus dependent on the "treadmill of production," the practices and economic framework of growth-oriented techno-industrial modernity. The sad prospect is that humanity may end up undermining the ecological conditions necessary for its own survival. Some cosmologists explain the lack of any (officially accepted) encounter with non-terrestrial intelligent life as resulting from the fact that other intelligent life forms inadvertently destroyed themselves for the same reasons that we might do so.

Having said all this, I would like to emphasize that humans go far beyond other life forms in terms of linguistic capacities and in terms of the ability to develop abstractions. Widespread recognition of these abilities provides good reasons for Western anthropocentrism, and Eastern anthropocentrism, for that matter. Such anthropocentrisms have often dissociated humankind from nature, however, as indicated in eco-sociology's critique of exemptionalism. Any successful reinterpretation of humanity's relation to nature must abandon such dissociation, without simultaneously acknowledging that evolved human intelligence allows – even requires – us to *interpret* our place in nature and that such interpretations are *various* and *conditioned* by many complex factors. Nature will always be a human "construct" insofar as humans don't encounter nature immediately – except perhaps in certain circumstances – but always mediately, in terms of the personal, cultural, and social attitudes that predominate in any given historical moment. There is considerable truth to the notion, so important to modern philosophy and social theory, that humankind exists within historical "worlds" conjured up in part by human imagination, which of course is associated with and dependent on the human organism that has evolved over eons. Re-embedding ought not to become a reductionism that in emphasizing our kinship with other life ends up overlooking the *creative strangeness* of the human, a characteristic that figures into most ancient cosmologies and creation myths.

Second, Antonio counters Catton's observation that it would be a "sterile" exercise and a "distraction" to critically appropriate classical social theory – as developed by Durkheim and others – for the sake of developing an "effective, eco-friendly, alternative ecological paradigm." In re-conceiving the humankind–nature relationship, we must cast our nets widely, seeing insights from the wisest in past and present. Many eco-philosophers have sought to tease out eco-friendly views from a wide variety of Western thinkers, ancient and modern, as I once did – for good or ill – with the thought of Martin Heidegger. Similarly, Antonio maintains that the social theories of George Herbert Mead and John Dewey "framed a comprehensive starting point for *postexemptionalist theory and practice* that is worthy of re-engaging." Such an exercise is hardly sterile, unless we are willing to conclude that much of the work that social theorists and philosophers is sterile, given how often such endeavors engage the work of previous thinkers.

Is Catton right, however, in describing such endeavors as "distractions"? Here, he may have a point. The first three sections of Antonio's article lead the reader to expect the closing section to reach a rousing crescendo, including dramatic calls to action. In fact, however, the finale was somewhat anticlimactic: a summons to social theorists to develop a paradigm that re-embeds humankind in nature. Such a call is perhaps not surprising, given that the favorite academic weapon for contestation is the pen, or these days the keyboard. May such an endeavor properly be called a distraction, however? This may depend on what Catton – were he specifically addressing AGW – thinks people should be paying attention to. Were everyone to focus solely on slowing AGW society would grind to a halt. Even social theorists who recognize the importance of AGW and other environmental problems must attend to matters other than taking direct action to solve such problems. Decades before environmental problems – and AGW – came to prominence, social theorists were well aware of countless social problems, ranging from inadequate public health measures to socio-economic stratification. Notwithstanding such problems, most sociologists continued theorizing rather than leaving their academic posts to directly tackle multiple real-world issues. Of course, many social theorists have engaged in political praxis as well as academic theorizing.

We should be grateful, however, that most social theorists continued doing what they were trained for, namely, developing the theoretical and methodological tools needed to disclose the social structures – beneficial and harmful alike – that under gird contemporary societies. Without the results of such investigations, policy makers and people on the ground would be much less well equipped to make effective interventions. Analogous insights need to be drawn from academics focusing on philosophy, religion, cultural anthropology, psychology, history, and other perspectives that shed light on human beliefs, motivations, aspirations, and capacities for self-delusion. Climate change, energy sources, and other such problems are so dauntingly complex that we must draw on insights from all quarters, if humankind is to do a little better than muddle through the 21st century.

There is a great deal with which I can agree with the words with which Antonio concludes his article:

> We need to frame theories and practices that stress culture's embeddedness in nature and that, consequently, manifest actively our responsibilities to distant peoples, future generations, and other life forms with which we share the planet. Our futures are entwined. This Copernican turn calls for profound rethinking, redefining, reshaping, and redirecting of growth, not abandoning it.

Perhaps, surprising is the closing reference to "growth," which Antonio proposes to re-construe so as to allow continued adherence to it as a social goal. Presumably, in a future article, he will outline major features of such redefined growth, one consistent with the long-term well-being of human life and the rest of life on Earth. I, for one, look forward to reading what he will propose.

NOTES

1. See also Fromm (2005).
2. See Krugmanm (2009).
3. Those interested in details of my reservations about AGWH may refer my article, "Including More Perspectives," in press.
4. See Hulme (2008).
5. See Sagan and Turco (1990). In his January 6, 1991 review, published in *The New York Times*, Len Ackland notes the following: "The original analysis of nuclear winter followed a 1982 study by Paul Crutzen and John Birks, published in the Swedish journal *Ambio*, in which they analyzed the atmospheric effects of the enormous quantities of smoke that would be generated by nuclear war. To this analysis the Turco and Sagan team added the idea that the smoke would lead to a severe temperature drop on the earth's surface." http://www.nytimes.com/1991/01/06/books/chilly-scenes-of-nuclear-winter.html. Accessed on July 7, 2009.
6. *Ibid.*
7. Lomborg (2009). Some readers may already be familiar with Lomborg's controversial book, *The Skeptical Environmentalist* (2001), which used a variety of publicly available statistics to argue that many social and environmental problems are showing improvement, despite gloomy claims. The book was savaged by a number of scientists and eco-activists, who in effect called for his views to be banned. In Lomborg's home country, Denmark, he was subjected to similar attacks, but a high-level review committee rebuked claims that his work was shoddy or unsubstantiated. Although Lomborg's work is important, it is of course not the last word on this matter.
8. No page numbers are included. All subsequent references to Noble are taken from this essay.

REFERENCES

Begley, S. (2009). We can't get there from here. Political will and a price on C02 won't be enough to bring about low-carbon energy sources. *Newsweek*, Published March 14, 2009, from the magazine issue dated March 23, 2009. Available at http://www.newsweek.com/id/189293. Accessed on July 3, 2009.

Bryce, R. (2009). Let's get real about renewable energy. *The Wall Street Journal*, March 4. Available at http://online.wsj.com/article/SB123621221496034823.html. Accessed on July 3, 2009.

Fromm, E. (2005). *To have or to be?* New York: Continuum.

Hulme, M. (2008). The conquering of climate: Discourses of fear and their dissolution. *The Geographical Journal*, *174*(1), 5–16 Available at http://www3.interscience.wiley.com/journal/119390422/abstract?CRETRY = 1&SRETRY = 0. Accessed on July 3, 2009.

Krugmanm, P. (2009). Betraying the planet. *New York Times*, June 28. Available at http://www.nytimes.com/2009/06/29/opinion/29krugman.html. Accessed on July 5, 2009.

Lomborg, B. (2009). The climate-industrial complex. *The Wall Street Journal*, May 22. Available at http://online.wsj.com/article/SB124286145192740987.html. Accessed June 19, 2009.

Noble, D. F. (2007). The corporate climate coup. *Znet*, May 8. Available at http://www.zmag.org/znet/viewArticle/15472. Accessed on June 27, 2009.

Pradhan, B. (2009). India rejects any greenhouse-gas cuts under new climate treaty. *Bloomberg Press*, June 30. Available at http://www.bloomberg.com/apps/news?pid = 20601091&sid = aWs0Pts2Kxes.

Sagan, C., & Turco, R. (1990). *A path where no man thought: Nuclear winter and the end of the arms race*. New York: Random House.

St. James's Palace Memorandum. (2009). London, May 26–28. Available at http://extras.timesonline.co.uk/pdfs/sjp_memorandum_290509.pdf. Accessed on July 6, 2009.

'CHOOSE LIFE' NOT ECONOMIC GROWTH: CRITICAL SOCIAL THEORY FOR PEOPLE, PLANET AND FLOURISHING IN THE 'AGE OF NATURE'

John Barry

For *Current Perspectives in Social Theory*

> ...
> Choose a fucking big television
> Choose washing machines, cars, compact disc players, and electrical tin openers.
> ...
> Choose leisure wear and matching luggage.
> Choose a three piece suite on hire purchase in a range of fucking fabrics.
> Choose DIY and wondering who you are on a Sunday morning.
> Choose sitting on that couch watching mind-numbing sprit-crushing game shows
> Stuffing fucking junk food into your mouth.
> ...
> Choose your future. Choose life. (Irvine Welsh, Trainspotting, 1996)

> Anyone who believes exponential growth can go on forever in a finite world is either a madman or an economist (Kenneth Boulding)

Nature, Knowledge and Negation
Current Perspectives in Social Theory, Volume 26, 93–113

ISSN: 0278-1204/doi:10.1108/S0278-1204(2009)0000026006

INTRODUCTION

Robert J Antonio's 'Climate Change, the Resource Crunch, and the Global Growth Imperative' is an excellent starting point for examining the dominance of the ideology of 'economic growth' within contemporary societies and social thinking against the background of the growing and profound socio-ecological crises such as climate change, resource scarcity and rampant consumerism. His article integrates contemporary scientific evidence about climate change and the ecological devastation being wrought by contemporary patterns of globalised carbon-fuelled and climate changing capitalism with classical and contemporary sociological theorising and calls for a greater engagement of social theorising with these defining challenges of the 21st century. In particular, Antonio's article begins the process of sociological theory re-engaging and re-interpreting classical sociology (Weber, Durkheim, Mead and Dewey) to develop forms of social theorising, which overcome a simplistic nature–cultural dualism and can equip critical social theory as fit for purpose for analysing the ecological, climate and resource 'crunches' facing all societies in the 21st century.

In this article, I wish to use Antonio's article as a springboard to examine some of the issues raised in his article but not elaborated upon. Firstly, the outline of a viable and normatively attractive political alternative to a growth-orientated economy. Secondly, issues omitted from his analysis such as the invaluable insights from ecofeminist political economy; the wealth of alternatives to a growth-orientated economy one can find in green politics and the sub-discipline of green political economy; and the wealth of evidence demonstrating the well-being and equality detrimental effects of a growth-focused economy. While agreeing with most of what Antonio has to say, this article will be characterised throughout by taking issue with his conclusion that 'This Copernican turn calls for profound rethinking, redefining, reshaping, and redirecting of growth, *not* abandoning it'. This article will argue that we (in the industrialised North) need to abandon the *cornucopian* myth of endless 'material goods and services' central to which is a radical and profound rejection of growth in the 'minority' world of advanced 'affluenza-infected' consumer capitalism and its redirection to the majority world where economic growth is needed. 'One planet living', as a shorthand for sustainability demands this from a biophysical point of view (in terms of limited resources and ecological sinks) while global and egalitarian justice demands nothing less if a sustainable world is to be a just one.

GROWTHMANIA: THE CANCER STAGE OF CAPITALISM

John McMurtry in his wonderfully provocative book, *The Cancer Stage of Capitalism*, notes that

> the global market disorder ... has progressed by unseen money-sequence mutations and by systematically blocked connections to their life-destructive consequences. Throughout the planetary cancer's advance, it has found pathways of opportunistic invasion in an undetected 'pathogenic money code' whose mutating sequences have assumed forms unrecognised by economic theory and never seen by Karl Marx. *These life-attacking money sequences have typically invaded their social and environmental life-hosts by the non-living vehicles of corporate conglomerates.* (McMurtry, 1999, p. ix; emphasis added)

His analysis focuses on the issue of the vulnerability of 'life-conditions' and 'life-organisation', which includes environmental dimensions of human well-being. This 'cancer stage of capitalism' is the point where the global market and colluding states and transnational corporations are the most visible representations of an alien 'money sequence programme' or cancer that has invaded and is living off and destroying its human and non-human 'life host'. Essentially, this cancer stage of capitalism is characterised by what Marcuse (or Freud) would call the 'death instinct' (Thanatos) preying on the 'life instinct' (Eros) and in ecological terms the very life-supporting systems that sustain human and non-human flourishing. As McMurtry puts it, 'Increasingly we observe, the global market's investment cycles have no committed function to any life-organisation on the planet' (*ibid.*, p. 118).

The idea of the 'cancer stage of capitalism' was prefigured in many respects by the concerns of the green movement and ecological economists such as Herman Daly (discussed by Antonio) about 'limits to growth', the critique of 'growth-mania' in the late 1960s and early 1970s. The basic argument is well-established – beyond a threshold, beyond what we can call the 'sweet spot' according to Bill McKibben, author of *Deep Economy* (McKibben, 2007), growth turns from being a 'good', something that contributes to human and non-human well-being or life-sustaining, into a 'bad', something which threatens human and non-human long-term, that is, sustainable well-being. In short, economic growth or growthmania is a dominant ideology of 'growth for growth's sake', which is the ideology of the cancer cell. Growthmania as typified by debt-based and carbon-fuelled capitalist economic growth represents a form of life-denying and well-being undermining unhealthy development (or what Daly calls 'illith' – the opposite of health).

CLIMATE CHANGE, CULTURE AND SUBJECTIVITIES

Neoliberal American ideologues such as the American Enterprise Institute and others have been in denial about climate change and its profound implications for America and all societies – as outlined by Antonio in the first part of his article in revealing the 'truth' of right-wing think tanks and ideologues everywhere, namely, never let the empirical facts get in the way of one's preferred ideological agenda (see Antonio, this volume, p. 2). This has been in spite of the fact that 'orthodox' pro-establishment institutions such as the US military (pre-Obama) had already acknowledged the strategic and geopolitical importance of climate change (Pentagon study) or that the then UK Chief Scientist, Sir David King, had declared climate change a bigger threat than global terrorism.

Antonio's discussion of the science and likely impacts of climate change are comprehensive and underline a point made below, why we need to have a political economy based on science (and normative principles such as democracy and equality) rather than neo-liberal ideology. In short, the economy is far too important to be left to orthodox economists to design and manage. His discussion of the proposal for the 'Waveyard' water park for the Arizona desert is a vivid reminder of the ideological power of conventional economics to trump scientific and ecological rationality and why liquidating natural capital makes sense within the logic of a capitalist socio-economic order. The logic that presents this unsustainable project as not just viable but desirable is the same logic by which neo-liberal globalisers can justify US-style 'five planet living' as viable for the world's population as a whole. Neither ecological rationality nor scientifically informed arguments for 'one planet living' are on the ideological agenda of the cheerleaders for carbon and debt-fuelled consumer capitalism, confident as they are, as Antonio assiduously points out that their version of the Enlightenment project proves humanity can and should divorce itself from nature and its limits. As Antonio puts it, 'the prospects of [American middle class lifestyles] global diffusion … punctuates dramatically the depth of the *contradiction between growth and nature*' (see Antonio, this volume, p. 20; emphasis added).

Equally, the non-ecological and non-economic cultural dimensions of climate change also need to be acknowledged. As renowned climate scientist, Mike Hulme, has noted in a recent book on climate change:

> Apprehending the risks of climate change is a deeply cultural and social activity. If people perceive themselves as vulnerable to climate risks, either by prior experience or

> through falling into social or geographical categories deemed to be at risk, they are more likely to adopt the high-risk narratives of climate change espoused by some commentators. On the other hand, people living in cultures that are more risk-tolerant, or who fall into social categories that are deemed to have a high coping capacity for climate risks, may well be more sceptical about the quantitative risk assessments that emerge from science and that are subsequently amplified across society. (Hulme, 2009, p. 208)

This cultural and social dimension of perceptions of climate risk is of course central to any sociological analysis of climate change. An interesting example of a sociologically informed analysis of the cultural (and psychological) impacts of climate change from my own country, Ireland, is a 2008 report produced by the Irish American Climate Project (Sweeny et al., 2008). This report integrated scientific projections of likely climate change impacts on the island of Ireland with qualitative analysis of the cultural, political and social impacts in an innovative and interdisciplinary manner. The report included 'discussions of music and poetry because they explain the intense connections between the Irish landscape and Irish culture and how changes to one can affect the other' (*ibid.*, p. 2). A startling conclusion of the report is that due to water shortages as a result of climate change in the latter half of the 21st century, 'the potato ... may cease being a commercially viable crop on the island' (*ibid.*, p. 30). This is not simply a food/agricultural problem, but given the enormous icon and identity, constitutive symbolism of the potato is at the same time a cultural problem and a collective identity problem. What does Irish identity (related to ecological/climatic/agricultural practices) look like in a climate changed world? What identities, or which aspects, are to be re-negotiated or changed as a result of changed cultural–ecological relations? These cultural and psychological impacts of climate change (not all of which may be 'problems' or necessarily negative) also need to be factored into any sociologically informed analysis.

Thus, the 'powerful car culture' in America, which Antonio rightly points to as yet another element of the unsustainable lifestyle, needs to be equally seen as identity forming and not simply viewed from a resources and energy point of view (though important). Antonio also rightly calls for the need for a radical change in American middle-class lifestyles, but what sorts of alternative identities and alternative cultures do we need to consider in advocating this necessary step-change? Mat Paterson's recent book, *Automobile Politics*, offers a potential line of analysis bringing together the ecological/resource and energy dynamics and impacts of 'car culture' together with an analysis of its underlying growth and capitalist-orientated political economy of production and consumption, but more pertinent from a sociological perspective connects the power of this car culture to how 'automobile subjectivity' and 'automobile identities' are created and sustained. As he puts it, 'the rise of reproduction of automobility should be understood principally in terms of the particular way in which capitalist development was secured in the early twentieth century, on the one hand, and *the construction of particular sorts of mobile subjects, on the other*' (Paterson, 2007, p. 30). What alternative, non-car-based 'mobile identities' are possible or needed in the transition away from unsustainable car-based mobility? Paterson notes that 'the normative prognosis suggests, on the one hand, the need to attempt to detach automobility from dominant forms of subjectivity, but on the other, to accept that further 'greening' necessarily involves *governmentality* – a reshaping of what sort of people people are' (*ibid.*, p. 195). But of course while suggestive, this leaves a lot of theoretical and empirical work to be done in terms of fleshing out what these forms of 'green governmentality' may entail and how and who should reshape peoples' identities and senses of self. For reasons of space, it would seem to me that a place to start, not just in relation to mobility but the full range of sustainability issues, would be in moving from a critical analysis of the consumer and producer to that of the citizen (Barry, 2008). What are the 'citizen subjectivities' and associated practices and institutions compatible with or necessary for the transition away from unsustainability economic growth patterns and associated state, market and cultural formations?

At the same time, attention must be paid to the fact that sheer scale, rapidity and incontrovertible evidence about climate change is matched only by the prevarication of governments and other influential groups, especially key sectors of business and the academy (especially economics), and by the 'cognitive dissonance' displayed by millions of citizens who proclaim to know about and accept that their energy-intense, high-consumption and high-mobility lifestyles (and associated economic system and technological infrastructure) are the root causes of global and local ecological breakdown, but who either refuse or are unable to change their lifestyles to enable 'one planet living'. As Schellenberger and Nordhaus put it, 'while public support for action on global warming is wide it is also frighteningly shallow' (Shellenberger & Nordhaus, 2006, p. 9).

PEAK OIL AND A POST-CARBON WORLD

However, while climate change is doubtless the global socio-ecological challenge of the 21st century, and it has now achieved the media and policy focus it warrants, its causes, namely the burning of finite energy resources such as coal, oil and gas gains less media attention. That peak oil and gas is the flip side of climate change means that one cannot coherently discuss the latter without the former and vice versa. One of the issues not discussed in Antonio's article is the reality and implications of 'peak oil' and the implications of this for carbon-fuelled capitalism. It is interesting however to note that Antonio quotes the following from Julian Simon, 'The trend towards greater availability includes the most counterintuitive of all – oil' (see Antonio, this volume, p. 5). This is flatly denied of course by ecologists, greens and more recently the growing 'peak oil' and post-carbon movements and related movements such as the Transition Towns movement (discussed below) and what I call a growing 'ecological realism' or 'hard green' political perspective. This new ecological realism can be witnessed in the huge sales of books such as Jared Diamond's *Collapse: How Societies Choose to Succeed or Fail* (Diamond, 2006), Thomas Homer-Dixon's *The Upside of Down* (Homer-Dixon, 2006), Alister McIntosh's *Hell and High Water: Climate Change, Hope and the Human Condition* (McIntosh, 2008), Derrick Jensen's *The Culture of Make-Believe* (Jensen, 2002), the Odums' *A Prosperous Way Down* (Odum & Odum, 2001), James Howard-Kunstler's *The Long Emergency* (Howard-Kunstler, 2006) and Thomas Friedman's *Hot, Flat and Crowded* (Friedman, 2008). These books have complemented documentaries such as Al Gore's *An Inconvenient Truth*, Leonardo di-Capro's *Eleventh Hour* to less well-known ones such as *What a Way to Go: Life at the End of Empire*; *The End of Suburbia*; *Escape from Suburbia* and *A Crude Awakening: The Oil Crash*. These, and others, can be seen as articulating vital elements of the position Antonio rightly wishes to revive in his analysis of Dewey and Mead and their shared project of viewing human society as embedded in the non-human world and outlining the contours of not just ecologically literate social theorising but also one which goes beyond anthropocentric concerns alone. While neither classical sociological theory nor Antonio's article engages with non-anthropocentrism, there is, I would suggest a rich and fruitful research agenda for the development of non-anthropocentric social theorising, or at least for much more self-reflexive and critical anthropocentric social theorising (Barry, 1999).

While once confined to the margins, the peak oil position has (like climate change to which it is of course inextricably linked) increasingly moved centre stage in political and policy debates. In America, the Hirsch report released in 2005 was the first 'official' government recognition of the reality and geo-political and economic implications of peak oil for the United States. Its executive summary noted:

> The peaking of world oil production presents the U.S. and the world with an unprecedented risk management problem. As peaking is approached, liquid fuel prices and price volatility will increase dramatically, and, without timely mitigation, the economic, social, and political costs will be unprecedented. Viable mitigation options exist on both the supply and demand sides, but to have substantial impact, they must be initiated more than a decade in advance of peaking ... *In summary, the problem of the peaking of world conventional oil production is unlike any yet faced by modern industrial society.* (Hirsch, Bezdek, & Wendling, 2005, pp. 4, 7; emphasis added)

Given that modern industrialised societies are dependent on oil for transportation, electricity and heating, chemical fertilisers and food production, synthetic clothing and a whole vast range of other goods and services, the peaking of oil production is a hugely significant event, a paradigm shifting event. The peaking of oil does not mean that oil is about to run out, but rather we have passed the point of maximum production and therefore the geological fact of there being less oil available leading to the end of the era of cheap oil. And peak oil heralds nothing less than the wholesale restructuring of industrial civilisation as we know it, since, to put it crudely, that civilisation (enjoyed by a minority of the world's population) is 'addicted to oil'. None other than former U.S. President George W. Bush stated this in his 2006 State of the Union Address. According to him, 'Keeping America competitive requires affordable energy. And here we have a serious problem: *America is addicted to oil*, which is often imported from unstable parts of the world' (Bush, 2006; emphasis added). And from this flows a whole new 'national security' agenda for America and other nations in their pursuit of 'energy security' – one of the key objectives of current U.S. president Obama's 'Green New Deal'. This addiction and dependence on imported fossil fuels leaves countries vulnerable, and of course, the need to secure oil from the Middle East is one of the most compelling realpolitik reasons for the invasion and occupation of Iraq (given it has up to 17% of the world's proven oil reserves). If America and other Western countries are so dependent on this source of energy for everything from electricity generation to food production, is it little wonder we can see both current and coming oil and gas wars as well as the militarisation and securitisation of energy?

The reality of peak oil and the need to both 'detox' from carbon dioxide–emitting fossil fuels and begin the transition towards low and post-carbon energy can be seen in that this is no longer a fringe green position but now has found advocates within the establishment. An excellent example of this is the International Energy Agency's chief economist Fatih Birol who has declared that 'One day we will run out of oil, it is not today or tomorrow, but one day we will run out of oil and *we have to leave oil before oil leaves us*, and we have to prepare ourselves for that day. The earlier we start, the better, because all of our economic and social system is based on oil, so to change from that will take a lot of time and a lot of money and we should take this issue very seriously' (Birol, 2009; emphasis added). What are the sociological implications of peak oil and the dawn of post-carbon societies? As with climate change, but perhaps even more so given the potential for peak oil to mobilise governments, market actors and communities (as in the Transition Town movement discussed below, Quilley & Barry, 2008), we can see in peak oil and its potential societal-wide implications the 'return of nature'. As Thomas Homer-Dixon reminds us, 'The twenty-first century will in fact be the Age of Nature. We'll learn, probably the hard way, that nature matters: we're not separate from it, we're dependent on it, and when there's trouble in nature, there's trouble in society' (Homer-Dixon, 2006, p. 13). And of course this sentiment runs through Antonio's argument for a post-exuberant, ecologically embedded social theory that sees 'humanity and culture as part of nature, dependent upon it, and constrained by it' (see Antonio, this volume, p. 31).

SUSTAINABLE ALTERNATIVES TO GROWTHMANIA

While Antonio's article offers some excellent and well-informed lines of critique of conventional economic growth and what I would call the ideology of growthmania, we do not get a sense of what the alternatives are to the ecologically irrational model of globalised consumer capitalism. This section will briefly indicate what these alternatives might look like.

I begin this exploration by noting the following from green economist, Tim Jackson, 'Every society clings to a myth by which it lives. Ours is the myth of economic growth. For the last five decades the pursuit of growth has been the single most important policy goal across the world' (Jackson, 2009, p. 5). Thus, Jackson begins his recent report for the United Kingdom's official Sustainable Development Commission, *Prosperity without Growth: the Transition to a Sustainable Economy*. This seminal publication represents both the most eloquent and the evidence-based demolition of conventional growth-orientated economics and social theorising as well as a wonderful summary and précis of over 50 years of 'dissident' and heterodox critiques of neo-classical economic growth thinking and policy-making from a largely but not exclusively green perspective. The report outlines the necessity as well as desirability of moving beyond conventional economic growth as a public policy goal and myth or collective narrative towards a 'post-growth' position (which fits neatly with the post-carbon and climate change analyses above).

The critique of conventional economic growth has been a long-standing position of green thinking. It is interesting that while Antonio rightly highlights the work of Herman Daly as a pioneering ecological economist pointing out the negative ecological and socio-economic impacts of a political economic of 'growthmania', there is no sustained discussion of the 'Limits to Growth' hypothesis and school of thinking from which Daly and ecological economics emerges (apart from Antonio noting Daniel Bell's dismissal of the main hypothesis of the 'Limits to Growth' position). The 'Limits to Growth' debate of the late 1960s and early 1970s is the intellectual and political context from which Daly (Daly, 1974) and ecological economics emerges and indeed informs both the 'ecological realism' outlined above and the political objectives of most Green parties and large sections of the green/ecological movement, including the recent Transition Movement as indicated below. There are essentially two parts of this limits position – one is well covered by Antonio, namely the limits in terms of non-renewable resources (such as fossil fuels) and the limits in terms of the absorptive capacity of the biosphere to deal with human pollution (climate change being the most serious). In short, one-planet living means just that – if sections of the human population are living at levels of resource, energy and pollution levels beyond their 'fair share', this means either:

(a) this lifestyle is not sustainable in the long term and/or
(b) it is biophysically impossible for this lifestyle to be extended to the world's population.

The second part of the limits to growth debate relates to non-ecological/resource issues in relation to the impact of 'economic growth' on human well-being. Here, the argument is that while ecological limits means exponential conventional economic growth is simply *biophysically impossible*, the negative impacts on human well-being after a threshold means economic growth is *normatively undesirable*. This is for a number of reasons – here, for reasons of space, I will focus on the negative impact of economic growth on well-being, equality and global justice.

Growth and Well-Being

A major report by the International Labour Office *Economic Security for a Better World* found that 'economic security' (not economic growth) coupled with democracy and equality were key determinants of well-being and social stability. According to this report:

> People in countries that provide citizens with a high level of economic security have a higher level of happiness on average, as measured by surveys of national levels of life-satisfaction and happiness ... The most important determinant of national happiness if not income level – there is a positive association, but rising income seems to have little effect as wealthy countries grow more wealthier. *Rather the key factor is the extent of income security, measured in terms of income protection and a low degree of income inequality.* (International Labor Organisation, 2004, p. 23; emphasis added)

Such findings give empirical support to long-standing green arguments stressing the need for policies to increase well-being and quality of life, rather than conventionally measured economic growth and gross domestic product (GDP), rising personal income levels or orthodox measures of wealth and prosperity. Therefore, one possibility is for 'economic security' to replace 'economic growth' as a macro-economic objective.

Another is to highlight the empirical evidence we have in the industrialised world, which conclusively demonstrates that despite over 40 years of (intermittent) economic growth, greater wealth creation and so on, the populations of these societies are no better off in terms of well-being than they were in the immediate post-war era. As Gus Speth (Speth, 2006) in America and Tim Jackson (Jackson, 2009) in the United Kingdom point out, the statistical evidence is telling us the reality of the 'crocodile graph' demonstrating that while GDP per capita increases, well-being has plateaued out for decades (Graphs 1 and 2). This point has also been noted by official government bodies. For example, the UK's Cabinet Office published a report in 2002, which suggested that 'despite large increases in national income (and expenditure) over the last 30 years, levels of life satisfaction have not increased commensurately' (Donavan & Halpern, 2002, p. 2).

As Victor has eloquently put it, 'Americans have been more successful decoupling GDP from happiness than in decoupling it from material and energy' (Victor, 2008, p. 125). A Swedish Environmental Protection Agency report perhaps nailed the issue best when it stated, 'The ultimate question facing today's society in developed countries is whether consumerism actually contributes to human welfare and happiness ... Strategies are missing that

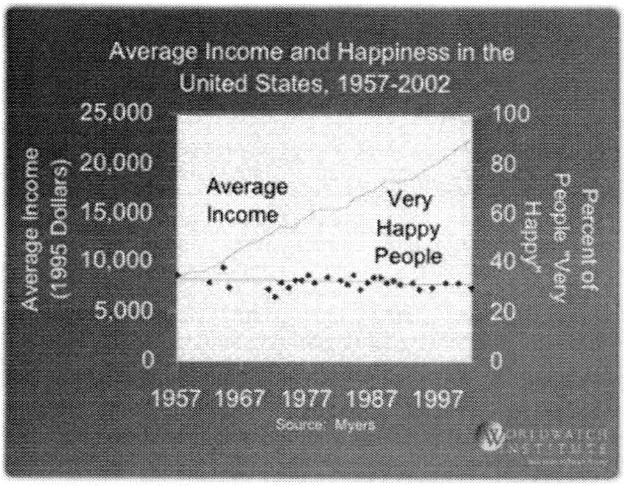

Graph 1. Average Income and Percentage of 'Very Happy' People (1957–2002), USA (Worldwatch Institute).

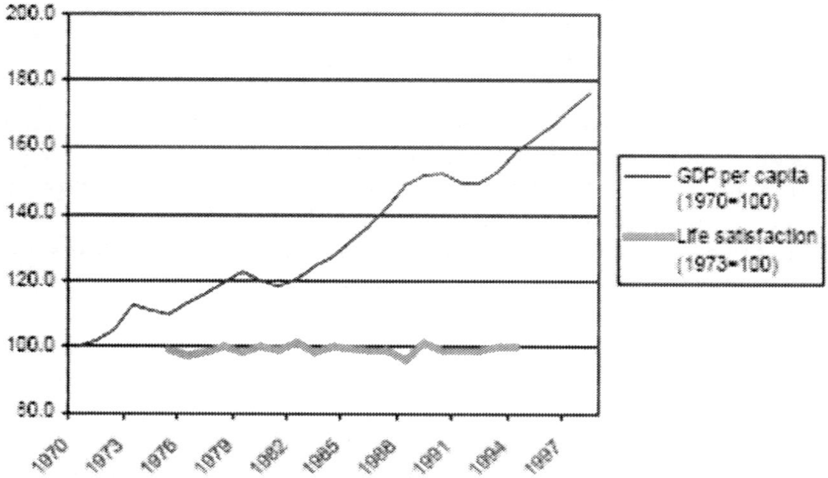

Graph 2. Life Satisfaction and GDP per Capita (1970–1997), UK (Donavan & Halpern, 2002).

would conceive ways of shifting from current culture of limitless consumerism to a society with less materialistic aspirations' (Mont & Plepys, 2008, p. 536).

While there are a variety of reasons for this disjuncture between income/GDP and well-being, including what Donavan and Halpern term 'hedonic adaptation' (Donavan & Halpern, 2002, p. 19) – namely the fact that the well-being benefits of increased consumption are temporary and soon wear off, requiring further consumption (and work and income) – the key question we need to ask here is this: if conventional economic growth isn't working to make us happier or contribute to our long-term well-being, why are we locked into a growth paradigm? This is particularly pertinent when we consider that as far back as the late 19th century, John Stuart Mill in a remarkably prescient passage was already raising the issue of the need for western capitalist industrial societies to move 'beyond growth'. As he eloquently put it,

> I cannot, therefore, regard the stationary state of capital and wealth with the unaffected aversion to it so generally manifested towards it by political economists of the old school. I am inclined to believe that it would be, on the whole, a very considerable improvement on our present condition ... It may be a necessary stage in the progress of civilization ... *But the best state for human nature is that in which, while no one is poor, no one desires to be richer, nor has any reason to fear being thrust back, by the efforts of others to push themselves forward* ... It is only in the backward countries of the world that increased production is still an important object: in those most advanced, what is economically needed is better distribution, of which one indispensable means is a stricter restraint on population. (Mill, 1900, p. 453; emphasis added)

Is economic growth and its associated modes of consumerism and overconsumption in the developed world under free-market capitalism – all of which require more use of natural resources, energy, human labour and creativity – rather similar to a drug so that we can say that what we need in the developed world is to detox from the addiction to economic growth and consumerism?

Inequality and Growth

While Antonio is correct to focus on the neo-liberal supporters of conventional economic growth – such as the American Enterprise Institute, Julian Simon, Friedrich Hayek and his followers – what is not more fully developed in his article is the crucial fact that these neo-liberal ideologues promote not just economic growth but *unequal economic growth*. Inequality is central (indeed functional) to the neo-liberal economic vision since it is argued that inequality creates the necessary incentives to reward innovation, hard work and greater productivity, which leads to greater economic growth. As the same time, economic growth acts as a substitute for demands or policies for greater socio-economic equality (or lowering levels of inequality) since the political and policy focus is on absolute rather than relative shares of income or wealth in a growing economy. Thus, it is necessary for inequality to increase or (at best) stay the same in a growth-orientated economy. The empirical evidence from the past 20 years of neo-liberal economic policies is of course that inequality has increased in countries such as America and the United Kingdom in that fewer people are now owners of more of the nation's wealth and productive assets. This indicates that what we have seen in those countries committed to neo-liberal economic growth policies is not a Keynesian 'trickle down' effect but the opposite a 'steam up effect' as wealth has been transferred from the poorer to the richer. This has also happened globally in that the world today is more unequal in terms of the distribution of income, wealth and productive assets (Pogge, 2007).

As Wilkinson and Pickett demonstrate in their recent book, *The Spirit Level*, unequal societies (such as the United Kingdom and United States) do worst when compared with less unequal societies whether one measures childhood or adult obesity, teenage pregnancy, rates of recycling, social capital and trust, political participation all of which mean that more unequal societies have lower levels of well-being when compared with less unequal societies (Wilkinson & Pickett, 2009a). As they put it,

> Not only has economic growth in the rich countries ceased to bring the social benefits it once brought (and continues to bring in poorer countries), but it now threatens the planet. We are therefore the first generation to have to find new ways of improving the real quality of life. The evidence suggests that we need to shift our attention away from increasing material wealth, to the social environment and the quality of social relations in our societies. For rich countries to get even richer makes little or no difference to the prevalence of health and social problems but, as other pages of this web site make clear, the social problems which beset many rich societies are much more common in more unequal societies. Societies with smaller income differences between rich and poor are more cohesive: community life is stronger, levels of trust are higher and there is less violence. *The vast majority of the population seem to benefit from greater equality.* (Wilkinson & Pickett, 2009b)

The bottom line here is that if one is an egalitarian or wishes to see less levels of inequality in society, then supporting conventional economic growth is counterproductive. Thus, egalitarians and progressives should support post- or degrowth positions since this directly addresses the distribution of work, income, assets and the relative distribution of rewards and benefits (as well as burdens) in society. This suggests that critical social theory if it wishes to normatively promote less inegalitarian socio-economic relations needs to move towards advocating a post-growth perspective (at least in the minority industrialised North).

Global Justice

One possible answer to Antonio's pertinent question 'How can China, India and other developing nations provide their rural and often impoverished masses electricity and still reduce carbon emissions?' (see Antonio, this volume, p. 23) can be found in the 'degrowth' or 'post-growth' perspective of green political economy in which the key to global sustainability is for the minority world to stop conventional economic growth and freeing up ecological and energy resources to enable the majority world to develop and progress. But this global perspective, based on the distribution of development opportunities globally, requires different forms of development than the 'one size fits all' one of debt- and carbon-fuelled, growth-orientated capitalism. And here again opens up a new research terrain for critical social theory – what are the sustainable socio-ecological-economic models necessary or desirable for 'global sustainability' and the goal of 'one-planet living'? If sustainability will be different in Boston, Bogotá, Berlin and Beijing, what will its localised embedding and institutionalisation look like?

At the same time, a non- or degrowth or post-growth economic strategy would have to shift its focus towards providing more free time for citizens in the North rather than focusing on productivity gains aimed at increasing more commodities. As Schor has succinctly put it,

> in the global North a successful path to sustainability must confront our commitment to growth and will ultimately entail a stabilization of consumption through reductions in hours of work. *Indeed, it is difficult to imagine a globally ethical, timely, and politically feasible resolution to the global ecological crisis in which populations in the North do not reduce the number of hours worked per capita.* (Schor, 2005, p. 38)

TRANSITIONS AWAY FROM GROWTH: CREATIVE DESCENT?

The 'Transition Town' movement in the United Kingdom, Ireland, Australia and America, beginning as it does from the premise of preparing communities to be resilient in the face of peak oil and climate change, offers some very pertinent and suggestive ideas and practices of what post-oil, climate-changed societies might look like. A key aspect of the Transition Movement is that it strongly indicates that, ruling out nuclear power and large-scale biofuels, if we wish to create a renewable energy economy, *this means an economy and polity with less energy*. This is why a, if not the, key aspect of the Transition process is for each community to create what is called an Energy Descent Action Plan (EDAP). Planning for a future with less energy represents an enormous transformation of high-energy, consumer societies in every sector and level – indicating a 'full spectrum' transformation of society and the economy as it were. Hopkins, the founder of the Transition movement, describes energy descent as follows:

> the continual decline in net energy supporting humanity, a decline which mirrors the ascent in net energy that has taken place since the Industrial Revolution. It also refers to a future scenario in which humanity has successfully adapted to the declining net energy availability and has become more localised and self-reliant. It is a term favoured by people looking towards energy peak as an opportunity for positive change rather than an inevitable disaster. (Hopkins, 2006, p. 19)

Energy descent planning is a key feature of this movement, arguing that creativity and innovation can be released by the power of a community joined together to create a more resilient and positive vision of their future in a carbon-constrained and climate-changed world. Transition as a movement is therefore, in part, a politics of hope in the face of the geological and economic facts of the end of the era of cheap oil and the dawning of a post-oil economy and the facts of a climate changed world in which *adaptation* to inevitable climate change will be as important as *mitigation* of future climate change based on reducing CO_2 emissions and the transition to a low and post-carbon energy economy.

One of the strengths of the Transition perspective is that it locks climate change and peak oil together as delimiting the major biophysical horizons of collective human nature in the 21st century. As Shaun Chamberlain (2009) points out,

> On reading the peak oil literature it is clear that the concerns revolve around depletion and shortage. However, it is also clear that there is more than enough carbon in the fossil fuels we have discovered to bring about catastrophic changes in our climate and environment. There is the clear necessity of leaving existing fossil fuels in the ground. It is rarely considered that this turns the scarcity paradigm in its head, and we could find ourselves in a situation in which unexploited fossil fuel reserves are regarded as valuable but dangerous resources which need careful supervision or even guarding by governments to prevent their extraction. We are used to considering resource depletion

in terms of having too little of a good thing. *We may need to start thinking that there is effectively too much of a bad thing.* (p. 128)

The implication of leaving oil and other fossil fuels in the ground has major economic, political and cultural as well as ethical import. An indication of this is the statement by NASA climate scientist James Hansen, one of the leading climate change scientists in the world, that we need to abandon the burning of coal. In a lead article in the UK-based *Observer* newspaper in February 2009, Hansen declared the following:

Coal is not only the largest fossil fuel reservoir of carbon dioxide, it is the dirtiest fuel. Coal is polluting the world's oceans and streams with mercury, arsenic and other dangerous chemicals. The dirtiest trick that governments play on their citizens is the pretence that they are working on 'clean coal' or that they will build power plants that are 'capture-ready' in case technology is ever developed to capture all pollutants. The trains carrying coal to power plants are death trains. *Coal-fired power plants are factories of death.* (Hansen, 2009; emphasis added)

This explicit political engagement by a scientist, which has characterised the 'science wars' around climate change for the past 20 years, is of course a pertinent and pressing topic for analysis and explanation by social science. The politicisation of science can now be seen as the counterpart to the scientisation of politics as critical social theorists such as Ulrich Beck and Anthony Giddens and others working in the sociology of science and technology have long suggested. This politicisation of science not only occurs with climate change but also debates around 'peak oil', thus giving such established approaches to the study of knowledge (especially so-called expert and authoritative) production, dissemination and consumption of a new impetus. However, another significant point, and one raised above in relation to what an economy would look like if designed by a scientist rather than a neo-liberal economist, is that social movements such as the Transition one see themselves as firmly informed by science (and while not dismissing as 'non-scientific', local and indigenous forms of expert knowledge) and see themselves as offering creative and innovation localised experiments in fusing science, ethics and new models of economic activity to ensure community resilience in the face of adapting to a climate-changed and carbon-constrained world.

While of course not containing a complete solution to the inter-linked triple crunch of climate change, peak oil and the financial/economic crisis, the Transition perspective deserves serious study by social theorists as a grassroots, 'viral' response to these challenges and opportunities. In particular what peak oil, climate change, the economic recession and the continuing degradation of the planet's life support systems seems to call for is not just a post-exuberant, post-growth and ecologically embedded sociology and critical social theory, but even more than that. Rather provocatively, one could suggest that what these interlinked socio-cultural-ecological crises call for is a more 'mission-led' critical social theory, that is, one that takes these currently problematic interrelated nature–culture relations as the central or at least the starting point for any serious and progressive analysis of the current 'human condition'. That is, can we say that the issues raised in both Antonio's article and the current article suggest that any sociological account that does not foreground these issues is 'ecologically illiterate' and therefore of limited use in helping us understand, explain and analyse human (and non-human) societies in the 21st century? (Barry, 2007) Are forms of critical social theory 'fit for purpose' unless they integrate the human relationship with and utter dependence on the non-human world – through the material or metabolic throughput as given by resources, energy throughput and pollution, but equally through non-material relations between the human and the more than human world, as given by, inter alia, symbolic, cultural, identity-constitutive and normative dimensions?

CONCLUSION

Pierre Bourdieu notes the following:

> the strength of the new dominant order is that it has found the specific means of 'integrating' (in some cases you might say buying, in others seducing) an ever-growing fraction of the intellectuals, all over the world. These 'integrated' intellectuals often continue to see themselves as critical (or simply on the left), according to the traditional model. And that helps to give great symbolic efficacy to their work in rallying support for the established order. Bourdieu (1999, p. 65)

Now while I am *not* saying Antonio is one of these bought intellectuals, cheer leading for capitalist globalisation, it is unfortunate that his conclusion in defence of economic growth does at least indicate to me the cultural and geographical specificity of where he is coming from. His concluding sentence that 'This Copernican turn calls for profound rethinking, redefining, reshaping, and redirecting of growth, *not* abandoning it' is the most disappointing part of what was otherwise an excellent article outlining the empirical evidence and normative arguments against economic growth. Growth, more, maximisation, mobility, rapidity and speed are all so ingrained in the American cultural imaginary that perhaps even critical social theorists have to work within rather than against these imagined and societal desirables. And I do admit that from a political strategic point of view, speaking a language of 'redirecting growth' rather than directly debunking it has more potential to mobilise people – especially in America. After all, it is one thing to be one step ahead of your supporters, quite another to 5 or 10 steps ahead. But, at the same time, as indicated above, there are multiple and compelling ecological-, egalitarian-, well-being- and sustainability-related reasons why we need to abandon growth in the minority 'developed' world and allow what growth potential there is, in those countries where need is greatest.

If we, as a species, are to have any future, it will be one, I would suggest from the available scientific evidence and normatively mobilisable and attractive principled views of the future, based not on the principles of 'growth', 'more' and 'speed' but on the alternative imperatives of post-growth, an ethics of sufficiency, enough, and a focus on the slow and more localised solutions. Suggestive intimations of what that may look like can be found in social movements and mobilisations around the world which make 'livelihood', 'simplicity', economic security, protection of the commons and 'environmental justice' their key demands.

Growth beyond a human life sustaining and human well-being sustaining threshold can too quickly become the ideology of the cancer cell, that is, growth for growth's sake. We need to ask ourselves the most basic and most radical question at the beginning of the 21st century – what is an economy for? Surely one of the criterion we should use is its contribution to sustaining human and non-human life, and beyond that its contribution to human (and I would also include non-human) flourishing. Allied to that thought we must also consider that perhaps one of the most politically radical things we can do in a high consumer culture is very simple. Refuse to consume. Life-sustaining and life-affirming politics, an economy focused on flourishing and a politics of resistance based in part on 'the great refusal' to overconsume and to focus on creativity and sharing more equally the products of the sustainable metabolism between human economy and non-human ecology … now there's a research and political agenda for critical social theory for the 21st century! And above all, conceptualisations of 'progress' which are measured not by increases in economic growth, GDP and income or wealth, but how a society looks after its most vulnerable – the old, the sick and the young. And in this project for a post-exuberant critical social theorising, Robert Antonio has more to offer.

REFERENCES

Barry, J. (1999). *Rethinking green politics: Nature, virtue and progress*. London: Sage.

Barry, J. (2007). *Environment and social theory* (2nd ed.). London: Routledge.

Barry, J. (2008). Towards a green republicanism: Constitutionalism, political economy, and the green state. *The Good Society: Committee on the Political Economy of the Good Society, 17*(2), 1–12.

Birol, F. (2009). Leave oil before it leaves us, Energy Bulletin. Available at http://www.energybulletin.net/node/43604. Accessed on 24 May 2009.

Bourdieu, P. (1999). *Acts of resistance: Against the tyranny of the market*. New York: New Press.

Bush, G. W. (2006). State of the Union Address. Available at http://georgewbush-whitehouse.archives.gov/stateoftheunion/2006/. Accessed on 6 March 2008.

Chamberlin, S. (2009). *The transition timeline for a local, resilient future*. Totnes (UK): Green Books.

Daly, H. (1974). Steady-state economics versus growthmania: A critique of the orthodox conceptions of growth, wants, scarcity, and efficiency. *Policy Sciences, 5*(2), 149–167.

Diamond, J. (2006). *Collapse: How societies choose to succeed or fail*. London: Penguin.

Donavan, N., & Halpern, D. (2002). *Life satisfaction: The state of knowledge and implications for government*. London: UK Strategy Unit, Cabinet Office.

Friedman, T. (2008). *Hot, flat and crowded: Why we need a green revolution – And how it can renew America*. New York: Farrar, Straus and Giroux.

Hansen, J. (2009). Coal-fired power stations are death factories. Close them. *The Observer*, 15 February 2009. Available at http://www.guardian.co.uk/commentisfree/2009/feb/15/james-hansen-power-plants-coal. Accessed on 20 February 2009.

Hirsch, R., Bezdek, R., & Wendling, R. (2005). Peaking of world oil production: impacts, mitigation and risk management, US Department of Energy. Available at http://www.netl.doe.gov/publications/others/pdf/Oil_Peaking_NETL.pdf. Accessed on 1 August 2009.

Homer-Dixon, T. (2006). *The upside of down: Catastrophe, creativity and the renewal of civilization*. Toronto: Knopf Canada.

Hopkins, R. (2006). *Energy descent pathways: Evaluating potential responses to peak oil*. MSc Dissertation, University of Plymouth. Available at http://totnes.transitionnetwork.org/system/files/msc-dissertation-publishable-copy.pdf. Accessed on 1 August 2009.

Howard-Kunstler, J. (2006). *The long emergency: Surviving the end of oil, climate change and other converging catastrophes of the twenty-first century*. New York: Grove Press.

Hulme, M. (2009). *Why we disagree about climate change: Understanding controversy, inaction and opportunity*. Cambridge: Cambridge University Press.

International Labor Organisation. (2004). *Economic security for a better world*. Geneva: International Labor Organisation.

Jackson, T. (2009). *Prosperity without growth*. London: Sustainable Development Commission.

Jensen, D. (2002). *The culture of make believe*. New York: Context Books.

McIntosh, A. (2008). *Hell and high water: Climate change, hope and the human condition*. Edinburgh: Berlinn.

McKibben, B. (2007). *Deep economy: The wealth of communities and the durable future*. New York: Henry Holt.

McMurtry, J. (1999). *The cancer stage of capitalism*. London: Pluto Press.

Mill, J. S. (1848/1900). *Principles of political economy*. London: Longmans, Green and Co.

Mont, O., & Plepys, A. (2008). Sustainable consumption progress: Should we be proud or alarmed? *Journal of Cleaner Production*, *16*(4), 531–537.

Odum, H. T., & Odum, E. C. (2001). *A prosperous way down: Principles and policies*. Boulder: University Press of Colorado.

Paterson, M. (2007). *Automobile politics: Ecology and cultural political economy*. Cambridge: Cambridge University Press.

Pogge, T. (2007). Why inequality matters. In: D. Held & A. Kaya (Eds), *Global inequality: Patterns and explanations*. Cambridge: Polity.

Quilley, S., & Barry, J. (2008). Transition towns: 'Survival', 'resilience' and the elusive paradigm shift in sustainable living. *Ecopolitics Online*, *1*(2), 12–31.

Schor, J. (2005). Sustainable consumption and worktime reduction. *Journal of Industrial Ecology*, *9*(1/2), 37–50.

Shellenberger, M., & Nordhaus, T. (2006). The death of environmentalism: Global warming politics in a post-environmental world. Available at http://www.thebreakthrough.org/PDF/Death_of_Environmentalism.pdf. Accessed on 15 February 2007.

Speth, G. (2006). Money can't buy you love or happiness. Available at http://environment.research.yale.edu/documents/downloads/a-g/EnvYale-CantBuyYouLove.pdf. Accessed on 3 August 2009.

Sweeny, K., Fealy, R., McElwain, L., Siggins, L., Sweeney, J., & Trinies, V. (2008). Changing shades of green: The environmental and cultural impacts of climate change in Ireland, The Irish American Climate Project. Available at www.irishclimate.org. Accessed on 28 July 2009.

Victor, P. (2008). *Managing without growth: Slower by design, not disaster*. Cheltenham, UK: Edward Elgar.

Wilkinson, R., & Pickett, K. (2009a). *The spirit level: Why more equal societies almost always do better*. London: Allen Lane.

Wilkinson, R., & Pickett, K. (2009b). Equality not growth. Available at http://www.equalitytrust.org.uk/why/equality-not-growth. Accessed on 3 August 2009.

REPLY TO MY CRITICS: CHOOSING LIFE

Robert J. Antonio

"The earthquake reveals new springs."

– Nietzsche

Michael E. Zimmerman (p. 76) states that his comments are *occasioned* by my essay and that not every critical point that he poses is addressed to my "specific positions." For the most part, he expresses his own concerns. Although Professor Zimmerman chose not to engage my core argument, he challenges its underpinnings and intent – that is, that we have a global ecological crisis, that we need to make growth problematic and rethink it, and that public intervention and planning is needed. My response to his comments is intended to draw out our fundamental differences. I say less about our points of convergence. From a position much closer to my own views, John Barry engages my claims more determinately, but also uses the opportunity to articulate his own positions, which arguably, at many points, extend constructively themes that I have addressed in part or fill gaps in my perspective. I say less about his comments, because our views converge, even in regard to much of what he says I left out. Both thinkers add substantially to the overall discourse. I will not attempt to address all their points. However, I will respond to what I see to be their key objections, and further clarify my position. From contrary viewpoints, both thinkers draw attention to problematic facets of my conclusion and my all too brief final comment

Nature, Knowledge and Negation
Current Perspectives in Social Theory, Volume 26, 115–128

ISSN: 0278-1204/doi:10.1108/S0278-1204(2009)0000026007

about renewed growth. Their critiques here point to a missed opportunity in my essay.

Professor Zimmerman (p. 75) states that he shares my "major concern" about the need to employ humanity's intellectual, material, and other human resources to come to terms with "the many enormous socio-ecological challenges facing us."[1] He (p. 83) also says that creating "more sustainable and more equitable...social, political, and economic arrangements is a worthy and vital goal...." However, he stresses throughout his commentary the possible, or *likely*, policy failures and drastic political excesses that would occur should the greater sense of urgency about ecological problems, expressed in my chapter and more strongly and eloquently by Professor Barry, be translated into collective actions, public interventions, or new policy regimes to deal with problems that Zimmerman acknowledges very tepidly and skeptically.

Zimmerman (p. 79) holds that my assertion that "sweeping policy changes" are needed to avert "eco-catastrophe" and to curb consumerism might encourage imposition of a counterpart to the Stalinist efforts to create "a new socialist 'man'" to expunge capitalism's "destructive appetites." He asserts that other thinkers and activists, deeply concerned about climate change risks, have already "called for dissenters to be put on Nuremberg-like trials for crimes against humanity." Responding, in part, to my comments about J.K. Galbraith's critique of neoclassical economics and its naturalization of unlimited consumer wants, he suggests that leaders dedicated to instituting major ecological change and curtailing capitalist consumerism will find it hard to avert "resorting to a new Gulag to silence dissenters." He implies that state or public regulation of consumer choice will lead to repressive consequences, but he offers no argument why this *must* be the case. Professor Zimmerman (p. 76) surmises correctly my sympathy for Galbraith and other progressive liberal and left-leaning thinkers who hold that needs should be a matter of intellectual and public discourse. Although shaped by divergent historical and cultural contexts, needs have been a key, albeit usually implicit, part of debates, in modern liberal democratic and social democratic societies, about the balance between liberty and equality, justice in the distribution of the means of participation, and more mundane policy matters of welfare provision and tax burdens. For example, today's U.S. debate about health care provision and justice is all about needs and the most fanatical outcry has come from organizations, lobbyists, operatives, and politicians that profit from the current system and a hard right-wing that warn about socialist dictatorship and Nazi tendencies (e.g., posters of President Obama with a Hitler mustache and in a handshake with Hitler superimposed in a Nazi crowd [Luce, 2009; photograph and story]).[2]

Professor Zimmerman parses his words carefully about his criticism of me at the start and sets a genial tone, but I find his references to Stalinism and Nazism, and to the violent, hateful side of religious-like ethics of conviction to be hyperventilated and frankly to strain credulity in this context. What if the scientists and critics, who argue that a severe ecological crisis is at hand, are right? Does Zimmerman's line of argument mean that we are doomed to extreme political repression if we acknowledge a crisis and employ public means to deal with it? I reject the association, even if indirect, of such consequences as possible or likely outcomes of my views. However, I do not dismiss his concerns about apocalyptic claims and sensibilities and anti-modern aestheticized politics, which can favor authoritarian means. I am aware of Nazism's green side and ecofascist threads of the European New Right and in radical versions of Third Way politics. As a younger scholar I departed the *Telos* circle because of their discourse about "organic negativity," which led to Carl Schmitt and the French New Right. Thus, I appreciate Zimmerman's cautionary notes, about possible fanatical strains of environmental activism. However, climate change deniers sometimes express a religious-like fanaticism and friend-enemy politics in their charges that scientists, environmental activists, and ecological reformers conspire against humanity, the free market, and Christianity and ultimately will recreate a Soviet gulag if they succeed in attaining power. Extreme right anti-government currents have been active in the United States for decades and are, for me, much more worrisome than a nascent green revolution sparking what some critics of communism have called a "dictatorship over needs." I do not dismiss entirely the possibility of the latter, but fear that it is much more likely to arrive via anti-scientific or pseudo-scientific propaganda and grossly neglected ecological problems precipitating severe and possibly unmanageable crises.

Zimmerman makes numerous points about how public policy initiatives, especially costly ones, can be misdirected, produce unintended consequences, be manipulated by the powers that be, and crowd out other important, worthy agendas. This emphasis on political prudence I do not contest. Reasoned intellectual debate and policy decisions must include critics who point to such problems and present alternative strategies. Moreover, I acknowledge, in my essay, profoundly problematic trade-offs with regard to reducing growth or instituting new patterns of growth. My appreciation for thinkers such as Weber and Dewey, who stressed incisively the uncertainty of social action, universality of unintended consequences, and need for reflexive, measured responses to problems are one reason why I concluded my essay in less than "scintillating" fashion, as Zimmerman rightly notes in a semi-appreciative manner. But contingencies suffuse the private sector and civil society as well as states and public institutions. Inhering in all social action, uncertainty should not preclude taking stock of serious economic, social, and ecological inequalities and ecological risks and acting to mitigate them. In the absence of exhorting unmet needs and the costly consequences of inaction, how is it possible to bring such issues and risks to the public's attention? Critical public discussion of inequalities and socio-ecological damages are usually opposed by powerful entrenched interests with the material and political resources to get their messages out and to create institutional blockages. The neoliberal era has been a time of flourishing for Washington lobbyists and their paymasters. Balanced concern for prudence in policy formation and aversion to political excess should express serious concern about the massive lobbying machines, politicians, and well-funded think tanks that disseminate pseudo-scientific propaganda and dismiss claims about ecological risk as a liberal-left conspiracy.

Professor Zimmerman (p. 81) states that he "spent a year delving into criticism of and alternatives" to the anthropogenic global warming hypothesis (AGWH) and that his "research" led him to conclude that the climate related sciences are insufficiently advanced and contain too many uncertainties to merit the kind of urgent concern expressed by the Intergovernmental Panel on Climate Change (IPCC). He simply asserts that the science is not developed enough to reliably predict future temperatures or regional climate changes. This general claim about the inadequacy of climate-related science has been a pivotal facet of corporate sector, climate change deniers' propagandizing and lobbying efforts.[3] I am *not* asserting that Zimmerman is in this camp, but he provides no context for a sweeping claim that undercuts climate change related science and that implies the irrelevance of my paper's extensive discussion of its recent findings. Twenty years ago, in a now classic book that introduced the general public to climate change, Bill McKibben declared (2006 [1989]) that a chief characteristic of the "new nature" is its "unpredictability." Increasing climate uncertainty has been a central part of the scientific discourse about climate change from the start. Recently, scientists have argued that climate change uncertainties have ended the "stationarity" (i.e., that natural systems fluctuate within predictable limits) presumed by water managers.[4] Climate change's complex, contingent feedback loops, made more intricate by other growth related ecological changes (e.g., cutting and burning in the tropics) and today's great scale and speed of ecological change increase uncertainty in many domains. The type of predictive powers implied by Zimmerman, with regard to global temperatures and regional impacts, will never be achieved and, likely, cannot be in principle. His very high standard sets aside evidential and theoretical arguments about how climate systems operate and the large body of supportive evidence about climate change impacts, especially in areas where their effects are expected to be most extreme (e.g., regions close to the poles and in the high mountains).[5]

Acknowledging and coming to terms with uncertainties, which inhere in the climate system's exceptionally complex biophysical interactions and feedbacks, were major concerns of the 2007 IPCC, *Fourth Assessment Report* (*AR4*) and do not negate the body of science that supports the AGWH.[6] Zimmerman states, in passing, that the IPCC panel is a "political body" that excludes disconfirming evidence about the AGWH. Even the famous climate change skeptic, Bjørn Lomborg (who Zimmerman defends and apparently respects) has a much higher estimation of the IPCC panel than does Zimmerman. Lomberg (2009) praises the IPCC's "careful research," and asserts that its "report writing process is robust and custom-made to weather criticism."[7] Moreover, much of the science that I discuss came *after AR4*'s release, appeared in top refereed journals, such as *Science* and *Nature*, and holds that warming and melting has been proceeding faster, or much faster than *AR4* projected. As explained in my essay, the IPCC underestimated the rate of increase of atmospheric releases of CO_2, and political brokering tended to understate, rather than overstate the warming related problems. Although pleading for prudence with regard to climate change crisis advocates' claims and ambitious policy aims, Zimmerman's argument and tone does not favor curiosity about the current state of climate related science and its post-*AR4* warnings about the ramped up scale and speed of climate change. Given the possible consequences, his view seems *very imprudent* to me.

Zimmerman chose not to address my points about other growth related ecological and resource matters that are partially or largely independent of warming (e.g., loss of biodiversity, degraded ecological services, food shortages, over-harvesting) and about how globalization has changed the scenario with regard to resource depletion and environmental damages. He could not have addressed all the issues in my paper and, understandably, wanted to express his own concerns. By electing not to address "overshoot," however, Zimmerman (p. 75) opted out of engaging my main point that globalization's huge scale and speed and penetration beyond the old core capitalist nations to populous developing ones, especially China, has greatly accelerated the consumption of natural capital and production of waste and, thereby, has changed the context of the planetary overshoot and limits to growth debates. This move, along with his setting aside the science that I reviewed, dodges my core argument. His points about my critique of the growth imperative and linkage of my sense of urgency about ecological matters to climate change panic and risks of political repression might have had more traction had he addressed my argument. His prudence and skeptical attitude is one-sided for it did not cause him to pause and entertain the possibility that the many scientists who have warned about grave climate change risks and reports of already existent severe climate-related damages in some parts of the world might be right or to acknowledge the hysterical, anti-modern responses of key factions of the climate change opposition. He does *not* deny the existence of ecological problems, but his dismissive view of the climate-related science supportive of the AGWH and emphatic warnings about future policy failures and authoritarian or totalitarian risks undercut the serious concern about climate change and other major ecological problems needed to generate a national discourse about the issues. Professor Zimmerman (p. 87) asserts that he is open to diverse policy options to cope with growth related problems and advocates critical reflexivity toward all of them, which makes sense to me. However, he stresses the catastrophic risks of concerted public action and planning, while he is silent about the broader social and ecological damages from the last three decades of unrestricted capitalism and the risks of its continuance.[8]

John Barry's comment about my "post-exuberant" theorizing is well put. The less than buoyant tone reflects my pessimism about the state of U.S. civil society and about the powerful blockages exerted by the wounded, but still hegemonic neoliberal regime.[9] I appreciate Barry's exuberant standpoint, necessary for freeing the imagination to frame ecological and political-economic alternatives. His opening quotation of "Rent-boy's" classic lines from *Trainspotting*[10] and closing exhortation about "life-sustaining and life-affirming politics"and an ecologically aware, 21st century critical theory challenge us to question the taken-for-granted growth regime and neoliberal treadmill. Barry's sense of urgency about ecological catastrophe, which I share, impels him to evoke an aesthetic rupture, or Marcusean "great refusal" – precisely the type of sensibility that Zimmerman fears will produce political overreach and repression. However, even the cautious re-embedding of culture and nature, which Zimmerman endorses, will likely need to tap aesthetic sensibilities. Most Americans are not receptive to bearing the economic costs or life-style inconveniences of even modest ecological reform. Environmental writers, such as

E.O. Wilson and Bill McKibben, stress how our way of life severs, or at least dulls, the aesthetic connection to nature. Their views make problematic neoliberalism's untrammeled accumulation and growth at any cost and absolutist idea of property rights, permitting and even encouraging trashing the commons. Wilson and McKibben imply that cultivating a sense of usufruct about our relation to nature and responsibility to future generations and other life on the planet is a compelling type of prudence as well as a worthy normative and aesthetic ideal. Barry's assertive points about "choosing life" and a "flourishing" of human well-being expresses a vital aesthetic trope and sensibility missing from my essay, but which helped motivate my writing it.[11]

Barry's critical points about missing themes in my chapter are mostly on target. Addressing issues such as peak oil, eco-feminism, and well-being would have enriched my argument. By focusing in a relatively detailed way on science and globalization, however, I had to exclude relevant issues that would have made the chapter even longer and more complex. I likely made some poor choices. For example, more commentary on the earlier "limits to growth" discourse would have likely strengthened my analysis of the post–World War II growth imperative, and lengthier treatment of inequality might have motivated a more robust conclusion and clarified my position about rethinking and reconstructing growth. However, I do not regret my omission of sustainable alternatives, which was deliberate; discussion of the topic would have added a highly problematic layer of complexity to the chapter. Lacking adequate knowledge of the issue, I could not have dealt with it nearly as intelligently as others who have long worked on it. My essay implies that the neoliberal growth regime is not sustainable and that a return to the post-World War II regime is neither possible nor desirable. Thus, although I share Professor Zimmerman's concern about tradeoffs, I am sympathetic to Barry's call for a "post-growth," or no growth regime, and "one-planet living." Barry (p. 111) is right – we need to ask "what is an economy for?" My point about radically rethinking growth poses a parallel question, even though I framed it in a tamer sounding fashion. How is growth ad infinitum possible, and at what costs?

The thrust of my essay questions the sustainability of modern capitalism per se (as we have known it); the growth imperative and treadmill of production inhere in its structure.[12] The chapter's missed opportunity refers to my failure to conclude my argument about this central issue and about the problematic trajectory of today's hypercapitalist, neoliberal globalization – its contradictions, erosion, and crisis. This type of denouement would not likely have been the "arousing crescendo," or "dramatic call action," that Zimmerman expected, but was pleased not to find. However, completing the argument about capitalism might have clarified my final comment about growth, making the conclusion congruous with the rest of the paper, terms of engagement more transparent to the critics, and conclusion a tad less "anticlimactic." Neoliberal deregulation, privatization, securitization, and financialization have sharply increased inequalities of wealth and income in the United States. Most of the income from increasing productivity gains and soaring stock-market and financial bubbles went to the wealthiest 1% of Americans, especially a tiny fraction of people at the very top.[13] The still developing 2008–2009, financial crisis and recession has opened serious questions about the substance and sustainability of neoliberal globalization's growth and, perhaps, neoliberal capitalism per se. Diverse analysts argue that the last several decades of debt driven, speculative financial growth is over, and some critics hold that capitalism faces a structural crisis as serious as, or worse than the Great Depression.[14] However, even if the problems are not as fundamental as pessimistically inclined thinkers project, the emergent consensus is that the current crisis resulted from neoliberal under-regulation. The recession's depth, duration, and outcome are now all matters of conjecture, but the neoliberal regime of accumulation is crisis ridden and may not be able to be sustained. Still its institutional and cultural foundations have not collapsed, and its leadership has not compromised. Formerly progressive California is now the test case of how far the safety-net and commons can be cut before even modest tax increases are entertained from those who can most afford to pay and, perhaps, of the limit to which the public sphere can be squeezed, reduced, and cut without a severe erosion of infrastructure, legitimacy crisis, and socio-political conflict or breakdown.

We face an economic crisis as well as an ecological crisis.[15] In the short term, efforts to curtail the recessionary downturn will likely aim at restoring the recent pattern of accumulation and growth. Recovery and continued neoliberal globalization will further accelerate the mounting throughput of natural resources and production of waste. Also, hundreds of millions of people in the developing world still eagerly await being brought onto the grid and integrated into the global capitalist system. The "bottom billion" poorest people, who have been unable to get on board the globalization train also await integration. Zimmerman is right to warn about the trade-offs of a no growth or low growth scenario. What happens to the poorest people? And how can the sliding living standards of poor and middling strata in better off parts of the world be raised? However, advocating growth as we have known it, without entertaining planetary overshoot and ecological devastation, is imprudent and self-defeating. Moreover, as I explained in my essay, substantial portions of the poorer parts of the world already have been suffering from climate change related ecological damages – floods, droughts, and conflicts (e.g., GHF, 2009). Barry is right that the growth imperative cannot be sustained without high costs. Continuous growth theories and policies and their treadmill of production and accumulation were fashioned when a much smaller proportion of the global populace was integrated into the capitalist system and, at least tacitly, assumed continued barriers to entry, dependent development, and limited pressure on resource consumption. Although uneven, neoliberal globalization's vast expansion makes this conventional view a dangerous illusion and produces a pincer between the economic crisis and ecological crisis.

Globalization has extended enormously the overall dependence on capitalist wage labor, and deploys ever more efficient labor saving technologies. Low wage work, contingent labor, underemployment, and unemployment, or Marx's "reserve army," is a taken-for-granted reality. Even in rich nations, the current regime provides no remedy for this problem. The growing horde of displaced people, mass of global informal sector laborers, and low-wage global diaspora challenge the old fiction that capitalism's expanding markets and "creative destruction" always will create, after temporary crises, sufficient work to sustain the vast majority of the populace. However, if this promised outcome were to eventuate today through the usual growth imperative, the increased resource consumption would make global carrying capacity an even more profound problem than it already is. Capitalism will likely be reinvented or transformed; its future is a contingent matter of communication, deliberation, struggle, and chance (i.e., new forms of free-market capitalism, social democracy, socialism, authoritarianism, and yet to be imagined regimes are all possible). Some public and corporate officials espouse green agendas, or entertain them as back-up strategies. However, most leaders from these sectors, with substantial public support, are rethinking ways to revive the growth imperative and capitalism as we have known it, with some new technical, organizational, and spatial-temporal twists For more than a decade, expansion of China's domestic consumption has been a topic of hopeful chatter among global firms, entrepreneurs, and financiers. Advocating the merits of such a strategy and tracking it, the McKinsey Global Institute (2009) holds that China growth could revive the entire global economy and, to do so, its leadership must act aggressively to increase this nation's private consumption to over 50% of GDP by 2025. China has already restored its meteoric growth, and may be on the path MGI touts.[16] A huge number of

China's and India's two and half billion people, still not integrated into global capitalism, want the benefits of inclusion and cannot return to a traditional economy overtaken by capitalist monetarization and property relations. Although advocating growth to address these unmet needs, Zimmerman (pp. 84–85) says he agrees with me that Julian Simon's infinite growth is implausible. However, does not expansive global growth presume precisely a Simonian scenario? If so, Barry's appeal for "one-planet living" ultimately calls for ecological realism and political prudence, not merely critical conversation about growth.

However, I do not retreat from my closing point about radically rethinking, reshaping, and restoring growth. Speculating that my affirmation of growth is rooted in an "American cultural imaginary," Barry (pp. 110–111) draws on the old linear European image of Americana (which has some veracity) – the ethos of bigger, faster, and, especially, *more*. But how could I call for a Copernican turn if I wanted to restore the growth imperative? We are cultural beings who must draw on our lived experiences and contexts (hopefully not always parochially). Most Americans of my generation have been influenced by the mythic, "golden age" narrative and memory of the postwar expansion and construction of suburban middle America. Younger generations of Americans and new immigrants readily assimilate this historically-based complex of sensibilities about growth and abundance. They also have been exported nearly world-wide with American entertainment culture. Could this U.S.-rooted postwar mythology be redeployed, with parallel energy, resources, and vision to carry out green reconstruction, curb the excesses of the old regime, and build the material and cultural bases of Barry's post-growth society? Fundamental redesigning of energy sources, transport, agriculture, built environments, and other facets of material culture and consumption would renew "growth" and *could* avoid eco-catastrophe, preserve democracy, and begin a transition from a capitalism as we have known it to a new regime. I expect that Barry's Institute for a Sustainable World and other such centers are already working toward parallel ends. I cannot imagine how a no growth rhetoric could be adopted now, or in the near future, in the U.S., except as a desperate reflex to cataclysm, imposed in a highly unequal and unjust way.[17]

In his *Critique of the Gotha Program*, the elder Marx warned that a new regime cannot be imagined or built effectively *de novo* all at once. Critical theory's ideas about the need for "determinate negation" had new meaning after the Holocaust and other 20th-century political disasters. Michael Zimmerman's critique suggests parallel concerns. However, I doubt that sustainability discourses, of people like Barry and his counterparts in my rural state and college town, will be hijacked for authoritarian purposes. The most pressing danger, in my view, is failing to get the ecological issue on the national agenda and to entertain a multiplicity of strategies to cope actively with mounting economic and ecological crises. Repression is more likely when unattended crises erupt into chaos and conflict. Nascent crises open new possibilities for forging new ways of life. In this regard, John Berry's arguments for "choosing life," imagining and working toward a less workaday, more just, democratic, and sustainable culture, which deploys our natural capital in a much more responsible, prudent, and compassionate manner makes a great deal of sense to me, and I share his goal.

Robert J. Antonio
University of Kansas
anto@ku.edu

NOTES

1. Professor Zimmerman (pp. 88–89) states that much of what I say in the final section of my chapter "makes a great deal of sense" to him. However, it is unclear whether his cautionary note about re-embedding culture in nature (i.e., avoiding reductionism and keeping in view humanity's "exceptional" linguistic and abstractive powers and consequent culturally mediated relation to nature) is directed at my argument. Zimmerman does not mention that my points on the relevance of Dewey and Mead stress exactly this type of balanced approach to re-embedding.

2. The petroleum industry appears to be adopting a similar strategy to the health care industry backed Town Hall "protests." Supported by the leading trade group, The Petroleum Institute, Energy Citizens has sponsored a protest and bused in oil company employees from their places of work. For this event and broader reportage on the topic, see Krauss and Mouwad (2009).

3. See, for example, the now defunct Global Climate Coalition (1996, pp. 1–2, 4–8). Also see the Union of Concerned Scientists' (2007) arguments about corporate funded "manufactured uncertainty."

4. The scientists acknowledge that climate systems always posed surprises, but that today's accelerated climate change has made matters much worse. See Milly et al. (2008), Pielke (2009).

5. Zimmerman's (p. 81) blanket claim that "just about everyone agrees that it is impossible to make any serious predictions about the effects of climate change on a regional basis" seems to me to be a gross overstatement, especially with regard to those regions where climate scientists' theories and models predict the temperature increases will be greatest and where empirical inquiry and observation have provided overwhelming support for the predictions.

6. The IPCC panel and other climate scientists that I read usually addressed directly and even dwelled on the uncertainties, imperfections, and possible gaps in their models.

7. Although stressing market-oriented, cost-benefit analysis, Lomberg (2009) sees anthropocentric climate change to be an urgent enough problem to merit exploration of risky geoengineering technologies to combat it. Zimmerman (p. 85) asserts that Lomberg has been "unfairly attacked" for providing environmental "good news" in his earlier *The Skeptical Environmentalist*. On this controversy, see Pielke (2007, pp. 116–134); see http://www.lomborg.com/ and http://www.copenhagenconsensus.com/CCC%20Home%20Page.aspx for Lomberg's views.

8. I do not agree with thinkers or policymakers who hold that ecological problems should, in principle, always be dealt with wholly or mostly by market-based strategies, but I do not deny the effectiveness of such approaches in certain contexts or suggest that they be abandoned.

9. The neoliberal regime is a widely shared cultural framework as well as a top-down political economic order; it includes attitudes, habits, and expectations that color the current phase of American individualism, impact people's receptivity to propaganda, and shape civil society's relations and discourses. Alternative visions meet resistance from below as well as from above.

10. This movie expressed forcefully the cultural wreckage of a neoliberal modernity gone awry. Strangely it was marketed in the United States as "hilarious," which I still cannot fathom.

11. My recreation and passion is "birding"; I have been recording the annual movements of birds in a tiny spot of riparian old growth, along the Kansas River in the heart of America, where migrant species have paused on their way north or south for millennia. My little space, Burcham Park, has been encroached upon by my college town in various ways (power company cuts, water company cuts, university boathouse cuts; city worker cuts to make the park look tidier etc.). The seasonal changes of the remaining old Cottonwoods and bird migrations continue. Every bird walk there brings ecological issues, aesthetics and problems, to mind and heart.

12. Anthropogenic alteration of natural environments and ecological damages can be traced back to primordial times, but the scale of these impacts and appearance of entirely new types of problems (e.g., anthropogenic climate change) bear the imprint of the current historical moment. Potlatch rituals and other forms of premodern ceremonial exchange, mentioned by Zimmerman, have nothing in common with capitalist possessive individualism and consumerism. State socialist regimes were dedicated to their own version of the growth imperative, which had harsh environmental impacts. Although having variable forms and impacts on environments, modern capitalism is the primary root of the growth imperative and of the ecological crisis. Neoliberal globalization's vast new built environments, and extension and consolidation of American-style consumption of goods and resources changes the scenario.

13. Charles Morris (2008, pp. 152–153) states that between 1980 and 2006 (the height of neoliberal dominance): "Almost *all* the top tenth's share gains…went to the top 1 percent, or the 'top centile,' who more than doubled their share of national cash income from 9 percent to 20 percent. Even within the top centile, however, the distribution of gains was radically skewed. Nearly 60 percent of it went to the top *tenth* of 1 percent of the population, and more than a fourth of it to the top one-hundredth of 1 percent of the population." The ending of the postwar capital-labor accord and decline of unionized labor, championed and executed by neoliberal

private sector and public leaders, have led to increased work hours, stagnant wages, and reduced benefits and security for most American workers (e.g., Mishel, Bernstein, & Shierholz, 2009).

14. I am speaking of the series of bubbles and bailouts, which have characterized neoliberal globalization and, especially, the latest wave of derivatives trading, backed by highly leveraged mortgage backed assets, and consequent massive financial bubble and financial crisis. The government served as the investor of last resort, bailing out the banks that have engaged in speculative practices when the bubble burst (e.g., see Morris, 2008; Foster & Magdoff, 2009).

15. Arguably the dual crisis is made even more problematic by the eroded cultural and social commons, which Tocqueville, Durkheim, Dewey, Putnam, and others have argued is needed to counterbalance modernity's economic and cultural individualism and moderate consequent fragmentation.

16. The Chinese government (GOV.cn, 2009) reports that GDP grew by 7.9% in the second quarter of 2009; industrial output is up 10.7%; and retail sales have risen by 15%. Even if this surge is not sustained and if the numbers are exaggerated, China's rapid growth continues to be a core facet of globalization and may continue to accelerate.

17. Rather than countering Barry, I am explaining roughly what I had in mind, when I made my closing comments about restored growth. I am not knowledgeable enough about Barry's position, or about sustainable practices, experiments, and visions, to be a competent critic of them.

REFERENCES

Foster, J. B., & Magdoff, F. (2009). *The great financial crisis: Causes and consequences.* New York: Monthly Review Press.

GHF. (2009). Climate change: The anatomy of a silent crisis (human impact report). Global Humanitarian Forum, May 29. Available at http://www.unep.org/ghfgeneva.org/Portals/0/pdfs/human_impact_report.pdf. Retrieved on June 10, 20009.

Global Climate Change Coalition. (1996). Primer on climate change science. Available at http://www.sourcewatch.org/images/8/82/GCC_Primer_Draft.pdf. Retrieved on August 21.

Gov.cn. (2009). China's GDP grows 7.9% in Q2. *Chinese Government's Official Web Portal,* July 18. Available at http://www.gov.cn/english/2009-07/16/content_1367091.htm. Retrieved on August 24, 2009.

Krauss, C., & Mouwad, J. (2009). Oil industry backs protests of emissions bill. *New York Times,* August 19. Available at http://www.nytimes.com/2009/08/19/business/energy-environment/19climate.html. Retrieved on August 22.

Lomberg, B. (2009). Engineering a cooler climate [interview by Jeremy Whipp]. *Financial Times* (August 13), 6.

Luce, E. (2009). Obama on drive to tackle healthcare anger. *Financial Times* (August 13), 3.

McKibben, B. (2006[1989]). *The end of nature.* New York: Random House.

McKinsey Global Institute. (2009). If you got it, spend it: Unleashing the Chinese consumer (executive summary), August. Available at http://www.mckinsey.com/mgi/publications/bbnewsletter/bb_unleashing_chinese_consumer.asp. Retrieved on August 21.

Milly, P. C. D., et al. (2008). Stationarity is dead: Whither water management? *Science, 319*(February 1), 573–574.

Mishel, L., Bernstein, J., & Shierholz, H. (2009). *The state of working America: 2008/2009*. ILR Press: Ithaca and London.

Morris, C. R. (2008). *The two trillion dollar meltdown: Easy money, high rollers, and the great credit crunch*. (Revised and Updated). New York: Public Affairs.

Pielke, R. A., Jr. (2007). *The honest broker: Making sense of science in policy and politics*. Cambridge, U.K., and New York: Cambridge University Press.

Pielke, R. A., Jr. (2009). Collateral damage from the death of stationarity. *GEWEX* (May), 5–7.

UCS. (2007). Smoke, mirrors & hot air. Union of Concerned Scientists, January. Available at http://www.ucsusa.org/assets/documents/global_warming/exxon_report.pdf. Retrieved on March 10, 2007.

DEVELOPING PLANETARIAN ACCOUNTANCY: FABRICATING NATURE AS STOCK, SERVICE, AND SYSTEM FOR GREEN GOVERNMENTALITY☆

Timothy W. Luke

1. INTRODUCTION

This preliminary survey begins to probe a few purposes and practices of "Earth System Science" to rethink the ways in which Nature is "taken into account" by this new power/knowledge formation. The workings of "environmentality," or green governmentality (Luke, 1999c), and the dispositions of environmental accountancy regimes depend increasingly on the development and deployment of such reconceptualized interdisciplinary sciences (Briden & Downing, 2002). These practices have gained much more cohesion as a technoscience network since 2001 Amsterdam Conference on Global Climate Change Open Science. Due to its brevity, this study is

☆An earlier version of the chapter was presented at "Nature's Accountability: Aggregation and Governmentality in the History of Sustainability," German Historical Institute, Washington, DC, October 10–13, 2008

Nature, Knowledge and Negation
Current Perspectives in Social Theory, Volume 26, 129–159

ISSN: 0278-1204/doi:10.1108/S0278-1204(2009)0000026008

neither an exhaustive history nor an extensive sociology of either Earth System Science or the new post-2001 Earth System Science Partnership (ESSP), which acquired new legitimacy during and after this professional-technical congress. Instead this critique reexamines these disciplinary developments to explore the curious condition of their rapid assembly and gradual acceptance as credible technoscience formations. This reevaluation allows one, at the same time, to speculate about the emergent interests hoping to gain hold over such power/knowledge programs for managing security, territory, and population on a planetary scale (Burchell, Gordon, & Miller, 1991; Foucault, 1991c, pp. 87–104).

The ordinary operations of environmental policy-making typically approach the Earth in one of three ways: first, as a site of accumulated material resources, which contains/holds "stock"; second, as a structure of vital processes, which dispenses/vends "service"; or third, as sites and structures of coupled artificial-and-natural events, which constitutes/delimits "system." In assessing the new risks of second modernity (Beck, 1992), many environmentalists have accepted these root metaphors to guide social practices as they institutionally map their new organizational initiatives for the sustained conservation of natural resources or the more secure preservation of the Earth (Cairncross, 1992; Chertow & Esty, 1997; Bliese, 2001). On the one hand, the imagination of Earth as a store of accumulated "material stocks" to be exploited judiciously and hopefully never depleted (WCED, 1987), or, on the other hand, as a structure of rising-and-falling vital processes with "ecological services" to be utilized on a more transparent cost/benefit basis (Westmen, 1978) already have served as productive policy stances (Gobster & Hull, 2000). Both of these conceptual frames have allowed experts to recognize the Earth's capacious, but still limited, carrying capacity to prevent any crippling degradation. Such analytical and axiological frames have underpinned the many resource mangerialist strategies employed in most American political jurisdictions for decades (Gore, 1992, 2006; Gottlieb, 1993). Having surveyed those practices extensively elsewhere (Luke, 2004, 2002, 1999b), this study will focus more on the growing importance granted in ecological policy studies to a notion of "system."

More recently, national scientific bodies, transnational scientific networks, and international nongovernmental organizations have augmented the spatial scope of their engagement as "environmental protection agencies" at local, regional, and national scales of analysis with a more global perspective, adopting a "planetarian" viewpoint that looks beyond stock and/or service to system (Luke, 2008b, 2005a, 2008a). No single jurisdiction has sovereign command-and-control of this planetary spatiality, but there are many organizations, firms, and individuals intent upon preparing to direct responses to such system-level challenges on a strategic basis rather than tactical action (De Certeau, 1988). Designs like these approach the environment and society as "coupled systems" – both artificial and natural – whose complexly tight, loose, or indeterminate couplings will be appraised by technical networks known as Earth System Sciences. Their role is to map, measure, monitor, and manage the "environmental sustainability" of complex social and natural systems generated from all the ecological services and natural resource stocks of the planet (Luke, 2005b). Consequently, the dispositif of Earth System Sciences opens the planet's workings to clusters of expert power and knowledge formations intent upon accounting for, and then perhaps administering, the "systems" of the Earth as coupled complex ecologies and economies (Briden & Downing, 2002).

The significance of Earth System Sciences often is neglected, but this veil of neglect can be easily pierced by turning to works like Alan Alan Wiseman's (2007) *The World Without Us*. In imagining "a world without us," Wiseman implicitly maps the concrete economic demands made by the elaborate materiel flows of human markets as they proliferate across the Earth in all of its ecological "systems." Even though he concludes that the planet can, and assuredly will, survive humanity, he also recognizes, like McNeill (2000) or even Lovins (1977), the weight, sweep, and duration of material consumption truly have made humanity through globalization (Beck, 2000) a systemic force of, or equal to, Nature.

In this vein, sociologists, like Mann (1986), trace out "the sources of social power," while Earth System scientists turn the tables around, tracking the "powers of social sourcing" to reconsider modernity's "society of power sources." Mann ironically maps traditional ecologies of resource nationalization (Luke, 1999b, 1999d) in the U.S. Environmental Protection Agency's and other state's ecological impact assessments when he claims, "domestically, the state is territorially centralized and territorially bounded. States thus can attain greater autonomous power when social life generates emergent possibilities for enhanced cooperation and exploitation of a centralized form over a confined territorial area" (Mann, 1986, p. 27).

Here, one sees disjunctures. The contemporary structures and subjects that Earth System Science assess are usually assigned a naturalized givenness, when they are, in fact, neither always natural nor already given (Foucault, 1970). Sassen observes – with a statement that could be an extended caption for decades of modernist social, political, economic, and cultural analysis – that the complex spatiality under surveillance here requires all humanity to understand,

> how territory, authority, and rights became assembled into the modern nation-state. It took work to make society national – from generals fighting for yards of territory, to lawyers inventing new juridical frameworks and instruments, to the work of merchants and capital owners that strengthened the national scale in economic operations, to the work or schools and other "disciplining" institutions in the forging of a national citizenry. There was nothing natural, easy or predestined about the national. (Sassen, 2006, p. 18)

Sassen sharpens this point in her excavation of the "organizing logics" behind transnational temporal/spatial "assemblages" to identify the denationalization created by globalization, but her unconscious characterization of ecological modernization nicely highlights the containerized materiality of economies and states in their local, national, and global expressions.

Given the constructedness of the economy and state in today's national statal forms, assessments of the materiality sustaining everyday life (Lefebvre, 1984), any country should be judged with regard to concrete spatial practices to track ecological underpinnings. As Lefebvre (1991, p. 38) claims, "spatial practice" at the local, national, and global levels in transnational contemporary capitalism is what "secretes that society's space; it propounds and presupposes it, in a dialectical interaction." In today's neocapitalist order, the concretion of spatial practices "embodies a close association, within perceived space, between daily reality (daily routine)" with all of "the routes and networks that which link up the spaces set aside for work, 'private' life and leisure" (Lefebvre, 1991, p. 38). Lefebvre also emphasizes that perceived spatial practices also disclose the attendant "representations of space," which often are the dominant ethical and political frames ordering materiality. Within those grids, he sees "conceptualized space, the space of scientists, planners, urbanists, technocratic subdividers and social engineers ... all of whom identify what is lived and what is perceived with what is conceived" (Lefebvre, 1991, p. 38). Finally, Lefebvre also suggests that studies of spatiality must appraise "'representational spaces,' or the materialization of spatial formations as they are directly *lived* through its associated images and symbols, and hence the space of 'inhabitants' and users ... this is the dominated – and hence passively experienced – space which the imagination seeks to change and appropriate" (Lefebvre, 1991, p. 39). By tracking the ecological flows of material production and consumption (McDonough & Braungart, 2002) across the world's many built and unbuilt environments as planetarian spatiality, Earth System

Science, in part, struggles to assay the interplay of practice, thought, and activity "which exists within the triad of the perceived, the conceived, and the lived" (Lefebvre, 1991, p. 39). In one sense, the struggles in the Earth System Sciences, as they take new organizational forms at various places, are shaped by their pursuit of these ends.

2. EXPERTARCHY, NATION-STATES AND "NATURE"

The new empirical scientific operations of Earth System Science "in action," like actually existing environmentalism, presume the expanses of Nature, more or less, are a site always at ready for continuous "resourcification" in global exchange by national economies and societies (Luke, 1999b). This intellectual arrangement for authority to morph into activity requires some scientific and/or technical expert(s) to shift from being professionally "an authority" in their disciplines to acquiring various institutional roles "in authority" for official, quasi-statal, or non-governmental organizations. As such expertarchic spaces are mobilized, the Earth is reduced, typically in experts' research programs and projects, to little more than a vast standing reserve, serving as a ready resource supply center and/or accessible waste reception site where "natural resources" for "local use" and "global exchange" are reshaped daily into the products and by-products of "national resources" consumption. From these patterns of activity, the engagements of environmental accountability emerge in the multiple practices of stocktaking, service-tracking, and system-tracing.

Once presented in this fashion, it is easier to see the productive power of experts linking together concepts like sustainability, resilience, or improvement to infiltrate Nature. Earth System Science aspires to scan and appraise the most productive use of these resourcified flows of energy, information, and matter as well as the sinks, dumps, and wastelands for all the by-products that commercial products leave behind (Virilio, 2000). Nature, therefore, "with the progress of capital accumulation and the expansion of economic development" is regarded as the material substratum of ecological stock, service, and system, which "is more and more the product of social production" (Smith, 2008, pp. 49–50). And, its clustered complexities always are already mobilized as political assets by clusters of experts at work for nations, states, enterprises, and individuals (Ong, 1999; Luke, 1999d). Hence, the environment's fungibilization, liquidification, and capitalization as resources cannot occur without expert intellectual labor to prep it, produce it, and then provide it for the global marketplace (Luke, 1999b).

The trick in natural resources management and environmental affairs education is to appear conservationist, while moving, in fact, very fast to disembed, liquefy, and commodify natural resources for more thorough, rapid, and intensive utilization by the nations producing them (Durning, 1992; Schmidheiny, 1992; Cairncross, 1992; Milani, 2002; Pinchot, 1910). Many currents in today's Earth System Science mostly aim at refining their theories and methods to expand the sweep and scope of such authority-in-action.

Planetary ecology and economy can thus become a unified spatiality of "the perceived, the conceived, and the lived" (Lefebvre, 1991, p. 39) for Earth System scientists' analyses of nation-states all work together in the world system. As environmental accountancy assignments, the intrinsic physical, operational, mechanical, chemical, or biological qualities of the environments and technological infrastructures intersecting in each nation-state, in turn, direct matter and energy toward particular paths of commodity creation (Van Kooten & Erwin, 2000). Green governmentality, as exercised by ecological expertarchy, is meant to guide these lines of flight and flows of life (Luke, 1999a, 2008b).

Despite Pinchot's (1910) sense that "the man who really counts is the plain American citizen" in Federal policy-making because only he or she can "set the common good of all of us above the private gain of some of us," and recognize "the livelihood of the small man as more important to the Nation than the profit of the big man" (pp. 3–4), this common good actually was framed above, or below, or beyond, all of us by many agents of a higher "executive power" working as "the steward of the public welfare" (pp. 3–4), or those professional technical experts who identify, manage, and transmorgify Nature into "national resources" (Gottlieb, 1993) through (pp. 3–4) their professional technical expertarchy (Luke, 1997).

The scientific scripts of ecosystem management, which are embedded in most approaches to environmental policy as green governmentality, are rarely made articulate by scientific and technical discourses. Yet, a Benthamite dream of optimality centered on creating "the greatest good, for the greatest number, for the long run" (Pinchot, 1947, pp. 261–262) is often their manifest goal. Over time, these practices of resourcification have woven a stock management model into mainstream environmental science and traditional natural resource policy-making (Cortner & Moote, 1999; Hays, 1959; Roosevelt, 1971; Mangun & Henning, 1999).

Against this policy horizon, Nature's stocks of materials do get treated as "resources," allowing the expertarchic presumptions of resourcification to become conceptually and operationally well-entrenched (Josephson, 2002).

The fundamental premises of such national resource managerialism, especially for key commodities, have not changed significantly over the past century (Virilio, 1995). Resource managerialism, as Cortner and Moote (1999, p. xi) suggest, is "in large part a social movement that embraces a new philosophical basis for resource management" by seeing nature as "stock." Consequently, the operational agendas of "sustained yield" in that stock's administration have directed the resource managerialism of the twentieth century to manage Nature's storehouses of stocks wisely. Good stock and service management repositions expertarchy to monitor the level of outputs, oversee the rate of increasing or decreasing demand, and manage the scale of sustainable use within an always emergent system of social production. Sustainability, if it means much of anything, is simply a softer, gentler style of sustained yield. Thus, the evolution of accountancy from the original visions of sustained yield to today's notions of sustainability is a win/win situation for both economic and ecological interests as they test out green governmentality for systemic application.

3. SITES OF GREEN GOVERNMENTALITY

Today's evolution of ecomanagerialism into planetarian environmental accountancy is a somewhat different project. New expert networks aim at explicating climate change research, biocomplexity analysis, and sustainability studies as they coalign in Earth System Science as "practices," or the material mobilizations of crucial positions, programs, and projects that constitute "places where what is said and what is done, rules imposed and reasons given, and the planned and the taken for granted meet and interconnect" (Foucault, 1991b, p. 75). Even though the Earth System Science project still has only limited traction in actual state policies or many corporate behaviors, the technoscience work of these expertarchic networks of/for/about green governmentality discloses, as Foucault (1991b, p. 75) argues, "programmes of conduct which have both prescriptive effects regarding what is to be done (effects of 'jurisdiction'), and codifying effects regarding what is to be known (effects of 'verdiction')."

The mediated, indistinct sovereign bases of these expertarchic efforts make them politically intriguing. Of the nearly 200 allegedly sovereign territories on the planet, maybe half, or even only a third, have any sort of continuous effective enforcement of their putative monopoly of legitimate force and legal authority within their borders as jurisdictions. Almost all of these efforts at arriving at some mechanism for environmental accountability are based in richest and most powerful of that minimally effective group, like the G-8 countries, the UN Security Council's permanent members of the EU nations where the key centers of verdiction also are found. Who is accounting what, where, when, why, and for whom, therefore, is an important question. In part, it is interesting because even those with the needed scientific assets cannot be certain; they have a sound accounting for their own territories much less those of others. And, in part, it is intriguing because all the remaining jurisdictions are under surveillance by these more verdictively mobilized nation-states, but what can this information mean for them and their peoples as it all is enfolded into Earth System Sciences? If sovereignty, as Schmitt argues, is the power to decide "the exception" (Schmitt, 2007), then who or what is aspiring to have this executive discretion for planetarian action, and is this aspiration ultimately unfulfillable in the current world order?

Foucault's wide-ranging account of the practices behind governmentality (Foucault, 1980, 1991c) stresses the formal conceptual apparatus needed to render both objects and subjects "visible" in some fashion. Systems of representation are necessary to map, measure, monitor, and then manage what must be governed. The technical practices of cartography, demography, geography, geology, hydrology, and statistics along with the organized groups of trained experts, who implement their operation, assess their findings and accumulate their discoveries as less as "type 1" academic knowledge and more as "type 2" applied knowledge (Gibbons et al., 1994) and become integral to the accountability regime responsible for documenting Nature.

One essential site of origin for the "visibilization" of Earth System Science is, of course, the International Geophysical Year (IGY). Launched on July 1, 1957, and running until December 31, 1958, this event was overseen by the World Meteorological Organization (WMO) and the International Council for Scientific Unions (ISCU). Using new scientific instruments developed to harness the atom and advance astronautical technology, the IGY initiated a series of experimental studies and monitoring efforts to develop new high-tech instrumental arrays as well as to develop big data sets about cosmic, solar, atmospheric, oceanic, hydrological, lithological, terrestrial, and geological processes. Aided by new equipment and strong financial support, these organizations began assembling and analyzing large complex data sets as part of an on-going global scientific service.

Within the United States, the Defense Department's and NASA's acceptance of this postulate about "the Earth system" provide a scannable object for a science ironically displaced into the domains of the geophysical sciences' new operational strategies that the U.S.S.R. matched with its own scientific scans. During the Cold War, two immense ideological blocs, each elaborating their own zone-regime expectations of coupled natural/social order, predictable aggregates of action, and shared values of control (Luke, 1989b), created different "Earth System" images in preparing to fight, and maybe win, a nuclear war. Aided and abetted by remote sensing satellite data, nuclear submarine oceanic monitoring, dispersed weather surveillance, and war game modeling, the military and intelligence services of United States and U.S.S.R. worked daily in ways that presumed the positive and negative coupling of human and nonhuman systems (Luke, 1989b) to maintain nuclear deterrence as well as anticipate nuclear attack.

Based on a perspective that regarded the Earth as an object of observation from astronautical, aeronautical, and nautical platforms, national command-and-control authorities often had to integrate their understanding of the world in pre-war, warring, and post-war scenarios in which human and nonhuman life worldwide would experience rapid coupled destructive events in a thermonuclear war (Luke, 1989b). Moving from such hypothetical strategic models to actual environmental monitoring in a more eco-managerial mindset was not a major methodological shift. Seeing the Earth as a composite of various thermonuclear battle spaces on a 24×7, 365×10, or 10×10 timeline speculated about plausible environmental damage zones on daily, weekly, yearly, or decade-long time horizons, while networks of experts tweaked their instruments of Earth surveillance to fulfill and improve upon new missions of ecological surveillance.

Earth System Science now seems verdictively ready to drive the next "order of things" that cogenerates order and things by propounding the environmental things to be ordered jurisdictively by science and government. To approach the Earth as systems capable of scientific study through greater systemic analysis, calculation, and organization already moves the project of systematic Earth sciences forward from theoretical speculation into a provisional starting point for a complex new political rationality. Indeed, planetarian power becomes a feasible project (Hoffman, 1997; Hawken, Lovins, & Lovins, 1999).

The nationality of such research is fascinating. On the one hand, all of these tendencies transpire within scores of national spaces, but the impulse to map, measure, monitor, and measure mostly draws its focus from a handful of wealthy OECD nation-states and transnational scientific networks. On the other hand, it is "national science foundations" that are defining these research programs, funding their practitioners, and circulating their results, but their mission statements are increasingly conceptualized in non-national planetarian terms that address biota as such, Earth systems as worldwide ecological services, and territories as biotic capacity carriers. Hence, the terms of account are often more rooted in abstract planetarian registers of assumption, inference, and modeling rather than determinate national criteria of economic output when the style of accountancy is assayed.

The Earth System Science problematic reverberates with significant influence from several intriguing sources. Inasmuch as any workable science requires the "immutable mobiles" (Latour, 1988) of experimental results, data observations, or shared metrologies, the establishment of the WMO and the International Telegraphical (later the Telecommunication) Union (ITU) was critical intervention going back into the 19th century to define, standardize, generate, and evaluate information about the weather worldwide. Once gathered, this information's rapid exchange, then continuous comparison, re-analysis, and chained prediction, which could all then be reprocessed and reused as well as stored and ignored, become instrumental foundation to define and document discrete global "systems," out of various local wind, current, weather, precipitation events, globally.

Even before the Cold War and the Space Race, proto-planetarian networks were forming. That coincident system of worldwide transport, trade, and travel required such globalized information often occludes the wellsprings of such representations and interventions. Nonetheless, they are real, and the sciences of systematizing the study of Earth-bound events, Earth-rooted trends, and Earth-constitutive processes all first arise in the world-spanning Victorian economy and society that arose with transnational traffic tied to steamships, telegraphy, telephony, radio, and aircraft travel (Luke, 2008b). Those technologies called forth their own patchwork of meteorological monitoring, globalization, and risk management. Often rooted in far-flung imperial domains around the world, the impulse to map, measure, monitor, and then manage natural events, perhaps as early as the 1890s, was already giving root to Earth System Sciences (Luke, 2008b).

A most intriguing and quite articulated organizational effort aimed at these ends, however, is far more recent. It can be observed in the workings of the ESSP that formed a decade after the collapse of the U.S.S.R. in 2001. As the terse self-characterization of its mission asserts, "The ESSP is a partnership for the integrated study of the Earth System, the ways that it is changing, and implications for global and regional sustainability" (http://www.essp.org/index.php?id = 10). While much is suggested in this one sentence, the ESSP also says even more by declaring unproblematically that the Earth indeed is a single system: "The Earth System is the unified set of physical, chemical, biological and social components, processes and interactions that together determine the state and dynamics of Planet Earth, including its biota and its human occupants" (http://www.essp.org/index.php?id = 10). Given this object of analysis, the science required to assay it is extensive.

Indeed, "Earth System Science is the study of the Earth System, with an emphasis on observing, understanding and predicting global environmental changes involving interactions between land, atmosphere, water, ice, biosphere, societies, technologies and economies" (http://www.essp.org/index.php?id = 10). Science here does not necessarily require any legitimating impulse beyond knowledge for knowledge's sake, but there is more to it than that. The ESSP openly asserts it represents a partnership formed for science-in-action, because its partners and many others see this now systematized Earth in a state of emergency. Today for its members, "the urgency of the challenge is great: In the present era, global environmental changes are both accelerating and moving the earth system into a state with no analogy in previous history" (http://www.essp.org/index.php?id = 10). While this claim appears to be true, what is implied as having greater truth is the need for some agent or structure to check such accelerating global environmental change to retard and/or reserve the earth system's entry into this "state with no analogue in history."

The ESSP formed in Amsterdam during 2001 at the first Global Change Open Science Conference where 1,400 participants representing over 100 countries signed "the Amsterdam Declaration on Global Change" (http://www.essp.org/index.php?id = 41&L = en). Essentially, this dense network of open science practitioners banded together the scientific communities of four major global change research initiatives, including DIVERSITAS (The International Biodiversity Program), the World Climate Research Program (WCRP), the International Human Dimensions Program on Global Environmental Change (IHDP), and the International Geosphere-Biosphere Program (IGBP). Each of these scientific technical bodies is a putatively self-organized collaborative effort to guide scientific research; yet, their respective individual and current collective agenda all point toward seizing the policy initiative in global environmental policy-making by doing a certain kind of "Type 2 science" rooted in Earth accountancy (Gibbons et al., 1994).

The presumption of Earth system scientists attached to this project to be granted legitimacy, if not actual authority to rule, is astounding. Charged by little more than their own personal anxieties, research agendas, and/or shared apprehensions, these transnational networks for conducting collaborative research projects on many different environmental phenomena pretend to know what must be studied to take charge in a state of emergency. Even though there is an emergent expertarchy forming in these enterprises, very little sense of how complex governance always will prove to be ever appears in their studies. Indeed, some naively believe their own certain faith in the nebulous notion of sustainable development is already operational through their current embedded structure, norms, and policies. Biermann, for example, suggests Earth Systems are a firm anchor for global environmental change studies, policy recommendations, and values. He even outlines an initial understanding of "Earth System" governance, asserting that it can be seen at work in "the formal and informal rule systems and actor-networks at all levels of human society (form local to global) that are set up in order to influence the co-evolution of human and natural systems in a way that secures the sustainable development of human society" (Biermann, 2007, p. 4). This profile for the Earth System Science Project as a governance engine says all at once that it is everything, nothing, and a certain expertarchic something is therefore well-worth critical analysis.

4. CENTERS OF ASSESSMENT AND ACTION

Three key nodes for observing Earth System Science at work are climate change research centers like the Tyndall Centre in the United Kingdom, the Pew Center in the United States, and PIK (Potsdam-Institut für Klimafolgenforschung) in Germany. Partly technoscience networks, partly green governmentality assets, partly environmental accountancy shops, and partly global governance collaboratives, these centers have multiple programs already in play (Lomborg, 2001). To simply disclose their verdictive aspirations as they display them in their web presence and then briefly discuss their would-be jurisdictive actions as points of environmentality is worth doing here.

At the same time, one must remember these centers of calculation and their scientific technical deliverables along with the policy work of their personnel, supporters, and stakeholders tend to, first, constrain conceptions of what must be known as well as, second, define directions for what must be done. At its limit, Earth System Science essentially expands the art of government by overlaying a planetarian accountancy of ecological processes with another "introduction of economy into political practice" (Foucault, 1991c, p. 92). Hence, let us turn to these cases.

4.1. The Tyndall Centre

The Tyndall Centre for Climate Change Research is an expertarchic ecological accountancy operation aiming to render global warming into a continuously accountable phenomena by conducting integrated modeling and policy analysis on energy use, social adaptation, international development, coastal damage, and urban endangerment. Officially, a research consortium of six British universities, including East Anglia, Manchester, Newcastle, Oxford, Southampton, and Sussex, it also receives considerable support from the Natural Environment Research Council, the Economic & Social Research Council, and the Engineering and Physical Sciences Research Council (http://www.tyndall.ac.uk/general/history/funding_inst.shtml) in the United Kingdom.

Its next phase organizational objectives are quite clear. The Centre aspires to become an internationally recognized climate change research center, while exercising significant influence on national British policies. As its "About Us" pages disclose, the Tyndall Centre presents itself as a key proactive site of intervention.

Tyndall Centre Objectives

Vision

To become an internationally recognised *source* of high quality and integrated climate-change research, and to exert a seminal *influence* on the design and achievability of the long-term strategic objectives of UK and international climate policy.

Purpose

To *research, assess* and *communicate* from a distinct trans-disciplinary perspective, the options to mitigate, and the necessities to adapt to, climate change, and to integrate these into the global, UK and local contexts of sustainable development.

Medium-Term Objectives

- *Advancing the science of integration* .... to develop, demonstrate and apply new methodologies for integrating climate-change related knowledge.
- *Developing responses* .... to seek, evaluate and facilitate sustainable solutions that will minimise the adverse effects of climate change and stimulate policy for the transition to a more benign energy and mobility regime.
- *Motivating society* .... to promote informed and effective dialogue across society about its ability and willingness to choose our future climate.

Emerging Strategies

- To *demonstrate*, by conducting integrated, trans-disciplinary, and frontier research, rather than through specialist studies in well established fields, that the Tyndall Centre can deliver insights about possible responses to climate change that could not have been achieved by the constituent parts working on their own.

- To *conduct* a selection of strategic assessments of complex problems created by climate change in a framework of both horizontal (across disciplines) and vertical (from problem to solution) integration.
- To *initiate*, and contribute to the development of, new trans-disciplinary research agendas relating to climate change and stimulate the necessary institutional, human and financial resourcing.
- To *engage* new elements of society in open and informed debate about the choices presented by climate change.
- To *foster* an international pool of talented young researchers who have the capacity to work at the interfaces of natural, social and engineering sciences and who can apply this creativity to further assist society in responding in sustainable ways to climate change.
- To *seek* matching funding during the initial five-year period and to secure expansion and diversification of financial resourcing over the following 10 years.
- To *evaluate* and pursue opportunities for founding fully or semi-commercial organisations that, while distinct from the Centre, will be able to apply new insights and tools generated by the Centre to the competent solving of climate-change management problems.

These strategies will require us to: i) engage widely with stakeholders in public and private sector organisations; ii) collaborate closely, and/or form strategic partnerships, with researchers outside the Centre in the UK, Europe and globally; and iii) listen carefully to the perceptions and aspirations of all peoples in relation to the risks created by climate change.

Output Performance Measures

These measures are defined by the Research Councils and are reported annually by the Centre. They include, for example, publications, external funding, research programme highlights, submissions to Parliament, etc. (http://www.tyndall.ac.uk/general/objectives.shtml)

Manifestly, the Tyndall Centre is an enterprise that "brings together scientists, economist, engineers and social scientists, who together are working to develop sustainable responses to climate change through transdisciplinary research and dialogue on both a national and international level" (http://www.tyndall.ac.uk/general/about.shtml). Along with such activities, this "development of sustainable responses to climate change" articulates itself through a series of global surveillance engagements that map, measure, monitor, and then maybe manage the patterns of contemporary global warming, if and when its accountancy is accepted by the general public, policy advisors, or business leaders.

This combination of policy analysis, scientific prognostication, earth monitoring, and a putative practitioners field guidance is easy to see as a set of practices in the Centre's "Latest Stories" on its "News," (http://www.tyndall.ac.uk/media/news/latest_news.shtml – pages) where the Tyndall Centre imparts its views – on climate change, for example, or its mitigation – to all of its publics:

Have a target of 2 degrees and plan for 4 -

Kevin Anderson and Alice Bows have today published a rigorous analysis of where modern-day emissions might be heading

Their paper, published by the Royal Society, is the first to look at global emissions since 2000 and so includes the rapidly industrialising economies of Asia and South America and incorporates recent deforestation, and spans the six 'main' greenhouse gases. They conclude "By focussing on long-term emission targets, such as 50% by 2050, climate policy has essentially ignored the crucial importance of current emission trends and their impact on cumulative emissions. As a consequence, although we should aim to reduce global emissions in line with a 2°C target, adaptation policy must focus on climate change impacts associated with 4°C or more."

Read more: http://journals.royalsociety.org/content/a7877169j7163rh2/
Further information: http://www.guardian.co.uk/environment/2008/sep/01/climatechange.scienceofclimatechange2

2008 Survey on attitudes to climate change & carbon offsetting -

This web survey asks about your attitudes to climate change and carbon offsetting. Your help in completing this web survey will be invaluable for the study.
Further information: http://www.tyndall.ac.uk/survey/survey.php?sid = 31

How do we create a sustainable coastline? -

The annual Tyndall Centre Forum is this year at the University of East Anglia on Wednesday 10 September and coincides with the annual Tyndall Assembly which brings together all Tyndall researchers from across the UK

Tyndall Centre researchers will open the afternoon by addressing climate change and coasts as a key sustainable development issue. Their brief presentations will highlight recent Tyndall Centre research and explore the science and management of a changing coast, linking the international scale to the case of Norfolk. Speakers from the Environment Agency and North Norfolk District Council will respond by giving the audience a practitioner's perspective on what climate change and other drivers mean for the present and future East Anglian coastline, and the substantial strategic and management challenges involved. To close, the panel of speakers and the audience will debate about the issues raised during the course of the afternoon.

First International Summit on Policy, Technology and Investment -

The Cambridge Climate Summit 'Entrepreneurship for a Zero Carbon Society' aims to step-up collaborations and understanding between policy, academia and business through enabling the latest research and ideas to be presented by international experts from the key fields involved in tackling climate change.

The Summit is designed to bring together cutting-edge energy research and innovative businesses to move the highly multidisciplinary clean energy innovation community forward on the scale required for a carbon free future. Relevant fields also include environmental tax reform and commercialization of low-carbon technologies. Pioneering technologies will be showcased by industrial participants as well as the latest student research in these areas.

All who are interested in politics, policy making, science, engineering, energy supply and green-technology are welcome. Held in the historic city of Cambridge amongst its impressive ancient and modern buildings, the meeting includes a banquet and plenty of opportunity for relaxed discussion and networking. We look forward to seeing you in September...

Read more: http://www.tyndall.ac.uk/media/press_releases/CambridgeClimateSummit 2608.pdf
Further information: http://www.cambridgeclimate.com/

The challenges of using market mechanisms to trade ecosystem services -

This is a summary of a recently published paper by Esteve Corbera and Katrina Brown in the leading development studies journal World Development

Their research exposes the difficulties in designing institutions to market forest-carbon and the constraints and challenges facing developing country governments if they are to effectively participate in global markets.

Read more: http://www.tyndall.ac.uk/publications/briefing_notes/worlddevelopment-summary.pdf

Measuring Impact: Highlights 2007–2008 -

This Briefing Note summarises the science and knowledge exchange highlights within this year's Annual Report to the UK's Research Councils

Read more: http://www.tyndall.ac.uk/publications/briefing_notes/tyndallimpact08.pdf

Adaptation and Mitigation for Europe -

The EU ADAM project, a consortium of 26 research organisations led by Tyndall UEA, has published its interim research results

ADAM is supporting the EU in the development of post-2012 global climate policies, in the definition of EU mitigation policies to reach its 2020 goals, and the emergence of new adaptation policies for Europe with special attention to the role of extreme weather

Read more: http://www.tyndall.ac.uk/publications/briefing_notes/ADAM_interimresults.pdf
Further information: http://www.adamproject.eu

This interoperation with the European Union's ADAM project illustrates how the Tyndall Centre's practices are cogenerative effort with another 25 research organizations. As Foucault argues, one can see here a politics of methodical action coalescing around questions of method. With its hybridized ecological and economic research, the Tyndall Centre becomes for ecological accountability devotees one of those "places where what is said and done ... and the planned and the taken for granted meet and interconnect" (Foucault, 1991b, p. 75) for environmentality actions. These otherwise ordinary web pages almost write the mission statement of general environment accountancy itself; that is, once their green governmentality basis is accepted as legitimate.

4.2. Pew Center

Likewise, the Pew Center on Global Climate Change in Washington, D.C. (http://www.pewclimate.org) pursues parallel objectives to the Tyndall Centre in the United Kingdom, namely, operating as a non-profit, non-partisan, and independent organization (under Internal Revenue Code section 501(c)(3) "to provide credible information, straight answers, and innovative solutions in an effort to address global climate change." Of course, the Pew Center also echoes the global state of emergency thematic, while suggesting its operations constitute "a forum for objective research and analysis and for the development of pragmatic policies and solutions" for "an issue that is often polarized and politicized" (http://www.pewclimate.org/about/history_and_mission). Declaring itself engaged across an array of spaces zoned as "International," "U.S. Federal," "U.S. States & Regions," and "Business," the more than obvious subtext is one of a pragmatic planetarian productivism pitched at national and transnational sites of action. As its headline reiterates in every web screen, the Global Climate Change project is simple: "Working Together ... Because Climate Change is Serious Business." And, to that end, much of its self-disclosure uncritically uses this language. To underscore its global reach, much of its on-going Anglophone scientific work also is scannable in Japanese, Mandarin, German, Spanish, and French.

The Pew Center is an enduring, embedded, and artfully engaged environmentality engine aimed at bringing together "business leaders, policy makers, scientists, and other experts," whose approach Pew claims "is based on sound science, straight talk, and a belief we can work together to protect the climate while sustaining economic growth" (http://www.pewclimate.org/about). Once again, the expertarchy offers itself as an objective, pragmatic, and sensible force dedicated to vital goals, like "producing first-rate analyses of key climate issues"; "working to keep policy makers informed"; "engaging the business community in the search for solutions"; "and reaching out to educate the key audiences" (http://www.pewclimate.org/about). As Foucault's (1991a, 1991c) examinations of governmentality suggest, the study of the Earth in such discourses can give all new jurisdictive responsibilities in the verdictive practices of establishing orders of things.

As advocates of environmental accountability, the Pew Center's personnel also mobilize their expertarchic forces in an openly liberal capitalist register. For the Center and its backers, it underscores the "Key Issues" (http://www.pewclimate.org/about/key.cfm). That is:

- **Risk of Inaction.** What are the consequences if we fail to act and meet the challenges of climate change?
- **Global Equity.** How do we establish a system that is fair, yet recognizes that different countries are at different stages of development and have varying levels of greenhouse gas emissions?
- **Developing Countries.** What actions can and should we expect of developing countries, and when?
- **Industry Equity.** How can we plan and implement programs to address climate change in ways that are fair to different industry sectors and members of the workforce?
- **Market Mechanisms.** How can we design market mechanisms and incentives that will result in timely, meaningful actions at the lowest possible cost?

The empowering logics in these accountability practices are both self-centered and diverse.

The "we" who are aware and acting are "experts," or all of those "top-tier researchers," "4,000 opinion leaders throughout the world," "business leaders and policy makers," and "the general public" who are allied to act on global climate change (http://www.pewclimate.org/about). Yet, they are ready to design mostly market mechanisms to intervene appropriately in the developed and less developed countries, to address the more and less greenhouse gassing industries, and to mitigate the greater and lesser suffering sites of climate change consequences. The message is unmistakable: (1) the adverse climate change impacts in, on, and around the United States and Canada are pre-imminently important, but they "may not be immune from impacts experienced elsewhere" (http://www.pewclimate.org/about/stake.cfm) and (2) moving forward to adapt now for the inexorable arrival of a changing climate is imperative as expertarchic authority steers "governments of the world," "business" as such everywhere, and "citizens throughout the world" to learn more about the issues and make changes in food production, transportation, energy use, and consumption "that will help lay the groundwork for change" (http://www.pewclimate.org/about/moving_forward.cfm).

4.3. The Potsdam Institute

In a different vein, the Potsdam Institute for Climate Change Research (Potsdam-Institut für Klimafolgenforschung) is a focused effort in Germany for the European Union and global NGO community to mobilize "systems and scenarios analysis, quantitative and qualitative modeling, computer simulation and data integration" from both the social and the natural sciences "to generate interdisciplinary insights and to provide society with sound information for decision making" as those decisions relate to the challenges of "global change, climate impacts and sustainable development" (http://www.pik-potsdam.de/). Launched in 1992 with five disciplinary departments, PIK's researchers believe that "science has today out grown the stage of simply analyzing climate change and needs to face the challenge of forging solutions" largely by "entering a new transdisciplinary phase" in its research program (http://www.potsdam.de/research/index_html/view?-set_language = en). These domains are indeed transdisciplinary, namely, "Earth Systems Analysis-Research Domain I; Climate Impacts & Vulnerabilities – Research Domain II; Sustainable Solutions – Research Domain III; and, Transdisciplinary concepts & Methods – Research Domain IV" (http://www.pik-potsdam.de/research/research-domains).

PIK's funding comes from a mix of international agencies, German government bureaus, private foundations, and NGOs, and the projects it has conducted range from updates to climate change research since the IPCC Fourth Assessment Report, global warming and world poverty analyses, a Europe-wide Stern Report-style study, and global CO_2 markets scans (http://www.pik-potsdam.de/research/research-domains/sustainable-solutions/externally-funded-projects). Driven by complex system modeling methods, PIK's mission statements about "research domains" disclose its wholly systemic dedication to create a domain of its own out of research by making on-going reappraisals of global climate change with all of their impacts and vulnerabilities. Mapping these coupled dynamics could never be an impossible task as far as PIK's research teams are concerned.

For example, the Earth System Analysis domain of research starts off with clear operational assumptions:

RD I Mission

Our goal is to study the coupled dynamics of the geosphere, biosphere and anthroposphere under natural and human forcing. Results should shed light on the implications of future human perturbations to the Earth system.

Earth as a planet is currently in an unusually long and stable interglacial period. The dominant force for change in the present anthropocene is human activity, which is altering the atmosphere, oceans and land surfaces. This activity consists not only of the emission of greenhouse gases but also of direct large-scale impacts on the land and marine biosphere.

The analysis of Earth system dynamics focusses on four lines of research:

- Analysis of past climate change in order to understand the processes that amplify or dampen driving factors, leading to gradual or rapid change in system components.
- Analysis of data and scenarios of current and future climate change and its system-wide consequences.
- Analysis of the impacts of anthropogenic actions on the environment, most notably on climate, oceans, ecosystems and the spatial structure of the land surface.
- Analysis of the role of human activities as major drivers of global climate and environmental change.

Research Questions

Which feedback processes explain extent and timing of past glaciations?

- What is the role of ice sheets in global climate?
- How much and how fast will the world warm and sea level rise under given greenhouse gas emission scenarios?
- How will the ocean circulation respond in a changing climate?
- Which are the tipping points in the Earth system under human forcing and what are their thresholds?
- How will climate change alter the spatial patterns of the biosphere?
- How will global food production and freshwater use change in view of climate change, population growth, life style changes and economic globalisation?
- What are potential future patterns of land use?
- What is the world potential for sustainable bioenergy production?

(http://www.pik-potsdam.de/research/research-domains/earth-system-analysis).

These operationally stabilized and methodically mapped predictions and prior trend documentation efforts lead to what PIK's researchers regard – as expertarchic advisory groups – as plausible vulnerability assessments, particularly for Europe:

Background and Mission

Vulnerability to the impacts of climate change on nature and human society is given when and where the extent of damage exceeds the capacity for adaptation. Most impacts are manifest at the regional level, but the continental/global scope of vulnerability is gaining importance due to the sheer size of the problem, aggravating the consequences. There is a widening gap between poor, vulnerable regions and regions with high capacity for adaptation. Since vulnerability depends on natural and on social conditions, we must integrate the biogeophysical aspects of climate change with other changes in society, such as demography, economy and politics. The integrated assessment of regional vulnerability now receives high priority in research and is expected to deliver findings suitable for synthesis in national and global assessments.

Key Research Questions

To advance the knowledge base of climate impacts and vulnerability, we will study a broad range of potential damages (sensitivity and exposure) as well as the capacity of society to cope with damages (adaptive capacity). Specifically, this implies the following questions:

- How sensitive are specific sectors (in selected regions) to climate and global change? Past research has been focussed on few sectors with expected high sensitivity or readily available models (e.g., agriculture and forestry) – and studies of other sectors (e.g., health, tourism) are comparatively sparse.
- Can the practical knowledge from stakeholders and other experts on adaptation be 'mined' in order to produce a new quality of regional vulnerability assessment?
- Can regional integration of vulnerability be achieved by systematic comparison of different sectors? Sensitivity and adaptive capacity differ between sectors not only in economic terms but also along social and/or ecological dimensions. New valuation techniques need to account for these differences, as well as for the interactions between sectors.
- How do the findings from regional vulnerability assessments scale up to broader spatial scales (e.g., national, EU) and to the globe?

Expected Outcomes

- A climate change damage function library. For some sectors (e.g., agriculture and forestry), advanced tools for the estimation of sensitivity exist, but these are usually inadequate for cross-sectoral integration. In order to overcome this limitation, and to cover more sectors, a damage function library will be constructed from a broad range of sources. We will classify observed factors affecting the multiple impacts of global change, e.g. on human health, ecosystems, infrastructure and other systems. New factors will be identified.

- Monetary and non-monetary valuation techniques for the relative importance of damages in different sectors. These will be applied in regional settings, based on different socio-economic scenarios. New emphasis will be given to extreme scenarios in order to better sample the range of possible future risk situations. Current regional vulnerability studies (e.g., on the Elbe catchment) will be extended. Findings and tools will also be tested on selected regions elsewhere (e.g., China, Amazonia).
- A meta database of functions and indicators for vulnerability studies and adaptation strategies. The database would allow the assessment of multiple drivers of global change. It could serve as a basis for economic assessment and other valuation studies, as well as for delivery of policy support in private and public institutions.
- A toolbox for the generation of socio-economic and climate scenarios suitable for vulnerability assessment in different regions and at the continental to global scale (including techniques to study vulnerability to changing extreme event frequencies).
- Integration of findings from regional studies and the UBA Germany Study to a new study of German national vulnerability.
- Maps of vulnerability at the global scale, using variable resolution and multiple indicators. (http://www.pik-potsdam.de/research/research-domains/climate-impacts-and-vulnerabilities)

Granted these highly formalized Earth system mappings, climate change vulnerability assessments, and transdisciplinary research alliances, PIK then specs out its expertarchical visions of various sustainability solutions and adaptation strategies:

> Research Domain III maps out strategies to mitigate dangerous climate change and to adapt to the unavoidable climate change. We develop instruments to accomplish the necessary reduction of greenhouse gas emissions by the world's energy systems. To mitigate dangerous climate change we consider four options that can be combined:
>
> 1. energy efficiency improvement
> 2. substitution of fossil fuels by renewable energy sources or nuclear power
> 3. carbon capturing and sequestration
> 4. structural change in consumption patterns
>
> Mitigation can be achieved through different international regimes of offering inducement. These regimes, however, affect the global distribution of income and wealth in different ways.
>
> Research Domain III develops regimes for competing normative models under discussion now, but especially for those that do not intensify global wealth inequality. In research for an international climate regime economic justice has to be considered just as much as economic efficiency. Since it is to the largest part the developing world that has to adapt to unavoidable climate change, strategies for adaptation should be mapped out especially for these countries.
>
> All strategies have to be built upon the current knowledge about climatic, ecological and economic systems' dynamics. The production of biomass for example, intensifies

conflicts about land and water use. Resolving these conflicts is part of all sustainable efforts to deal with climate change.

The risks of technological options to mitigate climate change, where so far no process-based representations exist, have to be weighed systematically against the risks of climate change.

Here, the "research domain" shows itself evolving into a "domain for researchers" from which its would-be expertarchs hope to assert strategic options, productive alternatives, or risk assessments. To admit as scientists that they are articulating "regimes for competing normative models" based on "current knowledge about climatic, ecological and economic system's dynamics," PIK's scientific technical workers implicitly outline their expertarchic agenda for governance. Knowledge is not for the sake of knowing; it is aimed toward the managerial ends of guiding green governance. This disposition is quite obvious in their sense of "central questions and products" for the environment coalitions of climate change accounting and navigation:

Central Questions and Products

- What are the options to mitigate dangerous climate change regionally, nationally and globally? Which of these are suited for short-term, medium-term or long-term application?
- What effect do technological change, urban dynamics and international trade have on mitigation and adaptation options? How easily can coalitions to mitigate dangerous climate change be formed?
- What kind of policy instruments can foster climate-friendly technologies and investments in adaptation?
- How do land use and climate change affect each other and what are the consequences for management strategies?
- How can it be ensured that the benefits of climate protection measures outweigh the associated risks?
- Which additional efforts (under which normative preconditions) should society make for climate protection to guard against possible failures of climate protection measures?
- Which measures of adaptation should be left to the market and which measures should be implemented by governments, where market is threatening to fail?

We develop climate policy scenarios that imply the current knowledge about the Earth system's determining components and that represent win-win solutions to stakeholders. Specific items are:

- Policy recommendations for the Kyoto-Plus process that are robust under climate, biosphere and socio-economic risks and uncertainties
- A prototypical adaptation fund
- Integration of adaptation and mitigation in cities
- To place climate policy within the dynamics of globalization

- To develop tools (e.g. REMIND model family) that provide and support climate policy scenario analyses (http://www.pik-potsdam.de/research/research-domains/sustainable-solutions)

At the end of the day, however, PIK's agendas are those of other expertarchic accountability assessments. Like the Tyndall Centre in the United Kingdom and Pew Center in the United States, PIK assures all that its work stresses how "the natural and social sciences work together to study social change and impacts on ecological, economic and social systems. They examine the Earth system's capacity for withstanding human interventions and devise strategies for a sustainable development of humankind and nature," which are aimed at providing "decision makers with sound information and tools for sustainable development" as needed by "national and regional authorities, and, increasingly, to global organizations such as the World Bank" (http://www.pik-potsdam.de/institute).

5. SUMMARY THOUGHTS

These scientific research centers are important inasmuch as they, first, represent exercises in epistemic re-engineering within the operations of the human and natural sciences and, second, prototype the discursive practices of the expertarchy functioning as green governmentality. Earth System Science might pretend to be a grand unifying theory of terrestrial processes, but, in fact, it is at best, as Foucault suggests, only another new epistemic prototype, or "a space of *dispersion*," which then opens, positions, or displaces existing conceptual formations in "an *open and doubtless infinitely describable field of relationships*" (Foucault, 1991a, p. 55). Those involved in such researches at the Pew Center, PIK, or the Tyndall Centre like to imagine they are capturing today's *Zeitgeist* for an "earthcare science." Yet, something else is moving with these practices.

Accountability for green governmentality resituates the discursive formations of natural and policy science into new interrelations rooted in, as Foucault asserts: (a) "the displacement of boundaries which define the field of possible objects ..."; (b) "the new positions and role occupied by the speaking subject in science ..."; (c) "a new mode of the functioning of language with respect to objects ..."; and (d) a new form of localization and circulation of discourse within society ..." (Foucault, 1991a, pp. 56–57). To postulate interrelations of an Earth System Science, therefore, displaces many verdictive means to work in the service of new jurisdictive ends.

As Foucault suggests, these epistemic enterprises bring forth new limits and forms for scientists and policy-makers alike who wish to identify the *sayable*, the *conservable*, the *memorable*, the *retainable*, the *valuable*, the *institutionalizable*, and the *controllable* (Foucault, 1991a, pp. 59–60). The *oikos* of ecology and economy converge in these discursive domains as the Earth is systematized and subjected to the operationalizations of Earth System Science whose accounting tracks "a machine which has to be continuously made, and not merely operated, by government ... in all its implications of possession, domestication, and controlling action" (Gordon, 1991, p. 11). It is made, it has completed prior accountable audits, and it stands ready for staging new departures today in the economy, society, and state.

Like accountancy, these environment accountability efforts, implicitly or explicitly, assume their mode of terrestrial reckoning must, or at least could, discover "irregularities" along with regularities in the economy of Nature and Society. The demarcation line between regularity and irregularity often is traced back to the 18th century when climate observers first noticed atmospheric warming and cooling variations. Increasing concentrations of greenhouse gases are a marker of the transition from agrarianism to industrialism, feudalism to capitalism, and localism to globalism. Once the turn is taken, climate change experts believe additional spikes mark acceleration in this transformation of the atmosphere's geophysical dynamics.

Here, the more common, and continuously debated, set of irregularities was discovered in the 1950s and 1960s in the rising tides of new data from the IGY and other research undertakings. Greenhouse gases, whose point-based combustion and exhaust were quite focalized, were discovered to be increasingly accumulating in parts per million of atmospheric gases as causes of generalized increases in overall atmospheric and oceanic temperatures. Once found in such acts of accountability, some forum for holding agents and structures accountable for these disturbances was regarded as being imperative. The back-and-forth over who is responsible, however, leads as often to calls for reassessment, rereckoning, or recalculation as well as better, more frequent, closer, more continuous measurement of the disruptions behind the observed irregularities.

The prospect of seeing Earth as a coupled human and natural system, which could be plagued by a sudden deep disruption, non-linear change, and chaotic complexity, is real. It first became a definite material possibility in a very decisive fashion on August 6 and 9, 1945. During this 96-hour period over six decades ago, it became quite evident – as various thinkers foresaw how atomic weapons technologies could proliferate and then different thermonuclear stockpiles would grow – that humanity immediately could be a profoundly earth-altering force in 60 minutes or less on any given day. Acknowledged openly in the MAD doctrines of the 1960s, and the nuclear winter debates of the 1980s, a full-blown nuclear war very quickly would prove human beings' powers to alter the planet on a historical and geological timescale (Luke, 1989a, 1989b). And, no matter what Earth System Science does to imagine the planet as a system ready and able to be closely controlled for the best environmental outcome through expert planetarian ecomanagerialism, this game-changing alternate type of planetary change also will sit on the sidelines as long as nuclear weaponry exists and nuclear power is used to abate the greenhouse gassing of caused by fossil fuels.

While part of the environmental science being mobilized here and now is concentrated upon improving its practitioners' instruments, laboratories, theories, and data, all of the current findings they offer come with the proviso that the data typically are incomplete, the theories are not entirely confirmed, the laboratories currently in operation are too few, and the instruments already are not fully reliable. Ordinary principles of good scientific practice, then, should stress the need for constant caution, determined doubt, and stiff skepticism about the findings of Earth System Science. Some producers and consumers of their findings do display this professional wariness about the entire enterprise. Yet, too many other consumers and producers of Earth System Science tend to have an unwarranted certitude about what is well-founded scientific knowledge, and what might be only provisionally detected and/or incompletely confirmed observations about the planet's many complex coupled systems.

Regrettably, these less cautious centers-of-calculation often become involved in latently political, or even manifestly institutional, networks of policy debate about "what must be done." Expertarch jargon about market mechanisms, fair programs of industrial growth, vulnerability assessments, and inducement regimes are mystifications of their grab for greater green governmentality. The conduct of conduct is the target of such green governance. Moving directly from statements about what "is" actually or allegedly true to statements, on what "ought" to happen is rarely productive, and logically flawed as rules of ethical and political practice. Nonetheless, one sees these elaborate networks of scientific mapping, technical monitoring, and managerial modeling operating in principle as engines of alternative administrative approaches, platforms of shadow cabinet consultation, or foundries for protyped environmental regulation.

As centers of calculation, concord, and conceptualization for Earth accountancy, many sustainability studies also are an initial positioning of liberty against security as well as security for liberty. On the one hand, global climate change studies, conferences, or accountings are "modes of state intervention whose role is to assure the security of those natural phenomena, economic processes and the intrinsic processes of population: this is what becomes the basic objective of governmental rationality. Hence liberty is registered not only as the right of individuals legitimately to oppose the power, the abuses, and usurpations of the sovereign, but also now as an indispensable element of governmental rationality itself" (cited in Gordon, 1991, pp. 19–20). Ecological accountability sets the limits of liberty and the scope of security through the same mechanisms of continuous environmental monitoring, managing, and mapping. Like the prison, the hospital, or the asylum, the Earth System Sciences project is being positioned to operate as *programmes*, or "sets of calculated, reasoned prescriptions in terms of which institutions are meant to be reorganized, spaces arranged, behaviors regulated" (Foucault, 1991b, p. 80).

Like Foucault, this analysis of green governmentality examines the interplay of scientific discourses and governmental initiatives within environmental accountancy. So, this discussion has tried "to define how, to what extent, at what level discourses, particularly scientific discourses, can be objects of a political practice, and in what system of dependence they can exist in relation to it" (Foucault, 1991a, p. 70). In a context in which considerable reserves of scarce assets, like remote sensing systems, geospatial imaging systems, meteorological monitoring systems, oceanographic measuring systems, and economic modeling systems, all can be mobilized for interdisciplinary missions, a progressive politics of responding to green governmentality must work "to understand the manner in which diverse scientific discourses, in their positivity (that is to say, as practices linked to certain conditions, obedient to certain rules, susceptible to certain transformations) are part of a system of correlations with other practices" (Foucault, 1991a, p. 70).

Generating these scientific studies, and circulating them as widely as possible, is a useful intellectual activity. Having better information is preferable to having no information, misinformation, or outdated information. And, there often is no authoritative force between these recommendations. Well-intentioned individuals, companies, and governments can try to work bits and pieces of the Earth System Science centers' research into their own behaviors, but there is no collective coordinated and cohesive response forthcoming simply because "the data" are there.

The on-going collaborations that produce Earth System Science are, as Gottweis (2003, pp. 258–262) might suggest, instances of policy-making as "ordering activities" that are precarious, relational, deliberative, and representative in the sense that environment accountability initiatives,

> Require attention to various forms of re-presentation, such as narratives, indices, computations, and procedures of assessment. Policymaking, then, must be studied as a material practice which does not simply react to its environment, but inscribes itself into its texture, and creates/rewrites order by drawing from a multitude of discursively available narratives, modes of representations, artifacts and technologies, a process which entails intermediation and transition between different social and non-social worlds and realms of perception. (Gottweis, 2003, pp. 260–261)

This characterization of inchoate policy-making as ordering activity is most likely too generous. Nevertheless, it captures the good intentions of environmental accountancy centers, like the Pew Center, the Tyndall Centre, and PIK. Whether or not, their representations of the rampant environmental degradation swamping the world today will gain a deeper hold is a political question well-worth waiting to hear answered. In the meantime, this initial examination of such centers illustrates a few prominent features in today's practices for taking Nature into account, the economies their diverse accounting practices tap into, and the normative values of sustainability behind how Nature can be brought "to account." In these relational ties, the expertarchic networks behind them accumulate new power and knowledge for would-be planetarian projects of control.

REFERENCES

Beck, U. (1992). *Risk society: Towards a new modernity*. London: Sage.

Beck, U. (2000). *What is globalization?* Oxford: Blackwell.

Biermann, F. (2007). Earth system governance as a crosscutting theme of global change research. *Global Environmental Change*, *17*, 326–337.

Bliese, J. R. (2001). *The greening of conservative America*. Boulder, CO: Westview.

Briden, J. C., & Downing, T. E. (2002). *Managing the earth: The Lineacre lectures*. Oxford: Oxford University Press.

Burchell, P., Gordon, C., & Miller, P. (1991). *The Foucault effect: Studies in governmentality*. Chicago: University of Chicago Press.

Cairncross, F. (1992). *Costing the earth*. Boston: Harvard Business School Press.

Chertow, M. R., & Esty, D. C. (1997). *Thinking ecologically: The next generation of environmental policy*. New Haven: Yale University Press.

Cortner, H. J., & Moote, M. A. (1999). *The politics of ecosystem management*. Washington, DC: Island Press.

De Certeau, M. (1988). *The practice of everyday life*. Berkeley: University of California Press.

Durning, A. (1992). *How much is enough? The consumer society and the future of the earth.* New York: W. W. Norton.

Foucault, M. (1970). *The order of things.* New York: Vintage.

Foucault, M. (1980). *History of sexuality, Vol. I.* New York: Vintage.

Foucault, M. (1991a). Politics and the study of discourse. In: G. Burchell, C. Gordon & P. Miller (Eds), *The Foucault effect: Studies in governmentality* (pp. 53–72). Chicago: University of Chicago Press.

Foucault, M. (1991b). Questions of method. In: G. Burchell, C. Gordon & P. Miller (Eds), *The Foucault effect: Studies in governmentality* (pp. 73–86). Chicago: University of Chicago Press.

Foucault, M. (1991c). Governmentality. In: G. Burchell, C. Gordon & P. Miller (Eds), *The Foucault effect: Studies in governmentality* (pp. 87–104). Chicago: University of Chicago Press.

Gibbons, M., Limoges, C., Nowotny, H., Schwartzman, S., Scott, P., & Trow, M. (1994). *The new production of knowledge: The dynamics of science and research in contemporary societies.* London: Sage.

Gobster, P. H., & Hull, R. B. (2000). *Restoring nature: Perspectives from the social sciences and humanities.* Washington, DC: Island Press.

Gordon, C. (1991). Governmental rationality: An introduction. In: G. Burchell, C. Gordon & P. Miller (Eds), *The Foucault effect: Studies in governmentality* (pp. 1–51). Chicago: University of Chicago Press.

Gore, A., Jr. (1992). *Earth in the balance: Ecology and the human spirit.* Boston: Houghton Mufflin.

Gore, A., Jr. (2006). *An inconvenient truth: The planetary emergency of global warming and what we can do about it.* Emmaus, PA: Rodale.

Gottlieb, R. (1993). *Forcing the spring: The transformation of the American environmental movement.* Washington, DC: Island Press.

Gottweis, H. (2003). Strategies of poststurctural policy analysis: Toward an analytics of government. In: M. A. Hajer & H. Wagenaar (Eds), *Deliberative policy analysis.* Cambridge: Cambridge University Press.

Hawken, P., Lovins, A., & Lovins, L. H. (1999). *Natural capitalism: The next industrial revolution.* Boston: Little Brown.

Hays, S. P. (1959). *Conservation and the gospel of efficiency: The progressive conservation movement, 1890–1920.* New York: Athenaeum.

Hoffman, A. J. (1997). *From heresy to dogma: An institutional history of corporate environmentalism.* San Francisco: New Lexington Press.

Josephson, P. R. (2002). *Industrialized nature: Brute force technology and the transformation of the modern world.* Washington, DC: Island Press.

Latour, B. (1988). *Science in action: How to follow scientists and engineers through society.* Cambridge, MA: Harvard University Press.

Lefebvre, H. (1984). *Everyday life in the modern world.* New Brunswick, NJ: Transaction.

Lefebvre, H. (1991). *The production of space.* Oxford: Blackwell.

Lomborg, B. (2001). *The skeptical environmentalist: Measuring the real state of the world.* Cambridge: Cambridge University Press.

Lovins, A. (1977). *Soft energy paths.* New York: Harper.

Luke, T. W. (1989a). *Screens of power: Ideology, domination and resistance in informational society.* Urbana: University of Illinois Press.

Luke, T. W. (1989b). On post-war: The significance of symbolic action in war and deterrence. *Alternatives: A Journal of World Policy*, *14*(3), 170–195.

Luke, T. W. (1997). *Ecocritique: Contesting the politics of nature, economy and culture*. Minneapolis: University of Minnesota Press.

Luke, T. W. (1999a). *Capitalism, Democracy and Ecology: Departing from Marx*. Urbana: University of Illinois Press.

Luke, T. W. (1999b). Training eco-managerialists: Academic environmental studies as a power/knowledge formation. In: F. Fischer & M. Hajer (Eds), *Living with nature: Environmental discourse as cultural politics* (pp. 103–120). Oxford: Oxford University Press.

Luke, T. W. (1999c). Environmentality as green governmentality. In: E. Darier (Ed.), *Discourses of the environment* (pp. 21–151). Oxford: Blackwell.

Luke, T. W. (1999d). The wilderness society: Environmentalism or environationalism. *Capitalism Nature Socialism*, *10*(4), 1–35.

Luke, T. W. (2002). The practices of adaptive and collaborative environmental management: A critique. *Capitalism Nature Socialism*, *13*(4), 1–22.

Luke, T. W. (2004). Ideology and globalization: From globalism and environmentalism to ecoglobalism. In: M. Steger (Ed.), *Rethinking globalism* (pp. 67–77). Lanham, MD: Rowman & Littlefield.

Luke, T. W. (2005a). Environmentalism as globalization from above and below: Can world watchers truly represent the earth? In: P. Hayden & R. El-Ojeili (Eds), *Confronting globalization: Humanity, justice and the renewal of politics* (pp. 154–171). New York: Palgrave Macmillan.

Luke, T. W. (2005b). Globalisierung als planetarisches Oekomamangement: Eine Kritik globaler Biokomplexitaetsmodelle. In: I. I. Schroeder & S. Hoehler (Eds), *Welt-Raeume: Geschichte, Geographie und Globaliserung seit 1900* (pp. 282–302). Frankfurt: Campus Verlag.

Luke, T. W. (2008a). The politics of true convenience or inconvenient truth: Struggles over how to sustain capitalism, democracy, and ecology in the 21st century. *Environment & Planning A*, *40*, 1811–1824.

Luke, T. W. (2008b). Climatology as social critique: the social construction/creation of global warming, global dimming, and global cooling. In: S. Vanderheiden (Ed.), *Political theory and climate change* (pp. 121–152). Cambridge, MA: MIT Press.

Mangun, W. R., & Henning, D. (1999). *Managing the environmental crisis: Incorporating competing values in natural resource administration* (2nd ed.). Durham, NC: Duke University Press.

Mann, M. (1986). *The sources of social power*. Cambridge: Cambridge University Press.

McDonough, W., & Braungart, M. (2002). *Cradle to cradle: Remaking the way we make things*. New York: North Point Press.

McNeill, J. R. (2000). *Something new under the sun: An environmental history of the twentieth-century world*. New York: Norton.

Milani, B. (2002). *Designing the green economy: The postindustrial alternative to corporate globalization*. Lantham, MD: Rowman and Littlefield.

Ong, A. (1999). *Flexible citizenship: The cultural logics of transnationality*. Durham: Duke University Press.

Pinchot, G. (1910). *The fight for conservation*. New York: Doubleday.

Pinchot, G. (1947). *Breaking new ground*. New York: Harcourt Brace Jovanovich.

Roosevelt, T. (1971). *The new nationalism*. Gloucester, MA: Peter Smith.
Sassen, S. (2006). *Territory, authority, rights: from medieval to global assemblages*. Princeton: Princeton University Press.
Schmidheiny, S. (1992). *Changing course: A global business perspective on development and the environment*. Cambridge, MA: MIT Press.
Schmitt, C. (2007). *The concept of the political: Expanded edition*. Chicago: University of Chicago Press.
Smith, S. (2008). *Uneven development: Nature, capital, and the production of space* (3rd ed.). Athens: University of Georgia Press.
Van Kooten, G. C., & Erwin, H. B. (2000). *The economics of nature: Managing biological assets*. Oxford: Blackwell.
Virilio, P. (1995). *The art of the motor*. Minneapolis: University of Minnesota Press.
Virilio, P. (2000). *A landscape of events*. Cambridge, MA: MIT Press.
WCED – World Commission on Environment and Development. (1987). *Our common future*. Oxford: Oxford University Press.
Westmen, W. E. (1978). How much are nature's services worth? *Science*, *197*, 960–964.
Wiseman, A. (2007). *The world without us*. New York: St. Martin's Press.

SOCIAL ACTION AND CATASTROPHE

Daniel M. Harrison

If you can keep your head when all about you are losing theirs... you'll be a man my son.
(Kipling, *"If"*)

1. INTRODUCTION

This chapter addresses the topic of social action in catastrophic situations. The main phenomenon to be explained is how social actors cope and adjust to a sudden and unexpected change in the ecological pressures around them, such as during natural and technological disasters, in war zones, in the aftermath of accidents, and so on.[1] Using interpretative and phenomenological methods (Bentz & Shapiro, 1998), this chapter builds toward a Weberian kind of "ideal type" model outlining key sociological variables that try to explain differential social responses from human beings as they experience catastrophic circumstances.[2] Human beings must act very differently during catastrophes than they do during normal life. Just why is it that some people just succumb to the elements, whereas others struggle to persevere? How do some social actors manage to rely on their intelligence, resources, skill sets and other assets to control an otherwise hostile environment while others fail to do so?

Many sociologists and most social theorists are fundamentally interested in the relationship between individuals and society. In catastrophes, such

Nature, Knowledge and Negation
Current Perspectives in Social Theory, Volume 26, 161–186

ISSN: 0278-1204/doi:10.1108/S0278-1204(2009)0000026009

relationships are severely strained, sometimes to a breaking point. They can involve the "destruction of families, entire villages, and traditional ways of life; the unraveling of preexisting social structures, social roles, social arrangements and rituals; the terrorization and 'silencing' of victims and onlookers; the massive disruption of trusting connections with others; and other massive upheavals in human relationships and activities and in culture itself" (Ehrenreich, 2003, p. 19). These are situations where nature prevails over individuals in brutal, terrifying, and unexpected ways, and human culture and activity recedes (if only perhaps just temporarily) into the background.[3]

This essay focuses on human actions that mitigate the consequences of such events – on the social practices through which culture gets (or fails to get) reestablished as human beings attempt to regain control of their immediate environment. In terms of the parameters of this volume, it attempts to delineate some of the more salient social dynamics and processes that occur in decidedly "dystopian" circumstances.[4] Catastrophes signal a definitive break from civilization and with some human beings regressing into a sort of pre-modern kind of existence.[5] These scenarios are obviously not voluntarily chosen by participants or willed into being by them. To the contrary, the types of environments discussed in this essay often are a person's worst nightmare and represent a complete "Otherness" of experience impossible for subjects to fathom.

Although it is important to note important differences across catastrophic situations, there are many parallels among them such as knocking people off balance, threatening immediate and lasting harm, and forcing actors to urgently reestablish identity (White, 1992) in the face of massive social disorganization. With Lanzara (1983, p. 71), the analytical focus of this essay "is on the micro-level, exploring the behavior small volunteers' or folk groups that emerge and take action in the aftermath of disasters." It seeks to understand human reactions to extreme events as they are happening. Taking seriously the notion that sociology should provide "a down-to-earth approach to the scientific study of group life and human conduct" (Blumer, 2007, p. 68) that explores and investigates social worlds in their "natural" state, that is, as people actually live their lives, the central idea of this work is that catastrophes and disasters offer theorists and other observers quasi natural experiments into differential forms of action after conditions of social breakdown; they allow us to consider the social factors influencing why some people manage to behave rationally, even humanely, and others less so. How and why do some human actors fail to cope with their rapidly changed ecological milieu, whereas others manage to improvise, adapt and survive them?[6]

As our world becomes increasingly dystopian (Diamond, 2006) and crisis ridden, there will be a great need for social observers to understand and analyze the social processes through which how humans act in catastrophic circumstances. This essay endeavors to make a contribution to uncover what occurs during experiences of these events, that is, how humans – in their "nature" as *Homo sapiens* – fare amidst catastrophe, and what sorts of insights such social environments may provide when society faces the next disaster.

Section 2 looks into these ideas about acting in extreme environments in more detail, after which we consider the sociology of anatomical social facts. A preliminary attempt at modeling action in catastrophes appears at the end of this work.

2. ACTION IN EXTREME ENVIRONMENTS

Although not an entirely neglected area of study (see, e.g., Prince, 1920; Kutak, 1938; Wallace, 1956; Barton, 1969; Janis, 1969; Erikson, 1976; Lanzara, 1983; Sztompka, 2000; Hoffman & Oliver-Smith, 2002; Clarke, 2003; Clarke, 2006; Alexander et al., 2004; Neal, 2005; Rodriguez, Trainor, & Quarantelli, 2006; Freudenberg et al., 2008) traumatic environments and catastrophic situations do not generally receive much intellectual attention by contemporary sociologists and social theorists. Given the increasingly risk-prone nature of contemporary societies, however, including a global recession, ubiquitous environmental calamities, permanent war zones across the world, etc., this seems a bit shortsighted. Indeed, with the persistent ongoing social turbulence in contemporary societies, it could be argued that it will become increasingly important to understand "how people respond to extreme events…to situations of social breakdown" (Clarke, 2003, p. 125). To better illustrate this mode of reasoning, consider the following vignettes drawn from two recent catastrophic events in American history.

In late August 2005, as New Orleans was still in the midst of coping with Hurricane Katrina, many survivors who could not evacuate the city languished in flooded homes or waited outside the New Orleans Convention Center and Superdome, hoping for aid and assistance. Other actors, however, were not simply passive witnesses to their own helplessness, but managed to engage in positive action to change their situation, and the situation of others, for the better. One such action involved Mr. Jabbar Gibson, a 21-year-old man from New Orleans, who in the wake of Katrina's destruction managed to commandeer an idle school bus, drive it to the

Convention Center and then take a load of passengers on "a wild 12-hour ride filled with breakdowns, mishaps and close calls with the law" (Perlstein, 2006, p. 1) to safe haven in the Houston Astrodome. Gibson's bus was the first to bring evacuees to safety. Sociologically speaking, just how are we to explain the social forces and dynamics that unfolded in this situation?

One could argue that the effort to provide solid explanations of incidents of this sort ought to be an integral part of the cultural narrative that sociologists provide in the 21st century. As another example, take the hijacking of United Flight 93 – the fourth hijacked plane on September 11, 2001, and the one that presumably was heading for the White House. As we know from the cockpit data recorder and telephone conversations passengers made to family members and friends on the ground, some passengers on board Flight 93 organized what turned out to be a robust (albeit tragic) form of social action. Determining strategy at the back of the plane away from view, the passengers mobilized against their hijackers. Using a concessions cart and fire extinguishers as battering rams, they stormed the cockpit, subdued the pilots, and forced a crash landing in Shanksville, Pennsylvania. Through their action, the passengers effectively prevented the terrorists from accomplishing their deadly ambitions (Longmon, 2003). As social theorists, how are we to make sense of this chain of events? What were the specific social (as opposed to psychological) forces that compelled the passengers to act in such a way? What were the key social dynamics and processes involved? In what ways was the passenger's collective reaction and action socially constructed?

As a final example of the sorts of cases that demonstrate the importance of studying social action in catastrophe, consider the case of six people stuck in an elevator of the World Trade Center on 9/11 just after the first plane hit the north tower at 8:46 a.m. Inside the elevator were Jan Demczur, a window cleaner, and five other men. Using the blade and handle of his window washing tool to pry open the elevator doors, the men took turns scraping at a plasterboard wall across from the shaft. When they struck ceramic tile, they surmised they had stopped adjacent to a bathroom and managed to smash a hole through the wall. After working for an hour, they had made a hole big enough for all of them to crawl through and they dropped into the bathroom and then ran for their lives (Bruinius, 2002; Dwyer & Flynn, 2005, pp. 154–156, 162). This is one of the most interesting stories to emerge from 9/11. Just what was its sociological basis? How should we explain its constituent features and salient dynamics to students in our classrooms?

When it comes to understanding social action and catastrophe, one must first consider that there will be events that deprive individuals from the opportunity to act with intent. Obviously some catastrophic situations come on very quickly, such as car accidents, plane crashes, tornadoes, bombings, or nuclear attacks, and actors have very little or no time to act. As Rousmaniere (1980, p. 165) puts it, these "are occasions when men can do little or nothing to help themselves." Yet other extreme events and catastrophic situations unfold more slowly and actually offer individuals opportunities to adapt or improvise in a "period of recoil" (Rousmaniere, 2002, p. 198) and provide the social transcript (Scott, 1995) for analysis and interpretation suggested by this essay.

It should be clear that such a research interest is in many ways quite different from other scholarly work on catastrophe and disaster. First, unlike much literature, the present work is interested mainly in human reactions to catastrophic or extreme environments as they initially occur and in their immediate aftermath, that is, during the "period of recoil." Although sociological and other scientific work on the causes and long term consequences of disaster is obviously highly important, that is not the primary focus of this work and is located elsewhere (see, e.g., Freudenberg et al., 2008). Second, this chapter strives to maintain the analytically necessary distance from narratives of heroism or horror that often emerge in the aftermath of catastrophic events. Rather than providing real insight into the human condition, more often than not such narratives often devolve into benign platitudes about people "doing the right thing" or "stepping up to the plate". Other narratives describe how people turn into anti-social "monsters" or "animals" during the catastrophes.[7] Although such insights may be pertinent with regard to particular instances, they do not facilitate rigorous sociological analysis or understanding. The third distinctive feature of the present research project is its reluctance to focus simply on the various psychologically maladaptive or "dysfunctional" responses to catastrophes, for example, depression and other forms of mental illness, alcoholism, drug use, or socially traumatic responses to catastrophe described by researchers such as Erikson (1976). Although such research is not only legitimate but also important and necessary, it does not follow in the sociological current of trying to understand extreme events and catastrophes as I am pursuing it here.

Instead of following these approaches to disasters, this chapter is oriented toward a tentative model of the key sociological variables and social processes at work in catastrophic situations. By employing and adding to the "ideal type" of *catastrophic situations*, and comparing it in reference to empirical research of past catastrophes, we might be better able to identify predictive dynamics and develop action oriented typologies helpful in understanding catastrophic situations of the future. Before concentrating on the development of this model, however, a few remarks about the relationship of this research to the subject matter of sociology proper, will be in order.

3. SOCIOLOGY AND THE STUDY OF ANATOMICAL SOCIAL FACTS

Although at first glance it might seem that the main objective of this essay, namely, understanding differential human action in catastrophic situations may be somewhat marginal to the discipline of sociology and domain social theory, closer inspection should demonstrate that the former is central to the latter. It is one of the defining purposes of this essay to reclaim for sociological and social theorizing an aspect of their subject matter that has been strangely neglected and largely forgotten.

More than any classical sociologist, Emile Durkheim was adamant that sociology is and must be a true scientific discipline. Yet he argued it could only be a real science, if like other sciences, sociology had its own unique subject matter, one that is distinctive from the other disciplines. Durkheim proposed that the only proper subject matter for sociology was what he called "social facts" (Lukes, 1985, pp. 10–12), that is, social forces external to individuals that constitute constraints on human action. Although Durkheim's remarks on this concept are not always clear, we do know according to Lukes that "Durkheim saw social facts as lying on a continuum. At one end are structural, 'anatomical or morphological' social facts, making up the 'substratum of collective life'" (Lukes, 1985, p. 9). As Durkheim (2003, p. 78) puts it: "Social life rests upon a substratum which is determinative both in its extent and its form. It is composed of the mass of individuals who comprise the society, the manner in which they are disposed upon on the earth, and the nature and configuration of objects of all sorts which affect collective relations". Anatomical social facts represent "the tangible, material forms of societies…the nature of their substratum" (*Ibid.*). These facts vary "according to geographical disposition…[and are] the base from which social life arises" (in Giddens, 1972, p. 72). These excerpts make it clear that in referencing structural, anatomical, or morphological social facts, Durkheim had in mind some concept of the socio-ecological or environmental pressures and constraints that operate upon social organization and human society. Such social facts are foundational because without coping with them, societies would never be able to remain stable for long, let along develop the sort of highly differentiated systems of organization that Durkheim thought would bring about organic solidarity. Studying such classes of social facts, then, was one of the primary purposes of sociology according to Durkheim, and he also deemed it significant enough to constitute an independent discipline of its own, which he called "social morphology."[8] Durkheim writes that "this substratum directly or indirectly affects all social phenomena" (in Emirbayer, 2003, p. 77). He goes on to write that "this new field will study, not the forms of the earth, but the forms which effect societies as they establish themselves on the earth" (in Emirbayer 2003, p. 78).

According to Lukes, the second main category of social facts delineated by Durkheim is what he called institutional social facts, by which he meant social pressures and constraints found in the context of various (i.e., legal, political, religious, familial, and economic) institutions. This is the type of sociology that is probably most familiar to contemporary sociologists and social theorists and represents the bulk of sociological teaching and research. Durkheim's third main category of social facts was the least well defined (Lukes, 1985) and involves external social pressures and constraints that are not specific to any particular institutional domain. Durkheim referred to these types of non-institutionalized social facts as "social currents" (Lukes, 1985, p. 10).

The point in discussing this brief intellectual history is to remind readers how much the discipline and subject matter of sociology has become overly delimited and circumscribed in recent years. Many sociologists today would be hard pressed to provide a definition of the concept of anatomical social fact, and most probably would not use the term "social fact" in reference to the purpose of the discipline of sociology in general. The upshot of all of this is that many sociologists and social theorists are neglecting to adequately reflect upon and theorize a large portion of the social world and a significant portion of their subject matter. Although many sociologists today spend a great deal of time and effort researching and teaching various institutional (e.g., poverty, religion, the media) and non-institutional social facts (e.g., fads, crazes, rock concerts, gun violence), there has been a corresponding neglect of the study of anatomical social facts in the discipline as a whole. This is highly problematic because the elucidation of anatomical social facts are no less important in Durkheim's terms for the discipline of sociology than the other two main classes of social facts, and arguably more so.

On a deeper level, the jettison of the analysis of anatomical social facts from the discipline of sociology[9] is even more peculiar since anatomical social facts are arguably the most constraining of any social facts facing human beings. As is quite clear to anyone who has had any experience with a car accident, for example, or has been lost in the woods, or even has experienced some sort of traumatic event (such as a fire) in the naturalistic conditions of home, the social constraints generated by anatomical social facts are quite tremendous. Many individuals in Euro-American and other contemporary societies successfully manage to enjoy a relatively high standard of living and through the cooperative and competitive workings of society manage to keep nature at bay. Yet in some important respects, societies are having increasing difficulty keeping nature removed from culture (Beck, 1984; Hazelrigg, 1995) and around the world anatomical social facts continue to exert a huge influence on how human beings live their lives. One need only reflect upon some of the tragedies of the early 21st century such as the 9/11 attacks, Hurricane Katrina, bridge collapses in Minnesota, tornadoes, forest fires in the West, not to mention micro-biological social facts of, for example, the swine flu virus and e-coli bacteria in our food supply, to realize that humans in contemporary societies at times are "overwhelmed by environmental forces [they were] never designed to withstand" (Kampler, 2004, p. 13). Given this reality, it is quite ironic, as well as worrying that much of sociology and social theory has seemed to have forgotten or neglected to theorize or think about this subject matter in recent years.[10] It appears that many social researchers and scholars no longer even recognize as legitimate such areas of study within their discipline and would prefer to embark on sterile analyses of global institutions, for example, or remain within the safe confines of established qualitative and quantitative research methodologies, which have little or nothing to do with Durkheim's conception of sociology. If Durkheim's delineation of sociology gets lost or forgotten, the whole discipline runs the risk of collapse. If, however, more sociologists would reflect upon the roots of their discipline with its own unique subject matter, and if they strive to be true to this call, perhaps more might become interested in once again understanding the profound impacts of anatomical social facts and human reactions to them. Extreme environments and catastrophic situations create social facts like none other and are arguably much more influential upon identity than many of the categorical variables (race, class, and gender, for example) that get bandied about with so much alleged significance in introduction to sociology classes and to which so many professional sociologists and social theorists unthinkingly devote so much of their time.

4. TOWARD AN ANATOMICAL SOCIOLOGY OF 9/11 AND HURRICANE KATRINA

Two recent case studies that offer extensive and compelling narratives for analyzing social action in extreme environments and catastrophic situations are the terror attacks of September 11, 2001, on the World Trade Center in New York City, and the devastation wrought on New Orleans, LA, and the other parts of the Gulf Coast region after Hurricane Katrina. According to the *New York Times* (2009):

> It was the day when the unreal became the unimaginable. Sept. 11, 2001, the crystalline morning when planes dropped from the skies and toppled the World Trade Center and punctured a hole in the Pentagon…shattered the security of the country and introduced a nebulous and virulent enemy previously unfamiliar to most citizens. Nearly 3,000 people died that morning, the vast majority of them in the gnarled rubble of the Lower Manhattan towers, others at the Pentagon and in a rural Pennsylvania field. A numbed country with red-rimmed eyes came to understand the ugly menace of terrorism.

The catastrophe of 9/11 began shortly before 9:00 a.m. when an American Airlines plane struck "1 World Trade Center, the north tower, at 450 miles an hour" (Dwyer & Flynn, 2005, p. 20). Instantly, "10,000 gallons of jet fuel ignited into giant fireballs, the biggest with a girth of 200 feet, wider than the building itself" (Dwyer & Flynn, 2005, p. 80) and temperatures in parts of the building reached in excess of 1,000 degrees (Dwyer & Flynn, 2005, p. 267). Hundreds of people died instantly, others started jumping from windows to avoid the extreme heat. By the end of the day both towers of the World Trade Center had collapsed and brought down adjacent buildings in their wake. The remains of the World Trade Center smoldered for a full 100 days (Wright, 2006).

Hurricane Katrina slammed into the Gulf Coast during the last week of August in 2005. According to the *New Orleans Times-Picayune* (2009):

> Hurricane Katrina struck the New Orleans area early morning August 29, 2005. The storm surge breached the city's levees at multiple points, leaving 80 percent of the city submerged, tens of thousands of victims clinging to rooftops, and hundreds of thousands scattered to shelters around the country. Three weeks later, Hurricane Rita re-flooded much of the area. The devastation to the Gulf Coast by these two hurricanes has been called the greatest disaster in our nation's history.

According to historian Douglas Brinkley (2006, p. 173): "Not since Atlanta had been burned to the ground had a swath of Dixie looked so wretchedly barren." 400,000 people in Louisiana, Mississippi, and Alabama needed immediate assistance (Brinkley, 2006, p. 249). Hurricane Katrina was responsible "for more than 1,600 deaths and tens of billions of dollars in damages" (Associated Press, 2009). A Slidell, LA police officer, Rob Callahan spoke for many when he said, in reference to the destruction, "Imagine your worst nightmare and quadruple that 100 times" (Brinkley, 2006, p. 357). The nightmare of Katrina was ubiquitous and many residents have yet to awake from it, now more than four years later.

Although 9/11 and Hurricane Katrina are obviously very different catastrophic events and generated their own unique extreme environments, they also share some important parallels and similarities. In their book detailing the last 102 minutes in the World Trade Center, Dwyer and Flynn (2005, p. 247) describe the dramatic change in the American psyche after the 9/11 attacks: "All across the northeastern United States, people were essentially on their own, stepping into the first minutes of a new epoch without the protections of an old world order whose institutions and functions seemed to have turned instantly decrepit." Similarly in New Orleans and on the rest of the affected Gulf Coast, people were also abruptly cut off from the rest of the world. Survivors of the hurricane also did not know the "protections of an old world order," and like their fellow citizens in New York City on 9/11, they also found their social institutions to be "instantly decrepit."[11] In both environments, during and after the catastrophe, things just "didn't make sense any more" (Brinkley, 2006, p. 321). Irrationality became the new hallmark of our time.

In the immediate aftermath of the terror attacks and the hurricane, survivors found themselves in a profound state of radical autonomy, one of the most salient characteristics of all catastrophic situations. Extreme environments make all the quotidian aspects of daily life that seemed so important a few seconds ago, pointless and absurd as people are unshackled from the normal routines of everyday life. In advanced contemporary societies most of the time, most individuals believe that they can depend on others to help them meet their needs and in the best of times, a Durkheimian organic solidarity pervades the scene. If something goes wrong, one only need pick up the telephone, make a call, and possibly later take out a check book to solve the problem. In extreme environments caused by catastrophic events, however, this structure of mutual dependency vanishes – there is simply no one available for assistance, with the exception of perhaps other social actors who find themselves in the same situation. During both the 9/11 attacks and after Hurricane Katrina, "the expectations of protection, the habits of defense, and [a] sense of safety" (Dwyer & Flynn, 2005, p. 247) disappeared in a heartbeat and people suddenly had to fend for themselves.[12]

In attempting to cope with the awesome and terrifying anatomical social pressures of 9/11 and Hurricane Katrina, people reacted in a multitude of ways. Some people were struck dead instantly. Other responses spanned the range of "human valor and frailty and struggle" (Dwyer & Flynn, 2005, p. xxii). The many comforts and relative affluence of contemporary societies make it difficult for even people unharmed during the initial waves of destruction or shocks of extremity, to cope very readily with *any* change in their social environment. Experiencing a sort of shellshock as their senses are overwhelmed they experience a sort of learned helplessness, a particularly tragic form of Max Weber's affective or emotional action (Collins, 1981).

John Rousmaniere (2002, p. 199) in his study of shipwrecks notes that some passengers and crew "escape in mad humor, others…in an equally mad observance of regulations." He writes that, "of every ten people in a crisis, psychologists report, one or two behave heroically, another one or two are helpless, and the rest muddle along, trying to do their decent best. Who will be in those groups is not easy to predict, except that well-trained people who know basic skills and who have been through simulated disasters generally fall into the first group" (*Ibid.*, pp. 7–8). Kutak (1938, p. 67) describes a "holiday spirit" that can overcome survivors of a disaster. As he put is: "Immediate needs replace ultimate purposes as the criteria in terms of which behavior is determined. The inhibitions and the formalities of social life disappear…A crisis situation thus produces a state of mind closely approximating intoxication" (Kutak, 1938, p. 67).[13] More recently, sociologist Lee Clarke (2003) discusses responses to catastrophe in terms of relations of solidarity (e.g., helping or assisting other people) or relations of destruction (e.g., looting, violence).[14] While acknowledging that relations of destruction can be found in catastrophic situations, he argues that "altruism and cohesion, not self-interest and panic, are the central tendencies in response to extreme events" (*Ibid.*, p. 132).[15]

In both New York City and New Orleans after Katrina, one finds plenty of examples of such relations of solidarity. The account described earlier about the men who managed to escape from the elevator on 9/11 is one case in point, and there are many others. One of the most memorable accounts in Brinkley's history of Hurricane Katrina involves his discussion of "citizen first responders" known as the "NOLA Homeboys." With representatives of the Federal Emergency Management Association not yet on the scene, the New Orleans Police Department missing in action, and the whole city crippled by disaster, New Orleans residents who remained had to take matters into their own hands. Brinkley writes:

> The very first responders will never be known by name. When the levees broke, and the bowl started filling, hundreds of residents with recreational vehicles went into high gear. The city had people conducting rescues with yachts, dinghies, ferries, canoes, rafts, sailboats, scows, skiffs, sloops, tubs, catamarans, dories, draggers, baiters, and ketches.... They were known at Johnny White's Bar as the NOLA Homeboys, the oddballs who refused to evacuate, who were saving New Orleans from the ravages of Katrina...They had no official insignia or training. They were bartenders and insurance salespersons and clerks – regular folks. It never dawned on them to wait for the FEMA trucks or the Oregon National Guard. (Brinkley, 2006, p. 303)

Tales of similar actions and initiatives could be found from throughout the city. Larry Bradshaw and Lori Beth Slonsky, emergency medical workers from San Francisco who happened to be in New Orleans for a conference, observed that the "real heroes and sheroes [sic] of the hurricane relief effort" were not members of FEMA, the National Guard, or state or local government, but rather "the working class of New Orleans" who used various skills at their disposal to help people cope with the tragedy. In their account, they describe:

> The maintenance workers who used a fork lift to carry the sick and disabled. The engineers, who rigged, nurtured and the kept the generators running. The electricians who improvised thick extension cords stretching over blocks to share the little electricity we had...Nurses who took over for mechanical ventilators and spent many hours on end manually forcing air into the lungs of unconscious patients to keep them alive. Doormen who rescued folks stuck in elevators. Refinery workers who broke into boat yards, 'stealing' boats to rescue their neighbors clinging to their roofs in flood waters. Mechanics who helped hot-wire any car that could be found to ferry people out of the City. And the food service workers who scoured the commercial kitchens improvising communal meals for hundreds of those stranded (Bradshaw & Slonksy, 2005, p. 2).

Bradshaw and Slonsky were also part of the hundreds of people without food, water, or shelter who attempted to cross the New Orleans Bridge and seek save haven in Gretna, LA, but were turned back at gunpoint by the Gretna Police. Retreating after shots were fired over their heads, they formed an "an encampment in the middle of the Ponchartrain Expressway on the center divide, between the O'Keefe and Tchoupitoulas exits" (Bradshaw & Slonksy, 2005, p. 4). For a brief period, this encampment flourished as a sort of emerging civic space. Over time, the

> little encampment began to blossom. Someone stole a water delivery truck and brought it to us...A mile or so down the freeway, an army truck lost a couple of pallets of C-rations [MREs] on a tight turn. We ferried the food back to our camp in shopping

> carts...We organized a clean up and hung garbage bags from the rebar poles. We made beds from wood pallets and cardboard. We designated a storm drain as the bathroom and the kids built an elaborate enclosure for privacy out of plastic, broken umbrellas, and other scraps. We even organized a food recycling system where individuals could swap out parts of C-rations (applesauce for babies and candies for kids). (Bradshaw & Slonksy, 2005, p. 5)[16]

One important research agenda for contemporary sociologists might be to explain just how such instances of "active agency" (Paine, 2002, p. 67) during 9/11 and Hurricane Katrina emerged and what social dynamics and forces sustained them. It appears that language or communication (Habermas, 1989)[17] is a key component to regenerating social organization. People begin to reclaim their environments and reinstitute culture once they start to converse with each other again. In describing how the 1,000 or so people still trapped in the World Trade Center reacted immediately after the initial attack, Dwyer and Flynn (2005, p. 40) write: "the men and women stranded on the upper floors dialed cell phones, tapped the miniature keyboards on their pagers, and spoke into two-way radios, fashioning a bridge of voices." Similar stories were told from New Orleans; with no telephone and cell-phone service, text messaging for some became the only form of contact with the outside world.

5. TOWARD A MODEL OF HUMAN REACTION TO CATASTROPHES

The previous section used the examples of the 9/11 attacks and Hurricane Katrina to illustrate the sorts of social dynamics of interest for a research project that took the responsibility of analyzing anatomical social facts in the context of catastrophes and disasters seriously. There are a great many more examples that could be provided from these momentous events in American history, and plenty of others drawn from around the world. Using additional insights from these and other case studies, the present section attempts to specify in further detail some of the key variables, processes, and dynamics that might help to explain differential responses to catastrophic situations and extreme environments. Lanzara's (1983) work on ephemeral organizations in disaster zones provides fresh insight. In his research, Lanzara argues that in reacting to extreme environments, human beings very often do three things. First, they often "*associate* two differing contexts which on the surface share very little similarity" (e.g., using a concessions cart as battering ram; freeway exit as refugee shelter). Second, they "*extend* a private daily life activity into the domain of social services" (e.g., using ones professional, mechanical or organizational skills with the post-Katrina relief effort; using one's private boat to rescue people from rooftops).[18] Third, people demonstrate "*an ability to respond* to the enacted environment" (e.g., actually taking the initiative to act in the period of recoil). Lanzara (1983, pp. 77–78) says that it is through being "able to associate, extend, and respond" that a "concrete intervention" is produced. Although Lanzara's model provides a solid heuristic account for what happens during a disaster and offers researchers a starting point for what to look for in analyzing catastrophic situations, it can be bolstered by addressing a number of other variables that influence an individual's ability to act. The outline of such a basic model is presented first, which is followed by a discussion of each of its elements.

6. A MODEL OF SOCIAL ACTION IN CATASTROPHIC SITUATIONS

Internal (Self)	Agentic	External (World)
Instinct	Technology/tools	Nature
Cognition	Improvisation	Contingencies
Conditioning	Routinized progress	Social networks
Socialization	Willpower	
	Attitude	

The model comprises three basic dimensions, internal – corresponding to human variations in psycho-social development intrinsic to individual actors, external – which refers to activities and events in the external world over which an individual has no control, and agentic, having to do with social action or agency. The agentic dimension mediates between self and world. One key aspect to this dimension is the ability of human actors in a catastrophe to make use of *technologies or tools* (however primitive) to better cope with their situation. In extreme environments technologies that, in Martin Heidegger's parsing, are "ready to-hand" (Rouse, 1990) can be instrumental to survival and may make all the difference. As an obvious example, consider the difference of being lost in the wilderness with or without a book of matches or stranded in a river without a life-jacket. After the first plane struck the World Trade Center on 9/11, in

"Cantor Fitzgerald's northwest conference room on the 104th floor…people temporarily managed to ward off the smoke and heat by plugging vents with jackets." One of those struck inside, Andrew Rosenblum telephoned a colleague telling him, "We smashed the computers into the windows to get some air" (Dwyer & Flynn, 2005, p. 138). A few other examples can illustrate the basic point: If Aron Ralston (2004) did not have access to a pocket knife he would not have managed to self-amputate his arm and hike out to safety when his arm was trapped behind a boulder in the Utah desert. If mountaineer Joe Simpson (1989) had not have been carrying a sleeping mat in his backpack, he would have not been able to fashion a make-shift cast for his broken leg which had been injured while rock climbing in Peru, and nor would he have been able to crawl over frozen tundra and boulder fields to return to his base-camp after being left for dead.

Another central variable to action in extreme environments is *improvisation*. As Brinkley (2006, p. 250) writes: "improvisation [is] the fundamental *modus operandi* in a disaster, flexibility being the true guiding principle." Harrison White (1992) refers to this as the importance of what he calls "ambage" in social organization. Deconstructionists might refer to the necessity of "play" – in the sense of looseness – in social organization (Lemert, 2001). The general idea is that actors who manage to make it through catastrophes are more often than not better at making things up as they go along than people who stick to preset plans and are immobilized by change.

Before attending to the final agentic variable, let us address some of the internal variations responsible for differential human action in catastrophic situations. One important one has to do with *willpower*. Kampler (2004, p. 118) writes: "What all survivors have in common is an energy flow from the top down…a spark in the cerebral cortex known as motivation." This spark is what keeps the will burning and may stem from a number of sources, including defiance, vengeance, sympathy, duty, loyalty, honor, responsibility, religious belief, etc. Obviously the concept of will is an important one for theorists and philosophers and a great deal has been written about it. Extreme environments and catastrophic situations are by definition not easy to cope with. It is reasonable to presume that strong-willed people more than others are more likely to act successfully when coping with them.

Another internal variable has to do with one's instinctual response. Dwyer and Flynn (2005, p. 64) suggest that in the wake of the 9/11 attacks, survivors relied on one or some combination of *instincts*, training, and instructions.[19] Instincts are a human being's first line of defense when assessing or coping with potentially hazardous conditions or events. As Langewiesche (2005, p. 148) writes, in a vivid description of the downing of the *Estonia*, "Survival that night was a very tight race, and savagely simple. People who started early and moved fast had some chance of winning. People who started late or hesitated for any reason had no chance at all. Action paid. Contemplation did not. The mere act of getting dressed was enough to condemn people to death." Experienced hikers can sometimes sense when a particularly bad storm is coming in. Before the Tsunami that devastated much of Indonesia in 2005, survivors witnessed animals and birds suddenly moving very quickly inland away from the ocean, trying to find higher ground. People who followed the lead of the animals were given a head start in trying to survive.

In recent years, the role of *cognition* in social life (see, e.g., Turner, 2002) has become increasingly apparent to sociologists and other social scientists. Variation in the capacity for knowledge, intelligence, and neurological functioning should increasingly be built into our models of social action. Kampler (2004, p. 147) states that humans have "a supreme capacity to learn by observation, to store and transmit information using language, and to integrate and apply their knowledge by turning survival skills into customs." These skills become even more necessary when negotiating extreme environments and catastrophic situations than they are in normal life. Jonathan Pincus' book, *Base Instincts* (2002) makes a convincing case for social scientists to think more about the neurological explanations for violence and murder. One can easily extend his argument to non-criminal forms of social action. It may indeed be the case that people who function more effectively during catastrophes have cognitive structures that are "wired" differently than people or are less successful, i.e. their mental chemistry is configured in a way that makes them easier to cope.

Extreme environments and catastrophic situations are obviously very physically demanding and require much from bodies of the actors involved. It should come as no surprise that people who are physically fit are, *ceteris paribus*, more likely to better endure situations of social breakdown compared to those whose bodies are not in such good shape. As Kampler (2004, p. 226) remarks, when facing extreme environments, the body should be "finely tuned, fully adapted, and highly motivated." *Physical conditioning* and strength are key to survival. Langewiesche (2005, p. 107) reports from the downing of the *Estonia*, "Of fifteen children under the age of 15, only one survived; and of 167 adults age sixty-five and older, every woman and all but two men had died." When the World Trade Center was attacked and its inhabitants had to attempt to flee the building, many people had to descend multiple flights of stairs to escape. Many who were ill conditioned or out of shape simply could not make the journey and died when the towers collapsed. It makes sense that an individual suffering from diabetes, heart palpitations, or who is morbidly obese, is not going to handle an emergency airplane landing as easily as someone who does not suffer from these conditions. Many other examples drawn from the empirical research of catastrophes (e.g., Klinenberg, 2003) bear this out and confirm that populations are differentially vulnerable to hazards and disaster.

As was mentioned in the quote by Dwyer and Flynn, above, another variable that can help us better understand action in extreme environments is the concept of *socialization*, in this case, preparatory training (either formal or informal) for coping with disasters and catastrophes. In the case of Hurricane Katrina, for example, many of the people who ended up making the most robust interventions and in some cases taking on more less institutionalized leadership roles in the community had previous training which influenced their propensity to act. New Orleans personality and radio announcer Garland Robinette, who remained on the air almost around the clock during the hurricane, notes that his strength for enduring that catastrophe came from his experience serving in Vietnam. As he stated,

> Knowing people were dying and hearing stories and talking to people who were in the process of dying…That reminded me a lot of Vietnam. When the people came out with me from our studio to the flooded cars, they were hysterical or shocked. I felt extremely calm. Vietnam…I think it was thirteen months of combat, and if you lose control, if you get emotional or you get afraid or too brave or just don't stay calm and think and be relaxed, you're going to die. So Vietnam was a gigantic plus…I have been to a bad place. I had that bad place to go back to, that helped me remain calm enough to get the job done. (Brinkley, 2006, p. 296)

Another person for whom socialization with disaster also played a significant role in the Hurricane Katrina tragedy was New Orleans resident Rick Matthieu, leader of a group of rescuers known as the "Soul Patrol" and who "saved at least five hundred New Orleaninians" (Brinkley, 2006, p. 308). Brinkley (2006, p. 308) reports, "What enabled Matthieu to act so efficiently was his childhood experience with Hurricane Betsy. He had been piloting boats since he was seven years old." As Matthieu put it: "With my dad during Betsy, we saved lots of folks…I prepared on Friday night before Katrina because my dad had prepared in the same way for Betsy" (Brinkley, 2006, p. 308).

As we know from the Thomas Theorem, the way situations are defined can have a dramatic impact on their outcome. Likewise, people's *attitudes* can affect their experiences of different situations. Although there is now a stigma associated with smoking cigarettes, at one point in time they were standard issue in the military partly because they were thought to improve one's attitude or outlook. It seems peculiar to us today, but Rousmaniere (2002, p. 193) writes that in the 1940s, "smoking was crucial to survival…Smoking was long known to relieve stress and provide boosts of energy. Few events were more comforting and symbolic of human connection than a buddy's offer of a cigarette or a match or the glimpse of a tiny bit or burning ash across a field or ship's deck."[20] Beck Weathers, the Texas doctor left to die on Mt. Everest in John Krakauer's (1999) book *Into thin Air* illustrates very well the importance of having a positive attitude on surviving extreme environments.

An important variable external to individuals facing catastrophic situations are one's *social networks.* This is what Rousmaniere (2002, p. xii) describes as "the power of personal connection to steer people through disaster." The extent of one's social ties can actually make "the difference between life and death" Clarke (2006, p. 133). The better developed one's network of social ties, the more people there are to offer assistance, and the more likely it is that one will act effectively. Without Simon Yates being on the mountain in Peru with him, there is no way that Joe Simpson (1989) would have survived his ordeal. When British United Nations officer Philip Ashby (2003) was trying to escape rebels in Sierra Leone he had to rely on the kindness of a non-rebel farmer who shuttled him and his men to safety; without this contact, he would have perished. The Underground Railroad from the American context offers another example of social networks aiding people in extreme situations.

One way that social actors can help themselves in extreme environments and catastrophic situations is through what may be described as *routinized action.* As Wiseman (2004, p. 38) writes: "When facing a disaster it is easy to let yourself go, to collapse and be consumed in self-pity. But it is no use giving up or burying your head in the sand, and hoping that this is a bad dream that will soon pass. It won't, and with that kind of attitude it will rapidly become much worse. Only positive action can save you." Doing something instead of nothing is an important theme in Ghinsberg's (2006) account of being trapped in the Bolivian jungle, on the Shackleton expedition (Alexander, 2000, p. 101), and in accounts given by Simpson (1989, 1993), Yates (2001), and Ralston (2004). To survive is to take things just one step at a time, to focus on small achievable goals which aggregate into larger success. Mitchell (1984, p. 17), for example, notes that the "joy of mountaineering" is to be found in the "moment-by-moment accomplishments" and "small successes along the way – picking a better route for the next few feet above, completing a difficult move, safely reaching the next ledge."

Our model of action in extreme environments would not be complete without a discussion of *luck*. As Clarke, 2006, p. 69 puts it, "We owe more to lady luck than we realize." Luck, also known as contingency (Rorty, 1989), serendipity, randomness, and so on, plays a significant role in all aspects of social life, extreme environments and catastrophic situations included. The opening minutes of the film *Magnolia* does well to illustrate just how much of an impact simply being lucky plays in our everyday lives. It is difficult to theorize issues of luck or randomness, in that it often appears as an intervention out of the blue, so to speak, which many attribute to God's will or acts of fate or fortune. Needless to say, people who have lived through catastrophes know the role of randomness in their survival. If Aron Ralston (2004) was not found by a couple of Dutch hikers after he had repelled down a 100 ft. rock face and jogged two miles in the desert, it is unlikely that he would have lived to tell his tale. Joe Simpson's (1989) book, *Touching the Void*, is filled with many examples of lucky chances. If Simpson had failed him just at one point a long the way, the whole social equation responsible for his ultimate survival would have fallen to pieces.

The opposite of luck is, of course, bad luck, which in extreme environments takes the form of additional impacts by other contingencies of nature, which may lead to a cascading chain of events completely outside of the actor's social control. As Hazelrigg (1989) points out, the experience of domination means precisely domination by nature. Social actors cannot withstand nature's bombardments forever. If conditions continue to deteriorate and some of the other variables in this model are muted or absent, the individuals involved will eventually fade from the scene.

7. CONCLUSIONS

Foucault once said, in reference to his investigations of power and knowledge, that it is important to persistently ask, "Who speaks?" This essay considers, in any extreme environment or catastrophic event, "Who acts?" The focus has been on how people manage to successfully respond to sudden, extraordinary changes in the anatomical social facts of their immediate environment. It suggests a model of some of the resources that people draw upon in these situations. As Rousmaniere (2002, p. 198) puts it, in the face of crisis, "a few people are calm and rational, a few others are irrational and even hysterical, [but] most are simply stunned." This work has tried to provide some insight into how "calm and rational" react to major threats facing their ecological milieu.[21] To the extent that contemporary society becomes even more dystopian than it is presently (see Gray, 2000) and the world becomes even more extreme, catastrophic, and turbulent, this work may help sociologists, social theorists and other social observers to better come to terms with how individuals act under catastrophic conditions. Actors facing such environments can be conceptualized as types of edge workers (Lyng, 2005). They have just stepped into a major breach, are no longer in control of their lives, and abruptly come to the realization that they must act (and do so effectively) to avoid disaster.

In contemporary society, the gap between normality and chaos, safety and disaster, rationality and irrationality, indeed between life and death itself, is becoming increasingly tenuous and difficult to discern. Under conditions of globalization (Barber, 1993; Friedman, 2005), the world is rapidly coming together and blowing apart at the same time. Owing to the nature of risks that are prevalent today, the world is seemingly a much riskier place (Beck, 1984) than at any time in recent memory. Natural and social environments are changing in dramatic and peculiar ways, and it is impossible to predict the outcome. As Hazelrigg (1989, p. 41) asks us, rhetorically, amidst all the hazards of our contemporary society and world, "Just where do we find safe territory today?" The answer, of course, is nowhere. The 9/11 attacks and Hurricane Katrina illustrate vividly that no one is immune from disaster and that risk is something with which all of us must be concerned. Avian flu and swine flu outbreaks, ricin in the post offices, not to mention the regular coterie of accidents, toxic spills, gas leaks, floods, tornadoes, wildfires, accidents and so on, constantly remind us that the postmodern world is no tea-party but instead rife with danger. Regardless of where or who you are, one's life can be ended or upended in a heartbeat. As the French radicals proclaimed during the events of May–June 1968: "Under the pavement: beach!" The wilderness is indeed never too far away, and because of this, as Wiseman (2004, p. 8) points out us, "the need for survival training has never been greater. Not only are we indulging in more exotic holidays, challenging expeditions, and foreign business ventures," but there is the continuing very real threat from radical Islamist movements (Sageman, 2003) for whom 9/11 marks the beginning of a new struggle against the West. Sociologists and social theorists also must do their part to come to terms with these developments. They need to devote more time and energy to studying individuals who in their engagement with anatomical social facts, as individuals, or as members of social groups, managed to engage in robust and positive social action. Social research on catastrophe needs to focus not just on the structural side of social reality (e.g., the causes of the catastrophe and all the utter devastation), but also on issues of human agency. In a Meadian sense, the analytic attention should be on social actors in catastrophes going beyond the "me" to become an "I." Through this process, identity (White, 1992) is developed in situations of conflict and chaos. It is hoped that further empirical attention paid to future catastrophes of this kind and the agentic responses of the actors involved, will elaborate some of these ideas in greater detail.

NOTES

1. This chapter assumes a certain structural equivalence in these crises where in reality there may be more differences among them. In 1937, Professor Margaret Wood, at a session of the Southern Sociological Society annual meeting, queried, in relation to "the absorbing topic of crisis behavior," whether "people behave differently in crises which are caused by so-called 'Acts of God' as floods, tornadoes, earthquakes, and fires" as they do "in crises which are a result of 'man's inhumanity to man'"? (in Kutak, 1938, p. 72). Social scientists and others are still developing an answer to this question.
2. The science of catastrophes spreads out across many disciplines and theoretical schemes (Jiobu & Lundgren, 1978).
3. On the culture/nature distinction, see Hazelrigg (1995).
4. Such a statement may seem absurd when so many members of contemporary societies have very comfortable middle class lives and never contemplate dystopia in the abstract, let alone experience anything like it in practice. Although certainly many individuals today live rather "safe" lives in society, the reality is that no one is completely immune from danger. Anyone's life can become dystopian in an instant and there is also a creeping dystopia (e.g., global warming; agribusiness, water shortages) that operate behind the backs, so to speak, of most social actors, but sooner or later will become more difficult to conceal. Catastrophe is becoming a more regular occurrence in contemporary societies and actors in these societies will experience (even if this experience is merely punctuated) extreme environments during their lifetime.
5. Remember August Comte's definition: "Sociology is the science of civilization."
6. It is hoped that the present work will provide a theoretical scheme which may be empirically tested and elaborated by others (including the present author) in future work.
7. For examples, consider all the references to looting, criminal behavior, widespread rapes, shooting at helicopters during Hurricane Katrina. Many stories reported as fact were found out later to be unsubstantiated or exaggerated by the media.
8. Maurice Halbwachs (1960) was the key figure to expound this aspect of Durkheim's thought. There were some important early parallels between the development of social morphology and the field of human ecology, but with

the emergence of these specialized disciplines, it appears sociologists as a whole largely lost interest in this important, socio-environmental, aspect to their discipline.

9. Despite a general paucity of interest in this subject matter, there are notable exceptions. Seventy years ago in his discussion of the Louisville flood at the Southern Sociological Society meeting of 1938, Robert Kutak expressed the essential idea which I am pursuing. He wrote: "man will not supinely by and let nature do with him what she will. It is because man has fought back at nature that civilization developed." Humans have an "age-old desire to prevent nature from controlling the conditions of his life" (Kutak, 1938, p. 70).

10. For a parallel argument, on the failure of sociologists to conceptualize environmental facts as social places, see Antonio's essay in this volume.

11. Indeed, four years after the Hurricane, many of these social institutions have still not recovered.

12. There is obviously "the need for leadership in any crisis" (Brinkley, 2006, p. 91). One would think that, in extreme and risky situations, people who are in leadership roles already would continue in that capacity. Sometimes this does happen (Dwyer & Flynn, 2005, p. 22), but many times it does not. In Brinkley's account, Mitch Landrieu, Lieutenant Governor of the State of Louisiana seems to have been a good leader in the aftermath of Katrina. But in the main, leadership was sorely lacking, especially by the New Orleans Police Department. Many in the force simply abandoned their posts and drove their police cars to safety. As Brinkley (2006, p. 203) writes, "It became sport in Houston [to] see how many NOPD cars you could photograph in town today. High-tailing NOPD officers had lost track of rules and regulations; many just drove their patrol cars straight out of the bowl to Texas." Heather Allan, a 52-year-old producer for NBC, offered this assessment of the New Orleans police during the tragedy. She said: "I saw a ninety-eight-year-old man paralyzed in a wheelchair ask for help and they just scoffed at him. They kicked three little huddled women with nowhere to go out of the Marriot because that was where the NOPD was sleeping…just tossed them on the street…They wouldn't even answer questions of people who asked which way the Superdome or the Convention Center was. They basically mocked all the homeless. I have never, ever, seen such a cold, I-don't-give-a-shit attitude from cops in my life" (Brinkley, 2006, p. 205). Brinkley (2006, p. 385) sums up the leadership vacuum: "For whatever reason – old tensions or storm-induced confusion – the police were often part of the problem after Katrina, not the solution." The US Coast Guard was arguably the only federal government agency to do well in the immediate post-Katrina aftermath. "Over the ten-day period following Katrina they evacuated more than 33,500 people using orange helicopters and flat-bottom boats" (Brinkley, 2006, p. 213). The Cajun Navy composed of members of the State of Louisiana Wildlife and Fisheries also performed exceptionally well after the storm.

13. Kutak (1938, p. 67) later refers to this using the term "crisis intoxication." A similar form of intoxication can be seen at "hurricane parties" in Florida, in the melee after the Rodney King verdict, and in people being snowed in for a few days in the Rockies. Sometimes this can lead to forms of hysteria. This is how quartermaster Hank Strauss observed his own ship, the *Pollux*, going down in the waves off the coast of Newfoundland. He wrote: "I went on deck with the other quartermaster

signalmen and when we looked up and saw the cliff we started laughing. There we were: forty-foot seas breaking over the ship; the deck smashed into kindling; heavy bunker oil pouring out of the ship everywhere and covering everything; officers telling us that the ship was likely to roll right off the shelf and sink at any moment. All that was happening and all we could do was laugh. It seemed so funny that *those stupid officers* had actually run the ship into a cliff!" (in Rousmaniere, 2002, p. 199).

14. In some ways, this seems to mirror Mead's discussion of the two movements of human behavior – toward social antagonism or social cooperation, which also is somewhat similar to the Freudian/Marcusian notion of life versus death urges (*eros v. thanatos*).

15. We are often surprised by the solidarity we find in catastrophic situations. People who heretofore were considered criminal and deviant banded together for the safety of the emerging population. One witness described what she saw at the New Orleans Convention Center: "These guys were criminals. They were. But somehow these guys got together, figured out who had guns and decided they were going to make sure that no women were getting raped…and that no one was hurting babies…. They were the ones getting juice for the babies… They were the ones getting clothes for people who had walked through that water. They were the ones fanning the old people" (Brinkley, 2006, p. 477).

16. Bradshaw and Slonsky's emergent organization, however, would come to a rather abrupt end. Quickly numerous media organizations were focusing on the camp and "officials were being asked what they were going to do about all those families living up on the freeway" (Bradshaw & Slonksy, 2005, p. 5). Not long after, the Gretna police returned to confront the group again, and after blowing away their structures with a helicopter, the group was, "at gunpoint, forced off the freeway" (*Ibid.*, p. 5).

17. In Harrison White's (1992) phrasing, "talk" or "stories."

18. This can be related to Clarke's (2003, p. 131) concept of the "role expansion" that often happens during catastrophes, "in which people assume a wider and more varied degree of responsibility for others. So the teacher becomes a nurse, the doctor becomes a rescue worker, the mayor becomes a counselor."

19. This raises the question of whether instincts are socially constructed or if they are more an innately given part of a human being's makeup.

20. It would be interesting to speculate on what, if anything, has taken the cigarette's place as a structurally equivalent aid to psychological recovery during catastrophic situations. What gap does this create in human relationships?

21. The more explicitly cultural or meaning-related aspects to this process, has obviously not been the focus of this essay and it is something that other researchers would do well to consider. As Oliver-Smith (2002, p. 41) notes, extreme events "challenge people's worldviews with profound existential questions for which meanings consistent with circumstances must be elaborated."

REFERENCES

Alexander, C. (2000). *The endurance: Shackleton's legendary Antarctic expedition.* New York: Alfred A. Knopf.

Alexander, J. C., et al. (2004). *Cultural trauma and collective identity*. Berkeley, CA: University of California Press.

Ashby, P. (2003). *Unscathed: Escape from Sierra Leone*. London: Pan Books.

Associated Press. (2009). Hurricane Camille's fury remembered 40 years later. *New York Times*, August 16. Available at http://www.nytimes.com/aponline/2009/08/16/us/AP-US-Hurricane-Camille-40-Years-ater.html?_r = 1 & scp = 1 & sq = hurricane%20katrina%20death%20toll&st = cse

Barber, B. (1993). *McWorld vs. Jihad*. New York: Harper Collins.

Barton, A. H. (1969). *Communities in disaster*. Garden City, NY: Doubleday & Company.

Beck, U. (1984). *Risk society*. Thousand Oaks, CA: Sage.

Bentz, V. M., & Shapiro, J. J. (1998). *Mindful inquiry in social research*. Thousand Oaks, CA: Sage.

Blumer, H. (2007). Symbolic interactionism. In: C. Craig, et al. (Eds), *Contemporary sociological theory* (pp. 45–67). Malden, MA: Blackwell.

Bradshaw, L., & Slonksy, L. B. (2005). *New orleans: Survivor stories* (pp. 1–8). Available at http://citypages.com/databank/26/1294/article13694.asp?page = 19

Brinkley, D. (2006). *The great deluge: Hurricane Katrina, New Orleans, and the Mississippi Gulf Coast*. New York: William Morrow.

Bruinius, H. (2002). A stunning tale of escape traps its hero in replay, *Christian Science Monitor*. Available at http://www.csmonitor.com/specials/oneyearlater/livesChanged_demczur.html

Clarke, L. (2003). Conceptualizing responses to extreme events. In: *Research in social problems and pubic policy* (pp. 123–141). New York: Elsevier.

Clarke, L. (2006). *Worst cases: Terror and catastrophe in the popular imagination*. Chicago, IL: University of Chicago Press.

Collins, R. (1981). *Max Weber: A skeleton key*. Thousand Oaks, CA: Sage.

Diamond, J. (2006). *Collapse*. New York: Viking.

Durkheim, E. (2003). Note on sociology morphology. In: M. Emirbayer (Ed.), *Emile Durkheim: Sociologist of modernity* (pp. 77–78). Malden, MA: Blackwell.

Dwyer, J., & Flynn, K. (2005). *102 minutes: The untold story to survive inside the twin towers*. New York: Henry Holt and Company.

Ehrenreich, J. H. (2003). Understanding PTSD: Forgetting "Trauma". *Journal of Social Issues*, *3*(1), 15–28.

Erikson, K. T. (1976). *Everything in its path: Destruction of community in the buffalo creek flood*. New York: Simon & Shuster.

Freudenberg, W. R., et al. (2008). Organizing hazards, engineering disasters? Improving the recognition of political-economic factors in the creation of disasters. *Social Forces*, *87*(2), 1015–1038.

Friedman, T. (2005). *The world is flat*. New York: Farrar, Straus and Giroux.

Ghinsberg, Y. (2006). *Jungle: A harrowing true story of survival*. Austin, TX: Boomerang New Media.

Giddens, A. (1972). *Emile Durkheim: Selected writings*. New York: Cambridge University Press.

Gray, J. (2000). *Straw dogs: Thoughts on humans and other animals*. New York: Farrar, Straus and Giroux.

Habermas, J. (1989). *The theory of communicative action*. Vol. 1. Boston, MA: Beacon Press.

Halbwachs, M. (1960). *Population and society: An introduction to social morphology*. Glencoe, IL: Free Press.

Hazelrigg, L. (1989). *A wilderness of mirrors*. Tallahassee, FL: Presses of Florida.

Hazelrigg, L. (1995). *Cultures of nature*. Tallahassee, FL: University Presses of Florida.

Hoffman, S. M., & Oliver-Smith, A. (2002). *Catastrophe and culture: The anthropology of disaster*. Santa Fe, NM: School of American Research.

Janis, I. L. (1969). *Stress and frustration*. New York: Harcourt Brace Jovanovich.

Jiobu, R. M., & Lundgren, T. D. (1978). Catastrophe theory. *Sociological Methods and Research*, *7*(1), 29–54.

Kampler, K. (2004). *Surviving the extremes*. New York: Penguin.

Klinenberg, E. (2003). *Heat wave: A social autopsy of disaster in Chicago*. Chicago, IL: University of Chicago Press.

Krakauer, J. (1999). *Into thin air: A personal account of the Mt. Everest disaster*. New York: Anchor.

Kutak, R. I. (1938). The sociology of crises: The Louisville flood of 1937. *Social Forces*, *XVII*, 66–72.

Langewiesche, W. (2005). *The outlaw sea*. Northpoint Press.

Lanzara, G. F. (1983). Ephemeral organizations in extreme environments. *Journal of Management Studies*, *20*(1), 71–95.

Lemert, C. (2001). *Social theory: Multicultural and classic readings*. Boulder, CO: Westview.

Longmon, J. (2003). *Among the heroes: United flight 93 and the passengers and crew who fought back*. New York: Harper Paperbacks.

Lukes, S. (1985). *Emile Durkheim: His life and work*. Stanford, CA: Stanford University Press.

Lyng, S. (2005). *Edgework: The sociology of risk taking*. New York: Routledge.

Mitchell, R. G. (1984). *Mountain experience: The psychology and the sociology of adventure*. Chicago, IL: University of Chicago Press.

Neal, A. (2005). *National trauma and collective memory*. New York: M.E. Sharpe.

New Orleans Times-Picayune. (2009). Katrina: The storm we always feared. Available at http://www.nola.com/katrina/. Retrieved on August 19, 2009.

New York Times. (2009). September 11, 2001. Available at http://topics.nytimes.com/top/reference/timestopics/subjects/s/sept_11_2001/index.html?scp = 1-spot&sq = sept%2011th&st = cse. Retrieved on August 18, 2009.

Oliver-Smith, A. (2002). Theorizing disasters. In: S. M. Hoffman & A. Oliver-Smith (Eds), *Catastrophe and culture: The anthropology of disaster* (pp. 26–47). Santa Fe, NM: School of American Research.

Paine, R. (2002). Danger and the no-risk thesis. In: Hoffman & Oliver-Smith (Eds), *Culture and catastrophe* (pp. 67–89).

Perlstein, M. (2006). N.O. man hailed as hero indicted. *New Orleans Times-Picayune*. April 14, p. 1.

Pincus, J. (2002). *Base instincts: What makes killers kill*. New York: W.W. Norton.

Prince, S. H. (1920). *Catastrophe and social change: Based on a sociological study of the halifax disaster*. New York: Columbia University Press.

Ralston, A. (2004). *Between a rock and a hard place*. New York: Atria Book.

Rodriguez, H., Trainor, J., & Quarantelli, E. L. (2006). Rising to the challenges of a catastrophe. *The annals of the American academy of political and social science*, *604*, 82–101.

Rorty, R. (1989). *Contingency, irony, and solidarity*. New York: Cambridge University Press.

Rouse, J. (1990). *Knowledge and power: Toward a political philosophy of science*. Ithaca: Cornell University Press.

Rousmaniere, J. (1980). *Fastnet, force 10*. New York: W.W. Norton.

Rousmaniere, J. (2002). *After the storm: True stories of disaster and recovery*. New York: International Marine.

Sageman, M. (2003). *Understanding terror networks*. Philadelphia: University of Pennsylvania Press.

Scott, J. (1995). *Domination and the arts of resistance*. New Haven: Yale University Press.

Simpson, J. (1989). *Touching the void*. New York: Harper Paperbacks.

Simpson, J. (1993). *This game of ghosts*. Seattle, WA: The Mountaineers Books.

Sztompka, P. (2000). Cultural trauma: The other face of social change. *European Journal of Social Theory*, *3*(4), 449–466.

Turner, S. P. (2002). *Brains/practices/relativism: Social theory after cognitive science*. Chicago: University of Chicago Press.

Wallace, A. F. C. (1956). *Tornado in Worcester: An exploratory study of individual and community behavior in an extreme situation (Publication 392)*. Washington, DC: National Academy of Sciences – National Research Council.

White, H. C. (1992). *Identity and control: A structural theory of social action*. Princeton, NJ: Princeton University Press.

Wiseman, J. (2004). *SAS survival handbook: How to survive in the wild, in any climate, on land or at sea* (New Edition). New York: HarperResource.

Wright, L. (2006). *The looming tower: Al Qaeda and the road to 9/11*. New York: Vintage.

Yates, S. (2001). *The flame of adventure*. Seattle, WA: The Mountaineers Books.

PART II
KNOWLEDGE

FORTY YEARS OF *KNOWLEDGE AND HUMAN INTERESTS*: A BRIEF APPRECIATION☆

Lawrence Hazelrigg

For those of us who are of a certain age as children of North America and the United States in particular, *Knowledge and Human Interests* was a little like the proverbial lightning bolt. Having endured entrapment however brief in the dustbin known as Parsonian sociology, we were eager to know that the discipline had a better future than taxonomic abstractions that seemed mostly to confirm a famous jibe about the relation of the abstract to the concrete – confusing the former as the latter to confirm the latter's existence.[1] Seeking alternatives of fresher air, we were often disappointed by

☆This chapter is written without attention to later installments in the massive corpus of Habermas' (1988) work (including the volume of articles entitled *Postmetaphysical Thinking*, in which he revisited and reworked – reduced, one might say – some of the landscape of *Knowledge and Human Interests*). I have written with a famous aphorism firmly in mind: Hegel observed, early in his preface to *Phänomenologie des Geistes* (§3), that "when considering a substantial, well-crafted argument, the easiest response is to judge it; more difficult is to grasp it; the most difficult, which unites both of those responses, is to produce an account of it." (My translation of "Das leichteste ist, was Gehalt und Gediegenheit hat, zu beurteilen, schwerer, es zu fassen, das schwerste, was beides vereinigt, seine Darstellung hervorzubringen" – which differs from Miller's standard only by interposition from the preceding sentence.) I have also written with recognition of Baxter's (1999) refuge: "In a culture afflicted with data-nausea, … everyone begins to hoard and value the autobiographical, a refuge safe from irony." Well, so we pretend.

Nature, Knowledge and Negation
Current Perspectives in Social Theory, Volume 26, 189–206

ISSN: 0278-1204/doi:10.1108/S0278-1204(2009)0000026010

regular offerings of intellectual ventilation. C. Wright Mills could be bracing on the surface, but closer inspection turned up more posture than rigor. *Character and Social Structure*, for which Mills was named as co-author with Hans Gerth, had seemed to hold promise, but pursuits during the intervening years suggested that the rapidly marginalized Gerth had been its musician, Mills mainly its stage director and promoter. Homans' *Human Group* retained some allure, yet the lack of any follow-up – or perhaps "follow through" would be the better phrase – left one wondering if its inspiration had really been beyond that of an introductory text. And there were other works that held one's attention at least for a while, without generating much stimulation and less enthusiasm. *Knowledge and Human Interests* was for us an exceptional event because of its capacity to weave us into *projects*, our futures, carrying strong threads of our discipline's heritage without simply (or oversimplifyingly) recapitulating them as in a theory textbook. And in the process, children of US culture that we were, *Knowledge and Human Interests* demonstrated to us that a specifically US heritage of social science deserved more and better scrutiny than it had usually received. I am referring to works collected as "American pragmatism" and in particular to the work of Charles Sanders Peirce and his social science heirs (loosely considered).

Although I had learned of Jürgen Habermas (1968) from discussions with Gerth, I had not read any of his work (or not that I now recall) until Jeremy Shapiro translated *Erkenntnis und Interesse* into English as *Knowledge and Human Interests*, barely three years after first publication in 1968. I should have come to it sooner, but my proficiency in German was less than stellar. (Here, obviously, I speak only for myself.) One of my mentors, Louis Schneider, convinced that I was too ignorant to understand Parsons properly, had set me to the task of reading parts of Max Weber in the original. That was sometimes a difficult task – and one compounded by the fact that in the view of another of my mentors, Joseph Lopreato, I needed much better Italian so that I could read Pareto in the original and thus avoid all the lapses of *his* Englishing translators. It seemed ludicrous that I, native to neither German nor Italian, could in principle do more than produce yet another translation – whether ever better than predecessors' an inherently arguable issue – although having much earlier enjoyed the experience of ancient Greek and learned to read a little of the tragedies of Aeschylus and others, I did understand the pedagogic as well as scholastic importance of making one's own understanding of an "original text," however incrementally.

In any case, I had only heard of *Erkenntnis und Interesse* before Shapiro's translation was announced, and by that time, I had convinced my mentors and their colleagues that they had been wonderfully successful in nurturing this lad's brain, even if it was still short of high proficiency in either German or Italian. I must confess I did occasionally wonder about Shapiro's skill, for I had known little or nothing of him at the time. But the English text was clear, coherent, and substantively challenging enough in its own right that I put aside any thoughts of how fine the line between "translating" and "traducing." Plus, Shapiro's "translator's footnotes" were reassuring. In fact, I later decided, and others more skillful than I confirmed, that Shapiro's work stands among the best.[2] For someone who had read philosophy, though mainly Hume, Kant, Hegel, Marx, and Nietzsche, before coming in a serious way to social science, reading *Knowledge and Human Interest* (hereafter, *K&HI*) was a very welcome bracing experience. There was so much to chew on, take apart, and re-assemble in varying sequences, argue with, engage other readers in, and sometimes just smile about. Even if his reading of Marx seemed oddly filtered and stifled ("re-Kantianized," it has been said, to my mind accurately), Habermas was demonstrating something of the range of restorative materials available in the ancestral heritage of the social sciences.

Another very large dimension to this initial reception of Habermas must be mentioned: the moment of *K&HI* was, at least for some of us, arriving in the wake of having just read Foucault's (1966) *The Order of Things*, the English premiere of which had occurred the previous year. Speaking again only for myself, this latter work, which I had read (indeed, was still re-reading) mainly as a restirring of part of Kant's *Critiques* with his later work (especially his *Anthropology from a Pragmatic Point of View*), made an unsettled context for the reception of *K&HI*. This must have become a matter of some frustration to Habermas, who was far more meticulous in his attention not only to Kant but also to the efforts of others who had engaged Kant, explicitly and at length, in efforts to move beyond. This meticulousness gives *K&HI* a somewhat faltered pace as compared to the panache of *The Order of Things*. One of the peculiarities of the latter book is that Kant is the great presence whose name is seldom written, and when it is, the acknowledgment and relief are gratefully felt, yet one also imagines that an air of regret attended the naming. Only after 241 pages (English edition) did Foucault manage to say that "the Kantian critique … marks the threshold of our modernity." Returning to the accord in his concluding pages, Foucault, associating some of his more explicit antecedents as heirs of the Kantian critique, made the contrast between implicit and explicit presence strikingly clear: "What occurred at the time of Ricardo, Cuvier, and Bopp, the form of knowledge that was established with the appearance of economics, biology, and philology, the thought of finitude laid down by the Kantian critique as philosophy's task—all that still forms the immediate space of our reflection. We think in that area" (p. 384). Habermas' *K&HI* is in that regard more straightforward. Indeed, Kant's explicit presence in this book is even greater than the name index counts, and there are more entries under "Kant" than under any other person name. Of course, Habermas surely felt no impulse to engage in a sort of cultural revanchism – to rebrand the terrain from Descartes to Kant to Kant's heirs, so to speak. What Foucault could well have said was that "we" – like Ricardo, Cuvier, and Bopp – "think in the area formed by the Kantian critical system." Indeed, if one had to name a single philosopher whose systematic thought has appeared in retrospect much like the template for main developments of "western" (i.e., European and Euro-American) societal-cultural formations during the next two centuries, that name would have to be Immanuel Kant.[3] Habermas was not wrong to return to him for beginnings, seeking to pry out and reshape timbers needed for framing his own construction. The very fact of the "re-Kantianization" of Marx, which proceeded long before Habermas, gives ample measure of that.

In any case, reading *K&HI* with fresh memory of Foucault's book was an exhilarating experience, and the memory remains remarkably emotive. That contextual reading elevated, and in a very specific way doubled, a sense of the principal motivation of *K&HI* as having been a project of recovery. There was also a sense of isolation and of worlds long gone or quickly passing. Some of this sense was very likely endemic to the text of *K&HI*, although present retrospection always must contend with the ongoing productivity of memory. But there is no (or little) novelty in the memory that reading *K&HI* along with *The Order of Things*, and then comparing imaginations born in/of those readings to the tenor and trajectories of the worlds around us during the early and mid-1970s, rekindled Humean skepticism without banishing Kantian hope, added another shroud around the promise/project of critical theory, and pointed toward escalating exhaustion of another, the current, order of intelligibility and judgment. When, writing during the 1970s, I urged rejuvenation of the utopist impulse, it was with that draft at my back (see, e.g., Hazelrigg, 1989, pp. 95–101).[4]

1. REFLECTION AND CRITIQUE

In his preface to the English edition of *K&HI*, Habermas described a main trajectory of his book as an exercise "to recover the forgotten experience of reflection," all the more difficult and crucial because of the stronghold that positivism had achieved. "That we disavow reflection *is* positivism" (p. vii). For some of us, the demonstrated message that followed this announcement was a signal to purgation; for, we had often heard the call to positivism's table but had rarely felt that its offering equated with dining well. For some of us, reflection was "a concept in philosophy" and, as more than one advisor had assured us, pre-scientific. For some of us, lacking any but a grammar school understanding of Narcissus, reflection had been only the superficiality of mirroring. Habermas' exercise showed that it was much more, even as we sometimes struggled to find our way through his negotiations of the unfamiliar depth.

Expressed in modern terms, the problem of reflection had been a topic at least since critiques of Locke's conception of mind and causal theory of perception. Soon, it would become identified almost solely with German idealism, however, both because of sustained treatments by luminary figures of that increasingly powerful tradition and because of neglect elsewhere. When Habermas announced his intention "to recover the forgotten experience of reflection," a US audience could be pardoned for puzzlement that they had apparently failed to remember this particular forgetfulness ("Does anyone know what he's talking about?"). But what was more jolting was the surprise that a work written by a young German in his native tongue primarily to a German audience would feature such intention. Had Ernst Mach and his fellow German positivists been *that* successful? Well, yes, though less so there, at more or less immediate proximity, than over here at some distance. Forgetfulness obscures its own criteria, often well enough that what is lost does not just *dis*appear but has already fallen into oblivion. That, of course, was part of what had to be understood, in understanding *reflection* – and part of why the very idea of "recovery," as Habermas proposed it, was daunting.

What, one might ask, is the condition without that recovery? That is to say, if people have forgotten the experience of reflection, in what condition do they live? Furthermore, since Habermas tied that experience to an "emancipatory interest," the previous question extends to another: emancipation from what? In a nutshell, from "the natural condition," from the naïve and dogmatic view that the object of reflection is external to consciousness (i.e., the thing in itself). As Habermas stated the point with reference to Fichte, "Self-reflection is at once intuition and emancipation, comprehension and liberation from dogmatic dependence" (p. 208).

It is of course this same "dogmatic dependence," the naïve attitude, which Kant sought to overcome with his critical philosophy. Reflection is the act by which the "subjective conditions under which we arrive at concepts" are revealed (Kant B316).[5] Indeed, for Kant, reflection is the condition of the principle of subjectivity (transcendental, not the empirical self but its condition of possibility and variety), which thus makes it the necessary precondition of any application of the understanding. In the process of reflection, consciousness catches sight of itself and its objects in the act of knowing and thereby gains not just a self-awareness but a self-knowledge that includes, in principle, knowledge of the limits of conceptuality, knowledge of the limits of reason (the antinomies). In Habermas' words, "It is in accomplishing self-reflection that reason grasps itself as interested" (p. 212).

Consider, in that light, what has often been described as the quintessential sociological act (such as it was emphasized by Max Weber, for instance) – namely, action with intent and anticipation of effect yet also with consequences, partially or wholly, neither intended nor foreseen by actors engaged in the action. For the specific actor in, and thus typified as, those conditions, self-consciousness is delusional, owing to dependence in the dogmatism of natural consciousness (the natural attitude). Thus, it could be said of Peirce and Dilthey, as Habermas did (p. 197, 212), that both were deluded in believing their respective grasps of a "logic of inquiry" because neither of them "discerned what they were actually doing" in their respective investigations of the sciences (to use Habermas' typology, the empirical analytic, for Peirce; the hermeneutic, for Dilthey), which was pursuing "methodology in the mode of the experience of reflection." This counted as a failure of "the emancipatory power of reflection," a power "which the subject experiences in itself to the extent that it becomes transparent to itself in the history of its genesis" (p. 197).[6]

Forty years later, that task is still daunting. The quick of human experience has become more hemmed in, thinned, and flattened, by equations of merchandising and consumption. The passivity of mirroring is if anything *more* regnant now than forty years ago. To the extent that an enterprise of theory in any of the social sciences has any remaining pulse, it beats to issues and questions that seemingly know nothing of the problem of reflection. But there are exceptions. One of the most notable of recent exceptions has occurred under the rubric of "information technology," where the traditional coding distinction between "data" and "instruction" (analogous to that between "substance" and "procedure," between "acted-upon" and "actor") is being relativized and virtualized in circuits of annealing reflection – or "reflective programming," as it is sometimes called.[7] But this activity is commonly bordered as a *technology*, thus as lacking more than tool value and implication, and as proper material for merchandising and consumption. What wondrous things we have learned to make, and with so little thought to feedbacks on/in/through the maker.

Reflection in this vein has departed from Kant and assumed an empirical cast as indeed a critical agency inasmuch and insofar as it interrogates with intent to make (i.e., with *intent*, and with intent to *make*). Understanding the import of reflection is to understand that *knowing* is itself an *activity*, integral to human being, integral to human doing. Here, the promise of reflection is an act of conscious being that takes consciousness as its own object, so to speak, but with awareness that this "objectness," the consciousness and that which is in consciousness, is a turning back on the actor, this "subjectness" of acting. The suppositions and conditions and methodical procedures of a knowing (i.e., a specific act of knowing) are not left behind as preparations for the completed act, which stands apart as a "know that" or a "know-how"; nor are they either external or ephemeral to the subjectness of the activity of that conscious being, or without intent (even if "behind the back of consciousness") beyond the completed act of the specific knowing. It is in that framework that we have a chance of catching sight of our own inconsistencies and contradictions of knowing and other activities. Dunn (1979, p. 56) supplied a chilling illustration years ago, when he drew attention to the twin facts that if human beings annihilated themselves in a nuclear holocaust it would not have been an act by collective choice; yet, most of those human beings would have been adherents in some degree to the tribalist loyalty of nation-state, as an act of collective choice, and the likeliest organization of activity resulting in nuclear holocaust would have been delineated and executed as effects of that very same tribalist loyalty. "Better dead than red": that sort of thing. Indeed, our activities of knowing, of citation and recitation and literate heritage (aural as well as, and more than, visual), are still remarkably nationalist, especially but not uniquely in the United States.

The possibility of reflection, this understanding of its absence because of a forgetfulness that might be overcome, has what should be obvious ramifications for theory – in particular, for what should count as *critique* and thus *critical* theory. As Dahms has stressed repeatedly and in a variety of venues, a critical theory and a theory that proposes criticisms are not the same organization of presupposition, standpoint, intent, and perspective. Thus, for instance, "many social theorists apply critical categories to decide what methods are best suited to the study of a specific social formation," but there is a difference insofar as "critical theorists endeavor to provide a framework for the study of contemporary society without necessarily reproducing its values" (Dahms, 2007, p. 198).[8] A method is not "value neutral" not because it is inherently value-free (i.e., unbiased tool and unbiased judgment), which it is not, but because value preference (and neutrality, insofar as this is not an empty category) is an action, whereas method, as such, does nothing. Method as such is only the promise of a characteristic, a particular methodicalness, of an activity – here, the activity of knowing. In the reflectedness of inquiry, "choice of method" is informed by intent of knowing, which is always a projection of value. Or, one might say, a hope. The framework makes explicit the criteria by which one decides that a parcel of evidence or an interpretation of meaning *is* that, because adequate to the intent of a claim to know.

For someone who had not been specifically schooled in trajectories of the reflection concept, the itinerary of *K&HI* offered selective instruction as Habermas built his argument in the interstices of a critical engagement with predecessors, weaving between the transcendental (Kantian reflection itself) and the empirical (ordinary actor recovering insight from reflection). It could not skate free of criticism by others, to be sure. Both in stature and in complexity of undertaking, this project advertised several virtually inevitable lines of attack by critics. The remarkably strong cognitivism by which he sought to guarantee a universal core of moral intuition while acknowledging the force of Hegel's critique of Kantian ethical doctrine was one such line. Another was the sometimes fuzzy distinction between the part of positivism that Habermas sought to reject and the part that he in effect adopted – inconcinnities and opacities that have prompted more than one critic (e.g., Geuss, 1981, p. 30) to wonder just how much Habermas had been captured by the object of his own critique. But these criticisms of *K&HI* are pushing beyond that work itself, partaking too much of later developments in Habermas' journey along the interstices. The peculiar vulnerability of *K&HI* itself, in the moment of its own situation, lay precisely in that situation: it was an offering that looked all too metaphysical in a recently declared "postmetaphysical age." Whereas Foucault's *Order of Things* gained an approving reception partly by assimilation to that declaration, *K&HI* looked more like the age being eclipsed, at least by declaration.

There was already a long trail of controversy about reflection, about what the concept precisely is (in particular, insofar as it is an active principle, not the passive "mirroring" notion associated with, say, Locke), about whether it is hopelessly metaphysical. Rorty (1989, p. 123) averred that it was an effort to overcome isolation of subjectivity, but that such effort amounted only to a "leap of faith," not logical argument: "This insusceptibility to argument is what makes 'the philosophy of reflection' ... the bête noire of philosophers who take public discussability as the essence of rationality."

While Habermas' development of a "discourse logic" was undertaken in part to facilitate a recovery of the experience of reflection within and as "public discussability," the acknowledged condition of success in that logic, later stipulated in such terms as "force of the better argument," is seldom realized empirically and rarely on a scale larger than a small group.[9] There was no sleight of hand in this motive, however. Intent – in particular the emancipatory intent by/in/through which an ordinary actor strives to recover insight of reflection – is an expression of the Kantian principle of hope. Habermas moves from Kant through Fichte, Hegel, Marx, and others to a position that, one might say, bears a family resemblance to later Husserl (along with Paul Lorenzen of the Erlangen School), paring and reforming and building on/with those materials, as he weaves his course of study/ instruction in *K&HI.* But for all the long trajectory of that motion, the memory of Kant never dims. Here, perhaps, it is important to be reminded that the structure of Kant's trilogy has at times been mistaken. The third of the three, *Critique of Judgment*, is neither an afterthought nor the least of the three, because dealing with aesthetics rather than logic or ethics, but indeed the pinnacle. Of the three questions stereotypically associated as emblematic of Kant's concerns (and stated toward the end of the first critique as manifesting the "whole interest of reason, speculative as well as practical"; cf. p. 203 of *K&HI*) – What can I know? What ought I to do? For what may I hope? – the third is reminder that human actions (including the cognitions of science, of philosophy, and so forth) are, when all is said and done, a matter of human *judgments*, not the pure rationality of an elegant logic or the practical rationality of a code of behavior. Choices among alternative courses of action are finally exercises of judgment, and these judgments are forward-looking, projective of intent (cf. Dunn, 2000, pp. 180–208). Habermas, insistently reaching for the universal, unfurls hope from an irremediably contingent stance in that projection. And do we no less, when we strive to understand one another (even if the understanding be for nefarious ends)?

2. DANCING WITH ANTINOMIES

An anonymous narrator begins "Stories of Our Lives" by introducing a main character –"a morally serious citizen as well as dedicated teacher of young people" – and telling us that she wonders

> whether people reflexively think of their lives as stories because from birth to death they are exposed to so many narratives of every sort, or whether, contrariwise, our notion of

what a *story* is, in every age and culture, reflects an innately dramatistic sense of life: a feature of the biological evolution of the human brain and of human consciousness, which appears to be essentially of a scenario-making character.

John Barth (1996, p. 96) brings a smile to one's face, and he does so without either sacrificing the complexity or mocking the seriousness of what is at issue. Indeed, a great deal is at issue (as was the case for an earlier morally serious citizen and dedicated teacher of young people): the contrast can be posed in terms of what belongs to "culture" and what to "nature," in terms of "being acted upon" and "acting," in terms of a "morality of freedom" and a "morality of determination," in terms of "reach" versus "grasp," and so on. Ultimately, one wonders, if the smile also reflects a weariness, perhaps even a resigned awareness of lasting incapacity. Is it finally just too much to deal with?

Barth is not alone in posing the contrast and its question, of course. In how many different sentences can the either/or statement be written? Stylists continue trying to invent new ones, sometimes with sadly hilarious effects (e.g., David Foster Wallace's *Infinite Jest*), and theorists (whether little "t" or big) continue trying to resolve the circularity, often to no effect at all. As with Habermas, it has been tempting to return to Kant and try again. Like characters of fictions, we hope for justice – at least the poetic sort if not something more palpable. It is excusable that when rounding a corner and being hit by some of the pustulating realities of life, we are tempted to infer cosmic significance to the onslaught – "fate" or "the hand of God," that sort of imagination of unstoppable movement. But though the experience can feel utterly cosmic, the events never are. Their place and reach are thoroughly of the quotidian. Such imaginations are of course tokens of experience straining to be more, the status of something that looks like a law, a determination by necessity. In that direction lay the antinomies of reason. These are, a sociologist would say, "structural," exemplars of what is distinctively *sociological* in social action (whether it be of an individual or of a group).

In returning to Kant for points of departure, *K&HI* implicitly refers to bounds inscribed in the doctrine of antinomies, but not all aporia are marks of one or another antinomy. Having developed powers of reason, human beings can ask questions they cannot answer, can invent relations of identity and difference that end in self-contradiction, have invented principles (rules, laws, etc.) of universal intent, each of them unquestionably good, yet the set (pairs, triplets, etc.) undeniably self-defeating. Because there has been a tendency to read these limitations backward, so to speak, into the antinomies of reason, let's be clear about the latter.

At any given time, there are lots of questions for which satisfactory answers have not been determined. Most of them do not count as antinomies. Consider this question, for instance: What is the mean lifetime of a photon (i.e., a quantum of light)? It might seem an esoteric question, but it is reasonable to ask. After all, doesn't everything have a finite lifetime? We have estimates of the mean lifetimes of various other particles – something in excess of 10^{29} years for a proton, for instance – so why not an estimate of the mean lifetime of photons? Well, according to the quantum theory, photons travel at limit speed (i.e., the speed of light); thus, because of time dilation (according to special relativity theory), there can be no reference frame in which photons decay. (Imagine, moreover, because a photon, unlike a proton or an electron or a boson, has precisely zero mass, is it even meaningful to ask about its rate of decay?) Or consider another current conundrum: according to the quantum theory, whereas sound, like other kinds of radiant energy, displaces three-dimensionally, light, also a radiant energy, displaces only two-dimensionally. Why? No one has come up with a satisfactory answer (or not the last I checked). Although so far deeply recalcitrant, these questions are examples of questions that have not been satisfactorily answered but that, so far as anyone can tell, are in principle *answerable* within the inventive framework by which the questions are logically implicated to begin with. Of course, it might yet be decided that the current framework is in some way severely flawed and thus supplanted by another that can answer all of its predecessors' questions, including those which had not previously gained satisfactory answers. That sort of event has happened before (e.g., an Einsteinian framework supplanting the Newtonian).

The antinomies are limits of reason (*Vernunft*) itself, and they are encountered when human understanding (*Verstand*) is extended beyond the sensible (that which is given by the senses), entering into various metaphysical entanglements.[10] For Kant, the boundaries of the understanding become clear as reason reflects upon itself, allowing apprehension of the logical condition of antinomies which are of reason proper. Therefore, imagine two principles or theses, each considered valid in its own terms and each appealing, but which when pursued through rational inquiry end in a mutual contradiction. One of the classic instances is that the universe must have finite limit in space and time, for everything has a beginning and end in time, a boundedness in space; yet, the universe must be infinite in space and time, for what could possibly be beyond the limit of space and the limit of time? Rational inquiry seems inevitably to lead to, or in some sense to uncover, inherent contradiction. Repeated experience of this phenomenon of inherent contradiction brings one to conclude that one has reached the limits of reasoned argument, and thus the limits of rational knowledge. But more, experience of the antinomies seems to point rational inquiry to something beyond itself. This "pointing" can be backward, toward an initiating rule or word (as in a religious doctrine of origins or in a secularized doctrine that substitutes, say, "laws of nature" for "word of God" as the initiating motive force and authority). However, the "pointing" can instead be forward, as in a projection of what might yet be made or achieved (as in Habermas' intent of communicative action).

It is "human interests," according to Kant, that impel us to expect systematically unitary, complete knowledge. These interests are practical and rooted in human reason. After all, the impulse "to know" is with expectation of accomplishment and with intent to eliminate error and eventually arrive at conclusive knowledge. But the impulse is a "transcendental illusion": we expect a necessary consistency and coherence in the fabric of our knowledge, but we mistake that expectation as something it is not and can never be – namely, representation of "objective necessity in the determination of things in themselves" (B354). Now, although Kant suggested that the illusion is unavoidable and even necessary to the integrity of an epistemology (B673), his own insight and indeed the system of his critical philosophy point to a project of maturation and responsibility whereby human beings can overcome what will be seen through effects of reflection as the "bad habits," so to speak, of succumbing to that illusion.

In the meantime, the experience of being able to think questions that seem to have no answers, create problems that seem to have no solutions, has been taken indiscriminately as the mark of something it usually is not, the antinomies of reason. From such experiences, some human beings have concluded that human beingness *as such* is deficient, that the essential condition of being human is deficit. The interest in completeness and systemic unitariness implies that the answer or solution is there to be had, if only "we" could do better.[11] That is to say, the experience is typically understood as recognition that human beings are capable of asking questions that *they* cannot answer. The implicit frame holds that a *complete* being, a non-deficient being, *could* answer any question available to posit. Universal intent is more demanding than that supreme force known by the ancient Greeks as Zeus. It is easy to see the resources that a careful observer such as Wallace or Barth can turn into an anguished hilarity, an amused knowingness, or some other reflexive irony.

One of the better lessons still available from those same ancient Greeks is (as Vernant, 1974, reminded us) that when their writers invented the tragic form and folded inherited myths into it, the seemingly free blanket of traditional security gave way (in part) to ironies of reflection. Prior to the tragic form, myths could be naïve; they gave answers and solutions before the questions and problems were explicitly formulated. Tacit understanding was powerful and nearly all. Greek tragedists taught how to use experiences of mythical tradition to formulate questions and problems for which there might well be *no* answers and solutions. The next (very short) step? A grand new template of self-reflection: being human, caught between tragedy and hubris. Habermas recognized this template, as he threaded his way through interstices, hoping to resolve the difference in a renewed project with emanicipatory intent. The insinuation of *Mangelwesen*, deficient being, is tenacious and profound, however, and it was reproduced in that project already in *K&HI*.

3. CAPACITIES AND INCAPACITIES

From the beginning, it seems, Habermas adopted a broadly defined standard discrimination of nature and culture, rather than treat that inheritance as itself crucially problematic. Some keen insights into that problematic were readily available to him – or so I have believed since before first encounter with *K&HI* (see, e.g., Hazelrigg, 1993) – but the filters through which Habermas approached Marx deflected those insights. For Habermas, nature is always inescapably "external," a reality that subsists in its own "ineradicable strangeness" (p. 86, 43). As a number of critics have asked, implicitly if not explicitly (e.g., Geuss, 1981, p. 28), how much can one relinquish and yet retain the capacity and capability of a critical theory?

When writing "Under the Thesaurus" during the late 1970s (see Hazelrigg, 1989, pp. 6–46), the evidence of climate change was already convincing (although estimates of rate of change were seriously short, mainly because complexities of so-called feedback loops were poorly modeled). The time series of densely spaced observations carefully made by Charles Keeling had demonstrated that atmospheric levels of CO_2 had surpassed any previously known historical record, and the recent rate of accumulation was accelerating.[12] At that time, it seemed too improbable to expect a cordial reception to argue that human beings produce nature – "palpably material" nature, like the stone against one's toe, "all the way down" – and so that thematic was kept mostly implicit (cf. Hazelrigg, 1995). But surely now, now that "global warming" has been registering in consciousness widely and perhaps deeply (Hulme, 2009), it is clear that *we* human beings, especially we of the northern hemisphere, have been producing a nature that might very well prove to be lethal to human being as well as to many other life forms. Can we now see that the labors that have been and are achieving such prospects are simultaneously "historical-hermeneutic" and "technical-empirical" – that an analytic division of interests between "work" and "interaction" perpetuates a lameness of human being? One of the greatest of our tragic ironies is that in our quest to find ever more that human being can be less than (justifying the "deficient being" posit, assigning ultimate responsibility elsewhere), we end up being less than our own inventions, less than what we ourselves have made. With Dunn (2000, p. 195), among others, surely we can agree the importance of starting inquiry from "the stakes which human beings have in what history does to them: from what is good or bad for them, in or against their interest." That history and the interests are of course human, even when humans pretend otherwise (e.g., the givenness of nature, and divine provenance), and the human making of human worlds (the part separated as Nature included) pushes its own history and interests ahead. Dunn is right to caution about the complications involved. Recalling Aristotle's *Politics*, he asked whether anyone can yet demonstrate definitively that human beings do have definite interests, and if they do, "what could be more evident than that most of them have little grasp of just what their own interests are?" There is something disheartening in the observation that today, more than 2,000 years hence, so much of *Politics* eludes the grasp of so many citizens who have ostensibly benefited from all the pedagogical activities, all the *Bildungsprozess*, of the intervening millennia.

K&HI was invigorating not only because of the stellar display of thoughtful heritage and an integrated process of argumentation that both built from and carried forward critical inquiry but also because the display carried much promise – regarding, for instance, capacities of self-governance, self-responsibility, and the like. Forty years later, *in*capacities seem to drag more inertially, more tenaciously, than they did.[13] One wonders whether young scholars today, insofar as they do read works of critical theory, imagine that these works must have been written in a language now dead. Perhaps they are a shade too young to share Axel Honneth's (2007) sense of the "great chasm" separating his current self from predecessors such as Horkheimer and Adorno, maybe Habermas as well. Evelyn Waugh once remarked that the generational gap between fathers and sons had never before been as deep and wide as it was between the combatants of the Great War and their fathers.[14] While the terms of generational change are different, one wonders with Honneth if the prospects that were visible to us in the advent of *K&HI* have retained their music – or still sound any music that newer ears can hear. Is there still an audience for anything like a critical theory? In a culture that advertises semblances of movement in fundamentally stagnant conditions, the (only slightly) camouflaged mirrors that re-present consumers' self-impressions and emotions so satisfyingly that little thinking – to say nothing of *critical* thinking – is ever stimulated obscure the action of semblance itself, for lack of contrast.

Since the 1960s, social science has had the advantage of ever larger volumes of data, combined with increasingly sophisticated analytic techniques for modeling their error structures so as to improve the quality of empirical conclusions based on presumably careful observations. Unfortunately, because almost all of the data have been cross-sectional in design, and so can support very little inquiry into dynamics, into process, the advantage has actually been a major disadvantage – namely, lack of practice in the ability to think in processual terms. Again, Dahms (2007, p. 192) was sharply on target when he pointed out that "the dominant model" behind virtually all development of empirical research during the twentieth century "has been resting on the paradoxical presupposition that research efforts should be directed at the refinement of concepts and methods that are inherently *static*, in the interest of illuminating a reality that is conspicuously dynamic." One of the strengths of critical theory was its insistence that reality is *process*. Granted, its analytic apparatus was sometimes crude, resulting mainly in descriptive accounts of discrete states arrayed loosely over long intervals of history with the benefit of two-, three-, or four-factor classifications of types and occasionally supplemented by the magical wave of a hand (see, e.g., Therborn, 2008). But critical theory did insist on dynamics, and that insistence remains current, one that might yet be converted into a critical inquiry into the dynamics of specific processes – a study not just *that* "things happen" but of *how* "things happen." *Knowledge and Human Interests* offered encouragement that this road was negotiable, that it could take us to better, more efficacious understandings of how we achieve knowledge, formulate interests, make worlds, and in the process further our emancipation from the dead hand of past times – or, as Kant put it, pursue maturation, responsibility, and self-autonomy.

Do we yet know what conversation is taking the place of that one? Or is the "place" itself being displaced, such that the next great conversation has yet to become recognizable?

NOTES

1. The authors of *The German Ideology* penned this quasi-syllogistic send-up (Marx & Engels, 1932, p. 481; with modification):

observation:	"The cat eats the mouse."
reflection:	"Cat - nature, mouse - nature; consumption of mouse by cat = consumption of nature by nature = self-consumption of nature."
discovered generalization:	"Devouring of the mouse by the cat is based upon the self-consumption of nature."

2. Not to say that I find it completely unproblematic. For example, his translator's note 23, to page 33, while compatible with the fact that in the long paragraph continued on that page, Habermas missed the point of Marx' critique of "nature's object," and so on, also contributes to that fact: "to give form in a symbolic system" applies just as much to the real sensuous object as to "communicative action." Similar cautions could be voiced with regard to notes 9 and 10 in the same chapter. In any case, however, Shapiro was generally overt in prejudicing toward, not against, Habermas' choices, and that one must expect of an honest translator.

3. And of course his works are still central to so much of academic philosophy, even when an author is writing primarily about another (e.g., McDowell's, 2009, essays on Sellars and Hegel).

4. Who could not be struck (as Amsler, 2008, pointed out) that the advertised theme of the 102nd annual meeting of the American Sociological Association was "Is another world possible?" Was the question meant to be ironic? One also wondered, "Other than what?" And, "On which side of that line sits the other world currently in production?" And, "Which of the 'other worlds' recently foreshadowed by Hulme (2009), for instance, will eventually be recognized as sufficiently 'other' to count as 'another world'?" And: "Whose theory makes convincing explanation of how, under what conditions, or even when, such recognitions do occur (rather than not)?" Compare Dunn (2005, pp. 352–353).

5. This is standard means of citation to the second edition of Kant's *Critique of Pure Reason*, published in 1787.

6. Imagine the consternation among theoretical physicists when, increasingly after the 1920s, they were forced to recognize (by results of their own work) that measured observation is not merely a tool, part of the technical apparatus of doing physics, but indeed a *basic principle* of physical reality. This story has been told several times now, with differing casts, but none more charmingly than in Louisa Gilder's (2008) new book, *Entanglement*.

7. Harrison White (1992, pp. 346–349) pointed to this nexus when he remarked on the analogy between a human actor and a Kalman filter (a concept in mathematics and a tool in engineering; see, e.g., Harvey, 1989). Apropos that, reflection can be thought of (as someone once said) as observing limits but dynamically.

8. See also, for example, Dahms (2002) and Dahms (2008) (the latter an extensive and intensive statement of the principled reliance of the idea of a science of society on critical theory).

9. Criteria of "public discussability" are hardly matters of concern only to philosophy or in the confines of the so-called historical-hermeneutic interests. See, for example, Laughlin (2008), who observes that the phrase "Knowledge Economy" acknowledges (as any economic formation does) the importance of a *scarcity* of knowledge, a scarcity mainly due to privatization (the rapid erection of "no trespass" signs around bundles of knowledge, skills, and embedded things – such as genomes and genomic variations – sanctioned by laws of patent as well as copyright).

10. Thus, one of the famous aphorisms from Kant's first critique (B75): "Thoughts without content are empty, intuitions without concepts are blind."

11. Think of one of several instructive examples from mathematics. Generation upon generation of mathematicians sought to establish mathematics as a complete system of formal theorems. Imagine the consternation, then, when in 1931 Kurt Gödel demonstrated that the quest could not succeed (his "incompleteness theorem"). There are of course antinomies other than Kant's "big three." Several antinomies in logic ensued from arithmetizations of the continuum, and new ones emerge. An example is known as Richard's paradox, often glossed (as Richard himself did) with the sentence, "Consider the smallest number not definable in English in less than twenty words."

12. These data are now known as "the Keeling curve." By the end of this decade, the Keeling number at his observatory will be at least 388 parts per million volume (ppmv), which is about 23 percent higher than in 1960, and the rate of increase has accelerated from 3.2 percent for the decade of the 1960s to about 4.9 percent for the current decade. Data from the Vostok ice core indicate that the highest CO_2 concentration during the previous 400,000 years was less than 300 ppmv.

13. Whereas one could argue (as has Therborn, 2008) that the state apparatus of governance in the "developed world" has been rather successful in public policy, it is reasonable to ask how much of that success has amounted to a reproduction of problems for which it can persuasively claim success and deflection or ignorance of problems for which effective response would require too much risk taking, with its electorate and/or with its congeries of special interests. Wolin (2008) has recently waded into these waters, as has Habermas in some of his most recent works (which, it seems to me, are less hopeful, by and large, that *K&HI*).

14. An awareness shared by Karl Mannheim and expressed in the impulse of his 1927 article on "the problem of generations."

REFERENCES

Amsler, S. S. (2008). Pedagogy against 'dis-utopia'. *Current Perspectives in Social Theory*, *25*, 291–325.

Barth, J. (1996). Stories of our lives. In: *On with the story* (pp. 147–179). Boston: Little, Brown.

Baxter, C. (1999). *The business of memory: The art of remembering in an age of forgetting*. St. Paul, MN: Graywolf.

Dahms, H. F. (2002). Sociology in the age of globalization. *Current Perspectives in Social Theory*, *21*, 287–320.

Dahms, H. F. (2007). Confronting the dynamic nature of modern social life. *Soundings*, *90*, 191–205.

Dahms, H. F. (2008). How social science is impossible without critical theory. *Current Perspectives in Social Theory*, *25*, 3–61.
Dunn, J. (1979). *Western political theory in the face of the future*. Cambridge: Cambridge University Press.
Dunn, J. (2000). *The cunning of unreason*. London: HarperCollins.
Dunn, J. (2005). *Setting the people free*. London: Atlantic.
Foucault, M. (1966). *Les Mots et les choses*. Paris: Gallimard.
Geuss, R. (1981). *The idea of a critical theory*. Cambridge: Cambridge University Press.
Gilder, L. (2008). *The age of entanglement*. New York: Knopf.
Habermas, J. (1968). *Erkenntnis und Interesse*. Frankfurt a.M.: Suhrkamp.
Habermas, J. (1988). *Nachmetaphysisches Denken*. Frankfurt a.M.: Suhrkamp.
Harvey, A. C. (1989). *Forecasting, structural time series models and the Kalman filter*. Cambridge: Cambridge University Press.
Hazelrigg, L. (1989). *A wilderness of mirrors*. Gainesville: University Press of Florida.
Hazelrigg, L. (1993). Marx and the meter of nature. *Rethinking Marxism*, *6*, 104–122.
Hazelrigg, L. (1995). *Cultures of nature*. Gainesville: University Press of Florida.
Honneth, A. (2007). *Pathologien der Vernunft*. Frankfurt a.M.: Suhrkamp.
Hulme, M. (2009). *Why we disagree about climate change*. Cambridge: Cambridge University Press.
Laughlin, R. B. (2008). *The crime of reason*. New York: Basic Books.
Marx, K., & Engels, F. (1932). The German ideology. In: *Collected works* (Vol. 5, pp. 19–539). New York: International Publishers (Reprinted in 1976).
McDowell, J. (2009). *Having the world in view*. Cambridge, MA: Harvard University Press.
Rorty, R. (1989). Is Derrida a transcendental philosopher? In: *Philosophical papers* (Vol. 2, pp. 119–128). Cambridge: Cambridge University Press (Reprinted in 1991).
Therborn, G. (2008). *From marxism to post-marxism?* London: Verso.
Vernant, J.-P. (1974). *Mythe et société en Grèce ancienne*. Paris: Librarie François Maspero.
White, H. (1992). *Identity and control*. Princeton: Princeton University Press.
Wolin, S. (2008). *Democracy inc.: Managed democracy and the specter of inverted totalitarianism*. Princeton: Princeton University Press.

PUBLIC SOCIOLOGY AND THE GOVERNANCE OF POSSIBILITY

Patricia Mooney Nickel

ABSTRACT

This chapter critically considers two conceptions of sociological labor as they have recently been articulated in two competing visions for public sociology. I use the contrast between Ben Agger's and Michael Burawoy's recent professions of public sociology as a lens through which to critically understand the way in which the narratives produced by sociological labor govern the emergence of knowledge, which would be the basis of transformation.

INTRODUCTION

As a tacit portrayal of the world and of the people who inhabit the world, sociology is an intimation of what exists, and, as such, it is also a statement about the possibility of transforming the circumstances that govern our lives. Sociological knowledge therefore always is a political statement that governs possibility in relationship to a public with whom sociologists communicate both directly and indirectly. From this perspective, governance is the narrative of legitimated knowledge that, in turn, results in a

Nature, Knowledge and Negation
Current Perspectives in Social Theory, Volume 26, 207–228

ISSN: 0278-1204/doi:10.1108/S0278-1204(2009)0000026011

legitimated practice that often masks the inaudible knowledge inherent to everyday life. It therefore matters whether sociologists conceive of their labor either as a depoliticized discovery of our experience in the world or as the search for a possible basis for transformation of the circumstances that result in such experiences. This chapter critically considers these two conceptions of sociological labor as they have recently been articulated in two competing visions for public sociology. I use the contrast between Ben Agger's (2000) and Michael Burawoy's (2005a) recent professions of public sociology as a lens through which to understand the way in which the narratives produced by sociological labor govern the emergence of knowledge, which would be the basis of transformation.

Given the near saturation of the discussion surrounding public sociology by Burawoy's division of labor, I am faced with the somewhat difficult task of shifting debate about a phrase that is already losing its ability to represent a critical stance and I do so within a division of labor that is already taken for granted by many who advocate Burawoy's reified portrayal of sociology as a legitimately governing body of knowledge in relationship to the state, publics, and possibility. In the short time since, with Burawoy's 2004 Presidential Address to the American Sociological Association, public sociology entered mainstream discussions of the discipline, the phrase "public sociology" has come to convey everything from NGO management (American University's MA Concentration in Public Sociology, 2007) to cosmopolitan state management (Beck, 2005; Delanty, 2006) to government policy consultancy (Perrucci, Ferraro, Miller, & Muschert, 2008). Although critically important, this empirical manifestation of public sociology practiced through the quiet mechanisms of the voluntary state (Nickel & Eikenberry, 2007) in concert with the market is not the type of governance with which I am concerned.

I am concerned with the ontological form of governance that *precedes* these incorrect assumptions about the inevitability of the collapse of the supposed boundary between the state, civil society, and the market (Beck, 2005) as it is now legitimated by the very public sociologists who claim to oppose such a collapse (Burawoy, 2006). I am thus concerned about the way in which representations of our experience in the world, such as organic divisions of labor (Burawoy, 2005a), and incorrect assumptions about the neutrality of knowing, such as the positivism and instrumental knowledge produced by professional sociology,[1] invisibly govern everyday knowledge about the possibility for transformation. My thesis is that Burawoy's public sociology is derived from the same assumptions that have produced the problems, which he argues that public sociology ought to oppose. I argue that Burawoy's organic division of labor is an ontological fiction stabilized by epistemological boundaries, which in turn, disguise the relationship between sociological knowledge and the neoliberal state, thus forestalling transformation.

BURAWOY'S ONTOLOGICAL FICTION

It has become difficult for those who advocate public sociology in the tradition of C. Wright Mills even to discuss public sociology without becoming bogged down in assumptions about the discipline of sociology with which they do not agree, including Burawoy's (2005a, p. 9) explicit rejection of the possibility of Mills' sociological imagination.[2] Since Burawoy's 2004 campaign, to say that one is an advocate of public sociology no longer conveys a critical stance toward the current practice of professional sociology. Given the dominance of Burawoy's conception, public sociology now conveys not only support for public intellectuals but also the division of labor through which Burawoy produces them. The result is that to speak on behalf of public sociology now implies that one agrees with the idea that sociological knowledge can be categorized according to two audiences (academic and extra-academic) and two epistemologies (instrumental and reflexive). Owing to the ontological fiction that Burawoy has written, public sociology as a phrase now reifies not only the idea that public problems exist *independent* of professional sociology but also Burawoy's entire fictional account of the discipline of sociology in relation to the way that knowledge boundaries govern possibility.

Public intellectuality is not a new idea, a fact that both Agger and Burawoy recognize[3]; however, its history has been rewritten by Burawoy as belonging to an organic evolution toward his taken for granted division of labor. The historical antecedents of the idea of a politically engaged intellectual date back at least to Karl Marx's oft-quoted statement in *Theses on Feuerbach*: "philosophers have only interpreted the world, the point is to change it." What is interesting about Marx's statement is not so much that – per Burawoy's (2005a) contention – intellectuals *ought* to change the world, but that, as Agger (1989a, 1989b, 1989c) recognized, intellectuals *already do* change the world. Marx did not suddenly inspire intellectuals to influence the world; intellectuals always influenced the world, whether they chose the label "public" or not. Thus, what is revolutionary about Marx's statement is not only its call for intellectuals to change the world, but its recognition that ideas circulated by intellectuals are politically functioning in relation to the possibility for transformation, albeit sometimes without our recognizing them as the social acts that they are (Agger, 1989a, 1989b, 1989c, 2000; Brodkey, 1987).

The theme of intellectuals in relationship with the public continued to be expressed in Antonio Gramsci's (1997) discussion of the formation of intellectuals, in C. Wright Mills' (2000) sociological imagination, in Alvin W. Gouldner's (1970) critique of professional sociology, in Michel Foucault's (1980) power/knowledge, in Russell Jacoby's (1987) criticism of the loss of space for public intellectuals, and Edward Said's (1994) statement that the role of the intellectual is to speak truth to power. I have inevitably left out innumerable statements on public intellectuality and the politics of knowledge, but my point is that thinkers at least since Marx have criticized the idea of professionalized knowledge with the hope that, through recognition that *intellectuals govern through the creation of knowledge about possibility*, a sphere of discourse aimed at improving the world might be achieved.

More recently, in 2000, Agger contributed to this tradition *Public sociology: From social facts to literary acts*. Agger's (2000) conception of public sociology is a critical one positioned in opposition to professional sociology's tendency to deny its authorship and thus to invisibly order the social world, resulting in a decline of discourse (Agger, 1989a, 1989b, 1989c, 1990, 1991, 1992). For Agger (2000), public sociology was a phrase used to advocate the transformation of sociological labor in such a way that writers would anticipate and value readers' criticism and be open to the public's revision of sociological ideas, thereby inviting public debate. Agger's (2000) argument, as the subtitle of his book indicates, was that professional sociology did not mirror inalterable social facts, but was always a literary practice that constituted a political stance. Agger does not call, as Burawoy does, for a division of labor that distinguishes public from critical from professional from policy sociology; Agger's call is for professional sociology to become public sociology through a transformation of hierarchies of labor *cum* knowledge to reverse the decline of public discourse (Agger, 1990).

In his 2004 Presidential Address, Burawoy used the phrase public sociology to describe something entirely different from Marx's initial critical stance and Agger's (2000) conception. As Paul Paolucci (2008, p. 382) has argued of Burawoy's mainstream public sociology program, "with its conflation of scientific discourse with political action, and its pronouncements on abstract principles such as democracy, freedom, and justice, public sociology is neither Marxist nor radical."

Although both Agger and Burawoy argue on behalf of sociological practice that addresses major social problems, this is the end of the similarity between their usages of the phrase, *public sociology*. In the first phase of his campaign for public sociology, Burawoy (2005a) set forth eleven theses "for public sociology."[4] These theses, unlike Marx's theses or Agger's critical public sociology, are *not* aimed at transformation of the intellectual endeavor known as sociology; Burawoy (2005a) very clearly indicates that his definition of public sociology is an addition to the field as he claims that it already exists. Indeed, Burawoy's (2005c, p. 318) argument is positioned *against* the radicalization of sociology: "Let me be clear, our disciplinary project cannot and should not be reduced to *critical sociology*, which makes no sense without a *professional sociology* to criticize or even without a *public sociology* to infuse with its commitments, just as all three find their compliment in a *policy sociology*[5] with its more instrumental deployment of knowledge."

My reading of Burawoy's text is that it is not primarily a call for public sociology, but is instead primarily an argument for the legitimacy of professional sociology and instrumental knowledge employed by the state. In his third thesis, the division of sociological labor, Burawoy (2005a) frames public sociology as belonging to an "organic" division of labor. Note well that Burawoy's use of the word organic in relation to labor is far more Durkheimian than Gramscian. Given that an "organic division of labor for sociology" is the basis of Burawoy's 2004 argument for public sociology, it is worth noting that in 1893 when Emile Durkheim originally wrote of the division of labor (specialization), he argued that it emerged not as a matter of choice, but as a matter of social law to which human choice is irrelevant. The division of labor, Durkheim argued, is necessary to maintain solidarity, or *order*. In other words, it is a necessary formation for the continuation of legitimate governing (Table 1).

What is curious about both the affirmative and the critical reception of Burawoy's call for public sociology is that no one seems to have recognized that he has called for much more than *public* sociology. Public sociology

Table 1. Burawoy's Organic Division of Labor.

	Academic Audience	Extra-Academic Audience
Instrumental knowledge	Professional	Policy
Reflexive knowledge	Critical	Public

Adapted from Burawoy (2005a, pp. 11–12).

seems an afterthought for Burawoy when one considers that this division of labor is a statement on behalf of:

- ending discussion about the transformation of professional sociology;
- bureaucratic intellectuals (Merton, 1968) producing instrumental knowledge;
- critical sociologists sequestered from the public;
- the uselessness of reflexive knowledge for public policy;
- the illegitimacy of the public's reflexive knowledge about the policy that governs their lives.

Public sociology, according to Burawoy's (2005a, p. 10) fictional account of what exists within the discipline of sociology, is an argument for professional sociology and instrumental knowledge: "there can be neither policy nor public sociology without a *professional sociology* that supplies true and tested methods, accumulated bodies of knowledge, orienting questions, and conceptual frameworks." Again, this is *not* an argument for public sociology; it is an argument for professional sociology. Public sociology is framed by Burawoy (2005a, p. 10) as a source of legitimacy for professional sociology, in exactly the same way that the discipline of *public* administration is framed as a source of legitimacy for the state, and professional sociology is said to be "the *sine qua non* of public sociology's existence." Indeed, Burawoy (2005a, p. 15) also argues that critical sociology is dependent on professional sociology: "Without a professional sociology, there can be no policy or public sociology, but nor can there be a critical sociology – for there would be nothing to criticize." *All* sociological labor is written by Burawoy to be dependent on the reification of professional sociology just as neoliberalism is dependent on the state's reification of instrumental knowledge to the exclusion of transformative knowledge. This division of labor smacks of what Paul Piccone (1978, p. 48) referred to as system-generated negativity:

> Counter-bureaucratic bureaucracies become one of the paradoxical expressions of artificially generated negativity. The problem with this system-generated negativity is that, *to the extent that it is itself bureaucratically sanctioned, it tends to become an extension of the very bureaucracy in need of control*...it simply extends the bureaucratic logic it was meant to challenge and becomes counter-productive. The organic negativity necessary to successfully sustain this challenge must develop *outside* the bureaucratic administrative framework (my emphasis).

For Piccone, it would be impossible for public sociology to emerge within a division of labor that legitimates artificial negativity based on the primacy of that which it opposes.

But it is not only the myth of negativity that we are faced with when considering Burawoy's division of labor. To say that Burawoy has generated an ontological fiction is to say that he has created the myth of a sociological division of labor to stabilize a world that simply does not exist; namely, a world in which professional sociology's instrumental knowledge does not govern possibility (Nickel, 2008). The boundaries between Burawoy's (2005a) role categories for sociologists give the false illusion that positivist and instrumental professional sociology and policy sociology do not *result* in a disengaged and discouraged public, thus preventing the successful emergence of reflexive critical sociology and public sociology. This emergence that, for Agger (2000), is dependent on the realization that American professional sociology and policy sociology are literary acts, open for revision, not professionalized facts, which people fail to recognize as being political and thus changeable. It is for this reason that Agger (1989d) rejects the idea that professional sociology based on seemingly authorless science can co-exist with public sociology: positivist and instrumental professionalism are destructive in relationship to public life because the public falsely internalizes "facts" and utility as inalterable truths (Agger, 1989a, 1989b, 1989c, 1989d, 2000). This was Agger's (2007, p. 270) point when, in his conceptualization of public sociology, he wrote that "[a] public sociology must want to change the world, and it *must recognize that it is already changing the world by intervening in it*" (my emphasis). When Burawoy (2005a, p. 104) wrote that public sociology promotes "dialogue about issues that affect the fate of society," he failed to recognize that it is also the practice of professional sociology, which he endorses, that affects the fate of society as it "secretly writes" the world (Agger, 1989d, 2000).

It is closer to the truth than is Burawoy's division of labor that when intellectuals produce professional sociology that fails to reveal that it is a political stance (Agger, 1989c, 2000), the public response is a decline in discourse because discourse depends on the assumption that *knowledge, including expert knowledge, is political*:

> We cannot comprehend, and thus remedy, the decline of discourse without considering the dominance of the university over intellectual life. Similarly, any story is incomplete without considering the marketplace in determining authorial choices and chances… *I am pointing to an absence* [of the public] *that is effectively a product of particular social and economic arrangements of knowledge* (Agger, 1990, pp. 33–34).

Burawoy (2004d, p. 126) has completely neglected Agger's earlier conception of public sociology except to say that "Ben Agger's *Public Sociology* is a minute analysis of the pathologies of professional sociology, but says curiously little about public sociology and *its* dilemmas." Yet, read through Agger's lens, Burawoy's (2005a) insistence that public sociology is dependent on professional sociology neglects that professional sociology exercises authority over the public sphere with consequences for the possibility of public sociology.

Agger's (2000) core argument in his conception of public sociology is that when positivist science dominates intellectual production, it dominates a public that, in response, fails to read science as resulting from questions that are themselves political. Thus, for Agger (1990), a reifying division of sociological labor such as Burawoy's (2005a) would be in part responsible for the decline in public discourse. This is so because when professional sociology produces the social as "facts" rather than as a literary acts it is exclusionary by virtue of secretly claiming authorial authority over the social world where such authority could otherwise be claimed by a public that realized that *they too could intervene in the world* (Agger, 1989b, 1989c, 1989d, 1990, 1991, 2000). The loss of the category *alterable* through the declaration of professional boundaries eclipses the possibility of the public participating in the narration of their social world and thus eclipses the possibility of public sociology. Politics scholars would call this governance without transparency. Professional sociology in this way is distinctly anti-discursive in that it excludes the possibility of change and in doing so excludes the public.

The current practice of professional sociology reified in Burawoy's table contributes not only to the decline of public discourse; it also contributes to the decline of public sociologists. Burawoy's (2004a, 2004b, 2004c, 2004d, 2005a, 2005b) claim to a division of labor for sociology reifies an *unnecessary* hierarchy of value imposed according to academia's reliance upon literary political economy,[6] which mirrors the neoliberal state's reliance on the market (Harvey, 2007). It is empirically untrue that appearing in public, even appearing intelligent, is the standard for determination of the value of one's academic labor, which is solely determined by the ranking of the academic journals in which one's writing appears: "[t]he commodification of academic writing structures quite differential career outcomes. Some people's work is 'less' worthy than that of others simply *by virtue of where it appeared, not what is says*" (Agger, 1990, p. 136). Academic survival is determined nearly entirely by a hierarchy of literary value, largely governed by sign-value in the service of academic rankings and tenure committees, as Timothy Luke (1999, pp. 349–350) explains:

> [T]he placement of articles and books become the blue book of one's career or the means for assaying the placement of one's labor in departments, between different universities, or within the discipline itself. Without any other stable measure of value, the systems of

> continuous normalizing judgment typically use obvious indicators of status, like institutional location or professional position to measure worth… 'Where' one publishes and 'how often,' then, clearly are accepted as a definition of 'worth' for ranking in these networks of production.

Academic networks of production are themselves self-legitimating, resulting in a downward spiral toward the decline of discourse: "instead of styling intelligent arguments decided in their own terms, academic writers attempt to clone whatever standards of literary presentation obtain in their own literary marketplaces" (Agger, 1990, p. 137). Within literary political economy of the academic career, the public has *no* influence on the determination of the value of academic texts, which *do* influence public life. This is so precisely because a reifying theory of sociological labor that is based entirely in the stabilization of professionalized discourse is not at all interested in the actually existing public[7] except to the extent that they are fodder for academic publishing, thus rendering the public itself a body of instrumental knowledge to be employed in the market of ideas.

Public sociology, if there is to be such an intellectual practice, would first have to acknowledge that writing and textuality are embedded not in the fictional account of sociological labor that Burawoy has produced, but in literary political economy. This requires that we reject the ontological fiction that Burawoy has imposed upon us and recognize that the extent to which sociologists are able to be public is largely governed by "hierarchies of value over valuelessness…" (Agger, 1989b, pp. 15–16). In contrast to Burawoy's (2004a, p. 105) call for professional sociology to "discipline" critical sociology, Agger (1993) calls for a critically feminist profession that expands the concept of value to include public labor. Public sociology would naturally emerge from such a profession because the hierarchy of valued over valueless (thus professional over public) would disappear, transforming not only sociology, but the entire realm of intellectual value.

An "organic" division of labor such as Burawoy's (2004a, 2004b, 2004c, 2004d, 2005a, 2005b), like a sexual division of labor, "conceals hierarchy underneath naturelike differentiation" (Agger, 1993, p. 142).[8] In Agger's (1993, p. 142) feminist conception, the logic of the division of labor is a hierarchical division, allowing for the valued to dominate the valueless. Burawoy's division of labor, which *differentiates* theory and *specifies* instrumental knowledge, is addressed by Agger in 1991 (p. 103):

> We must challenge the positivist definition of scholarly legitimacy – surveys, computers, grants, value-freedom – and instead insist on the validity of non-instrumental, non-quantitative knowledge. We must argue that theory *matters*, even if it defies easy domestication in the prevailing academic division of labor, which sets up rigid boundaries among specialties.

This is a key point in response to Burawoy's (2004a, 2004b, 2004c, 2004d, 2005a, 2005b) tacit statement that all sociological labor is equally valued; in truth, legitimated labor is valued more in sociology because the "open market of ideas," like the free markets that Burawoy (2008) is critical of, is an illusion (Agger, 1991, p. 103).

The academic world in which sociologists are employed values that which stabilizes the present over that which would require radical transformation (Nickel, 2008). Sociological labor, like all other labor in a capitalist state, is produced within the realm of valued and valueless activity (Agger, 1989b, pp. 60–61). This circuitry is as relevant to Burawoy's division of labor as it is to household divisions of labor. The discipline of sociology assigns value according to which authors it is willing to employ and what authorship it is willing to publish. As Luke (2005, p. 19) explains, "[b]odies of knowledge with perceived fundable potential, interdisciplinary fungibility or existing rankable luster typically are valued over those without high fundability, interdisciplinary openness, or potential for greater luster in national rankings." Public sociology cannot be transformative for any public unless it is also transformative for sociologists. If sociology is truly to go public, it must assign *value* to critical public sociology as something other than pro-professional sociological labor. We can easily rearrange Burawoy's (2005b) boxes to reflect how sociological labor typically functions in the day-to-day management of a junior scholar's career in North America (Table 2).

Neglect of this reality is damaging not only for the public, but for the sociologists themselves who are charged with stabilizing Burawoy's ontological fiction. The top row of Burawoy's table has more value in the academic market place than the bottom row does. This is so because that which is endorsed by one's peers has more value in the discipline of sociology than that which is endorsed by the public.[9] Burawoy (2004a, p. 105) assigns critical sociology a role as the "conscience of professional sociology." The "role" of the conscience of professional sociology is a secondary and servile one: professional sociology still *dominates*, especially given that Burawoy (2004a, p. 105) has assigned professional sociology the

Table 2. Sociology's "Actually Existing"[a] Division of Labor.

	Valued	Valueless
Functional	Professional policy discourse	
Transformative		Critical public intellectuality

[a]See Note 7.

role of *disciplining* critical sociology (Harry Dahms, 2008). Yet, as Richard V. Ericson (2005, p. 366) states quite rightly: "sociologists who do work what Burawoy labels critical – for example, feminism, queer theory and critical race theory – are as rigorously professional in their theories and methods as any others! Furthermore, being critical is a core element of professionalism."

Sociology as it is written by Burawoy no longer explains even the lives of sociologists, which are lived in the nexus of the ontological fiction he has created and the resultant myth that our own lives, careers, and voices exist in the supposed "ivory tower" (Burawoy, 2008, p. 358). As far as I know, sociologists buy their milk in "the real world," and they also attempt to find and keep jobs in a world where it is prohibitively difficult for most people to be critical public theorists and also employed in the production of instrumental policy knowledge. The reason that there are few, if any, *critical* sociologists employed as junior scholars in the field of public policy, or even in sociology, in the United States is that knowledge is political and thus critical knowledge is declared politically illegitimate by those who resist change.

It is naïve for Burawoy (2004a, p. 106) to write that "[g]raduate students may start out as critical sociologists, become professional sociologists, and then later turn to public sociology." Although this seems to have been Burawoy's path, in today's job market, dominated by professional sociology, most graduate students are told that they must start out as professional sociologists and that only when tenured may they may turn to critical and public sociology. Jonathan Imber (2001) specifically warns that Agger (2000, p. 354) gives bad advice when he recommends that graduate students carve out a career as critical public sociologists:

> Agger tries to imagine a world made safe for poetic sociology, even one that speaks of justice to the masses.... In the meantime, his hope for public sociology is potentially perilous advice for those risking a career in academic sociology. If you are lucky to have Ben Agger guiding (or Stanley Fish as a second letter), you might get promoted (or at least hired). The real lament is that so many of the unguided get stuck in what Alfred Whitehead rightly called the groove. And this is *after* tenure."

Although Imber (2001) was writing three years before Burawoy's (2004a) mainstream endorsement of public sociology, Burawoy's assumption that one may easily traverse the division of labor in sociology is specious. Junior scholars and future public sociologists cannot choose both the bottom row and the top row of Burawoy's (2004a) division of labor because the top row stabilizes the discipline (and thus the world) as it is, whereas the bottom row is aimed at transformation. Burawoy might reply that he has retained a space for critical public sociologists, but when academics choose to employ knowledge, they usually employ those whose stance is legitimated and legitimating and thus the top row. Evidence of this fact, the institutionalization of public sociology within the university in the United States has produced curricula that mirror public administration curricula (American University's MA Concentration in Public Sociology, 2007; NASPAA, 2008; Nickel, 2009).

Public sociology stands no chance of emerging from within a fictional division of labor that stabilizes an instrumental rationality that predetermines possibility. Professional sociology, like policy sociology, is based on the illegitimacy of particular forms of knowledge, specifically those that do not fall within its boundaries. Burawoy's institutionalization of knowledge and intellectuals thus involves political choice – what counts as knowledge is politically determined and, in the case of a reifying division of sociological labor, what counts as legitimate sociology is politically determined. As Michael Apple (2003, p. 11) explains, knowledge boundaries create a *constitutive outside*, which acts as a source of legitimacy for official knowledge. The pro-professional public sociology that Burawoy has constructed is based on concrete knowledge boundaries. It is through such exclusions, particularly the official exclusion of knowledge that power functions. "Out of the vast universe of possible knowledge, only some knowledge and ways of knowing it get declared to be legitimate or 'official.'… official knowledge is the result of conflicts and compromises both within the state and between the state and civil society" (Apple, 2003, p. 7). It is only through the declaration of public and critical sociology as *unprofessional* sociology that professional sociology can be maintained. Sociology's professionalism is constituted through a self-determined boundary of legitimacy, as Burawoy (2005a, p. 10) declared when he wrote that "there can be neither policy nor public sociology without a *professional sociology*." What Burawoy (2005a) has done is constitute the legitimacy of professional sociology through the determination of all other sociologies as somehow illegitimate by virtue of *not* being professional sociology.

From Fiction to Function

Like all dichotomies, Burawoy's boxes tell only half of the story. As Gouldner (1970, p. 12) pointed out, "[s]ociology has a dialectical character and contains both repressive and liberative dimensions," and as Foucault (1995) recognized, discipline is practiced exactly through binaries such as professional/unprofessional. If one reads between the lines, Burawoy has constructed a hierarchy in which critical sociology cannot be professional sociology, public sociology cannot be critical sociology, and policy sociology cannot be public sociology. Unpacked, Burawoy's boxes spill forth two more false assumptions: that there is a clean boundary between an academic audience and extra-academic audience and that there is a *necessary* boundary between the profession, criticism, reflexivity, and practice.

The false assumption of clean boundaries[10] stems from Burawoy's failure to recognize that the world is already influenced by positivism and instrumental knowledge. This was Agger's (1989a, p. 304) point when he argued that soci(onto)ology is "world-constitutive." Where professional sociology creates a positivist culture (Agger, 1989b), the extra-academic audience is *already constituted* by the academic audience. Burawoy incorrectly assumes that only knowledge that is *intended* for an extra-academic audience influences the extra-academic world. When Burawoy (2004b, p. 1607) writes that *"[p]ublic sociology* engages publics beyond the academy in dialogue about matters of political and moral concern," he fails to recognize that political and moral concerns are already determined by a public sphere that gives more value to instrumental knowledge than to reflexive knowledge. According to this assumption, policy intellectuals cannot be critical or reflexive, critical and public intellectuals are not policy intellectuals, and policy and the public are divorced.

The false assumption of a *necessary* boundary between profession, criticism, and reflexivity assumes that professional sociology is not already political. Knowledge and practice are political for professional and policy sociology just as they are for critical and public sociology. Read outside of their neatly packed boxes, Burawoy's sociological laborers risk falling into the following four categories:

- the unreflexive and unprofessional policy intellectual producing instrumental knowledge for an extra-academic audience;
- the unreflexive professional intellectual producing knowledge for an academic audience;
- the non-public critical intellectual producing reflexive knowledge for an academic audience;
- the unprofessional public intellectual producing reflexive knowledge for an extra-academic audience.

Public sociology needs to start with an understanding of how professional and policy sociology are *valued* in relationship to critical and public sociology and thus how power is hierarchically distributed within the field. Agger's (1990, p. 79) theory of literary political economy makes nonsense of Burawoy's division of labor: "[s]imply put, writers to write must make massive compromises with the profit and tenure imperatives of their employment in order to survive." I take from Agger that Burawoy is overly optimistic to think that most sociologists will dedicate their careers to "public labor" as long as he subjugates public and critical labor to "professional" and "policy" labor. No one does public sociology for tenure. Those who do public sociology do it for the public and often at cost to their "professional" career; as Derber rightly (2004, p. 119) points out: "Burawoy's assumption of an essential complimentarity among the four sociologies masks deep tensions among them." This is precisely so because public sociology, according to Burawoy (2004a, 2004b, 2004c, 2004d, 2005a, 2005b), is *not professional sociology* and thus does not "count" for advancement within the profession.

Burawoy's division of labor does a suspicious amount of what Abrams (1988, p. 76) calls "legitimation work" on behalf of professional sociology:

> what is being legitimated is, we may assume, something which if seen directly and as itself would be illegitimate, an unacceptable domination. Why else all the legitimation work? The state, in sum, is a bid to elicit support for or tolerance of the insupportable and intolerable by presenting them as something other than themselves, namely, legitimate, disinterested domination.

As we shall see in the following section, there is little distinction between Burawoy's legitimation of professional sociology and the legitimation of the neoliberal state.

PRO-PROFESSIONAL PUBLIC SOCIOLOGY'S CONTRIBUTION TO NEOLIBERALISM

The possibility of public sociology emerging from sociology's actually existing division of labor is only one part of the ontological fiction that Burawoy has constructed. The second phase of Burawoy's (2008, p. 354) campaign for pro-professional public sociology is an argument for a sociology positioned against "third-wave marketization"[11] in concert with civil society, by which sociology "lives and dies."[12] This stance against neoliberalism, though laudable, cannot emerge from Burawoy's ontological fiction, which stabilizes instrumental knowledge. If it is true that sociology lives and dies by civil society, it is also true neoliberalism practiced by the capitalist state lives and dies by the instrumental knowledge that Burawoy defends. Embedded within Burawoy's (2005a) division of labor are an instrumental theory of knowledge (professional and policy sociology), an affirmative theory of the liberal capitalist state (policy sociology), and several theories of intellectual labor; however, the relationship between knowledge, the neoliberal state, and intellectual labor and their amalgamated influence on public life is unproblematic for Burawoy. This is to say, public life is governed by instrumental knowledge produced by intellectuals in relationship with the neoliberal state, but Burawoy disguises this relationship through the creation of governing boundaries that are based on affirmation of the very instrumental knowledge that sustains the problems which he claims to oppose.

This is important because, given that Burawoy's call for public sociology is in fact a call for the legitimacy of professional sociology, it is not possible for us to consider his argument for public sociology independent of the ways in which professional and policy sociology – instrumental knowledge – govern through what Timothy Mitchell (2002) calls techno-politics and "rule by experts." Although Burawoy has separated professional and policy sociology in terms of their audience, he has maintained that they share a legitimate instrumental orientation towards public life. It is therefore necessary for us to consider the relationship between instrumental knowledge, which Burawoy supports, and the neoliberal state, which Burawoy claims to oppose. Instrumental knowledge stabilizes the logic of productivity that rules the market, the state, and, as I demonstrated earlier, academic life. Neoliberalism, what Burawoy has called "third-wave marketization" is sustained exactly by instrumental knowledge as the basis for the "naturalness" of productivity, what Lyotard (1997) called performativity.

My argument that instrumental knowledge governs possibility on behalf of the market builds on the Frankfurt School's critique of instrumental knowledge as the instrument of domination. Burawoy's attempt to achieve liberation from the market through the stabilization of instrumental rationality is untenable, as Max Horkheimer (1974, p. 21) argued in the *Eclipse of Reason*, "reason has become an instrument…completely harnessed to the social process. Its operational value, its role in the domination of men and nature, has been made the sole criterion." Burawoy does not understand that the domination of nature and the neoliberal state stem from the same thing: the instrumental knowledge that he defends, not once, but at least seven times (2004a, 2004b, 2004c, 2004d; 2005a, 2005b, 2005c). That Burawoy (2007, 2008) bases the second phase of his campaign in Karl Polanyi's *The Great Transformation* is specious given that Polanyi (2001, p. 41) himself viewed instrumental rationality employed as an industrial revolution as a "catastrophe." Burawoy's (2008) further criticism of the commodification of nature is similarly oblivious to the fact that, as Horkheimer and Theodor W. Adorno (1989, pp. 3–4) wrote, domination of nature is the outcome of instrumental knowledge, or enlightenment rationality, which results in totally administered society. Instrumental knowledge cannot coexist with liberation: "[t]he program of the enlightenment was the disenchantment of the world; the dissolution of myths and the substitution of knowledge for fancy…the concordance between the mind of man and the nature of things that he had in mind is patriarchal: the human mind, which overcomes superstition is to hold sway over a disenchanted nature. Knowledge, which is power, knows no obstacles."

The moral stance that Burawoy asks for public sociologists to take against the market is *impossible* in relationship to instrumental rationality, which requires all non-instrumental rationality to be "secondary, minor, or exceptional" (Mitchell, 2002, p. 300). In his study of development (instrumental rationality) in Egypt, Mitchell (2002, p. 15) finds that the "the binarisms fixed in place in modern politics open up the distance that requires and enables [expertise]…the place and claims of expertise are constituted in the separation that seems to open up, opposing nature to technology, reality to its representation, objects to their value…." A binary is precisely what Burawoy has achieved in his division of labor for sociology: a division that allows for a separation between public sociology's moral stance and professional sociology's instrumental knowledge. As Gouldner (1970, p. 13) argued of instrumental rationality, it requires "men to be moral cretins in their technical roles…choosing to ignore or to value other meanings and consequences of theories…in effect refusing to take responsibility for them even if they do exist…. The social function of such segmented role structure…is to sever the normal moral sensibilities."

Although Burawoy claims to oppose the market (2007) and neoliberalism (2008), the neoliberal state is fortified by his division of labor, which protects the validity of the instrumental knowledge upon which the neoliberal state is dependent for its emphasis on productivity, which is exactly the rationale used to support the shift from "welfare to workfare," which is just the recommodification of well-being which used to be decommodified by the welfare state (Esping-Andersen, 1990). As the welfare state was on the rise Gouldner (1970) identified the relationship between sociology and the welfare state as one of mutual fortification. As the welfare state is dismantled by neoliberalism, we are witnessing a deepening of this mutual fortification of sociology and the state's approach to well-being in Burawoy's ontological fiction, which, through the instrumentality of knowledge, absolves policy sociologists of responsibility for moral sensibility, now solely the responsibility of pro-professional public sociology.

Professional sociology is not autonomous from neoliberalism. In failing to recognize the politics of knowledge and its moral consequences, Burawoy fails to recognize that the very marketization of which he is critical is possible only through a particular orientation toward knowledge as it is produced by professional sociology, which he is *unwilling to criticise*. Every critical theorist since Marx has recognized the relationship between instrumental knowledge, the state, and the market (Aronowitz & Bratsis, 2002; Barrow, 1987, 1990; Foucault, 1978, 1980, 1995; Gouldner, 1970; Gramsci, 1997; Habermas, 1975; Harney, 2002; Horkheimer & Adorno, 1989; Lyotard, 1997; Mandel, 1972; Marcuse, 1991; Miliband, 1969; Mitchell, 1991, 2002; Offe, 1996; Polanyi, 2001; Poulantzas, 1968, 1969). Neoliberalism silences moral stances such as those advocated by Burawoy, through the assertion of the primacy of instrumental rationality, such as that advocated by Burawoy. Yet, Burawoy has asked us to base our moral stance against neoliberalism in the primacy of professional sociology and its associated instrumental rationality.

THE POSSIBILITY OF TRANSFORMATIVE PUBLIC SOCIOLOGY

Public sociology is not possible without insisting on the probability that professional sociology's emphasis on positivism and instrumental knowledge is contributing to the very problems that public sociology claims to want to end. A public sociology that argues on behalf of the possibility of transformation is possible, but it will not emerge from the ontological fiction that Burawoy has constructed because it is precisely through such fictions that the neoliberal state functions in concert with instrumental intellectual labor to govern public imagination. As Horkheimer and Adorno (1989, p. xiv) argued: "It is characteristic of the sickness that even the best-intentioned reformer who uses an impoverished and debased language to recommend renewal, by his insidious mode of categorization and the bad philosophy it conceals, strengthens the very power of the established order he is trying to break. False clarity is only another name for myth; and myth has always been obscure and enlightening at one and the same time…"

Critical knowledge cannot be transformative if is written to be just another incidental by product of an "organic" division of labor. A division of labor for sociology that stabilizes the production and employment of instrumental knowledge by professional and policy sociologists is not an argument on behalf of public sociology; it is an argument on behalf of the neoliberal rationale for legitimate governing. Public sociology must first tell the truth about the *possibility* for sociologists dominated by professional sociology to stand apart from instrumental knowledge and its employment by the state apparatus, demonstrating to people that it is *possible* to legitimate the knowledge gleaned from their own lived experiences in the world, which are increasingly out of synch with sociology's professional explanations.

NOTES

1. Despite Burawoy's global campaign, my understanding of professional sociology is based on mainstream American sociology. The phrase conveys very little in places such as New Zealand, where I teach a public sociology course in which Burawoy's argument struggles for relevance.

2. "C. Wright Mills (1959), and many others since him, would turn all sociology into public sociology. Mills harks back to the late 19th century forefathers, for whom scholarly and moral enterprises were indistinguishable. There is no turning back, however, to that earlier period before academic revolution. Instead we have to move forward and work from where we really are, from the division of sociological labor" (Burawoy, 2004a, p. 9).

3. The current disagreement between Agger and Burawoy can be traced to a 1989 debate between these two authors in the *Berkeley Journal of Sociology* about the "crisis of Marxism."

4. Burawoy's second phase of public sociology begins in 2006, with the publication of "A Public Sociology for Human Rights," followed by his 2008 publication of eight additional theses.

5. "Policy sociology" achieves mainstream use in the United States with Burawoy's (2004a) use of the phrase to describe consulting work or knowledge produced for the state. Of course, policy sociology has been in place for decades as an academic field known as public administration and policy. See Nickel (2009) for an overview of the confusion between public and policy sociology and public administration and policy.

6. Literary political economy refers to the commodification of writing and, by extension, the commodification of the academic career at the expense of the actually existing public: "Like any commodity form, academic writing is given value through exchange; in effect, academics themselves decide the value publication in

particular outlets is to have" (Agger, 1990, p. 136). Burawoy (2008, p. 359) recently acknowledged this point, although he does not attribute it to Agger or anyone else. However, he has only recognized it. He has offered no explanation of the implications for his division of labor.

7. I borrow the phrase "actually existing public" from Nancy Fraser's (1992) use of the phrase "actually existing democracy."

8. See Glenn (2007) for an outstanding feminist discussion of how Burawoy's division of labor through logics of exclusion cements hierarchy and marginalization.

9. See Judith Stacey (2007) for an excellent discussion of alternatives.

10. Although his overall argument is not necessarily compatible with my own, Orlando Patterson (2007, p. 176) rightly points out that Burawoy constructs "falsely crisp sets and categories."

11. Burawoy uses third-wave marketization and neoliberalism interchangeably. He first makes this argument as "public sociology vs. the market" in 2007.

12. It is important to note that civil society is a term that stabilizes the state, and it is deeply contested by critical sociologists (see Neocleous, 1995; Paolucci, 2008).

REFERENCES

Abrams, P. (1988). Notes on the difficulty of studying the state. *Journal of Historical Sociology*, *1*, 58–89.

Agger, B. (1989a). *Socio(onto)logy: A disciplinary reading*. Urbana, IL: University of Illinois Press.

Agger, B. (1989b). *Fast capitalism: A critical theory of significance*. Urbana, IL: University of Illinois Press.

Agger, B. (1989c). *Reading science: A literary, political, and sociological analysis*. Dix Hills, New York: General Hall, Inc.

Agger, B. (1989d). Is Wright wrong (or should Burawoy be buried)?: Reflections on the crisis of the "crisis of Marxism". *Berkeley Journal of Sociology*, *33*, 187–207.

Agger, B. (1990). *The decline of discourse: Reading, writing and resistance in postmodern capitalism*. New York: The Falmer Press.

Agger, B. (1991). *A critical theory of public life: Knowledge, discourse, and politics in an age of decline*. New York: RoutledgeFalmer.

Agger, B. (1992). *The discourse of domination: From the Frankfurt School to postmodernism*. Evanston, IL: Northwestern University Press.

Agger, B. (1993). *Gender, culture, power: Toward a feminist postmodern critical theory*. Westport, CT: Praeger.

Agger, B. (2000). *Public sociology: From social facts to literary acts*. New York: Rowman and Littlefield.

Agger, B. (2007). *Public sociology: From social facts to literary acts* (2nd Ed.). New York: Rowman and Littlefield.

American University's MA Concentration in Public Sociology. Available at http://www.american.edu/cas/sociology/maps/. Retrieved on November 22, 2007.

Apple, M. W. (Ed.) (2003). *The state and the politics of knowledge*. New York: RoutledgeFalmer.

Aronowitz, S., & Bratsis, P. (Eds). (2002). *Paradigm lost: State theory reconsidered*. Minneapolis: University of Minnesota Press.

Barrow, C. W. (1987). Intellectuals in contemporary social theory: A radical critique. *Sociological Inquiry*, *57*, 415–430.

Barrow, C. W. (1990). *Universities and the capitalist state: Corporate liberalism and the reconstruction of American higher education, 1894–1928*. Madison: University of Wisconsin Press.

Beck, U. (2005). How not to become a museum piece. *British Journal of Sociology*, *56*(3), 335–343.

Brodkey, L. (1987). *Academic writing as social practice*. Philadelphia: Temple University Press.

Burawoy, M. (2004a). Introduction, public sociologies: A symposium from Boston College. *Social Problems*, *51*, 103–106.

Burawoy, M. (2004b). Public sociologies: Contradictions, dilemmas, and possibilities. *Social Forces*, *82*, 1–16.

Burawoy, M. (2004c). The world needs public sociology. *Sosiologisk tidsskrift (Journal of Sociology, Norway)*, *3*. Available at http://sociology.berkeley.edu/faculty/burawoy/burawoy_pdf/PS.Norway.pdf. Retrieved on November 21, 2005.

Burawoy, M. (2004d). Manifesto for public sociologies. *Social Problems*, *51*, 124–130.

Burawoy, M. (2005a). 2004 American sociological association presidential address: For public sociology. *American Sociological Review*, *70*(1), 4–28.

Burawoy, M. (2005b). Response: Public sociology: Populist fad or path to renewal. *British Journal of Sociology*, *56*(3), 417–432.

Burawoy, M. (2005c). The critical turn to public sociology. *Critical Sociology*, *31*(3), 313–326.

Burawoy, M. (2006). Introduction: A public sociology for human rights. In: J. Blau & K. I. Smith (Eds), *Public sociologies reader* (pp. 1–18). Lanham, MD: Rowman & Littlefield Publishers, Inc.

Burawoy, M. (2007). Public sociology vs. the market. *Socio-Economic Review*, *5*(2), 356–367.

Burawoy, M. (2008). What is to be done? Theses on the degradation of social existence in a globalizing world. *Current Sociology*, *56*(3), 351–359.

Dahms, H. (2008). How social science is impossible without critical theory: The immersion of mainstream approaches in time and space. In: H. Dahms (Ed.), *Current perspectives in social theory* (Vol. 25, pp. 3–61). Bingley, UK: Emerald Group Publishing Limited.

Delanty, G. (2006). Cosmopolitan citizenship. In: J. Blau & K. E. I. Smith (Eds), *Public sociologies reader* (pp. 37–50). Lanham MD: Rowman & Littlefield Publishers, Inc.

Derber, C. (2004). Public sociology as a vocation. *Social Problems*, *51*, 119–121.

Ericson, R. (2005). Publicizing sociology. *British Journal of Sociology*, *56*(3), 365–372.

Esping-Andersen, G. (1990). *Three worlds of welfare capitalism*. New Jersey: Princeton University Press.

Foucault, M. (1991/1978). On governmentality. In: G. Burchell, C. Gordon & P. Miller (Eds), *The foucault effect* (pp. 87–104). Chicago: The University of Chicago Press.

Foucault, M., & Gordon, C. (Eds & Trans.), Marshal, L. (Trans.), Mepham, J. (Trans.) & Sopher, K. (Trans.) (1980). *Power/knowledge: Selected interviews and other writings 1972–1977*. New York: Pantheon Books.

Foucault, M., & Sheridan, A. (Trans.) (1995). *Discipline and punish: The birth of the prison*. New York: Vintage Books.

Glenn, E. N. (2007). Whose public sociology? The subaltern speaks, but who is listening? In: D. Clawson, R. Zussman, J. Misra, N. Gerstel, R. Stokes, D. L. Anderton & M. Burawoy (Eds), *Public sociology: Fifteen eminent sociologists debate politics and the profession in the twenty-first century* (pp. 213–230). Berkeley: University of California Press.

Gouldner, A. W. (1970). *The coming crisis of Western sociology*. New York: Basic.
Gramsci, A., Hoare, Q. (Trans.) & Smith, G. N. (Trans.) (1997). *Selections from the prison notebooks*. New York: International Publishers.
Habermas, J. (1975). *Legitimation crisis*. Boston: Beacon Press.
Harney, S. (2002). *State work: Public administration and mass intellectuality*. Durham and London: Duke University Press.
Harvey, D. (2007). *A brief history of neoliberalism*. Oxford: Oxford University Press.
Horkheimer, M. (1974). *Eclipse of reason*. New York: Seabury.
Horkheimer, M., & Adorno, T. W. (1989). *The dialectic of enlightenment*. New York: Continuum.
Imber, J. (2001). Public sociology: From social facts to literary acts (review). *Social Forces*, *80*(1), 353–354.
Luke, T. W. (1999). The discipline as disciplinary normalization: Networks of research. *New Political Science*, *21*, 345–363.
Luke, T. W. (2005). From pedagogy to performativity: The crises of research universities, intellectuals and scholarly communication. *Telos*, *131*, 13–32.
Lyotard, F., Bennington, G. (Trans.) & Massumi, B. (Trans.) (1997). *The postmodern condition: A report on knowledge*. Minneapolis, MN: University of Minnesota Press.
Mandel, E. & De Bres, J. (Trans.) (1972). *Late capitalism*. London: NLB Humanities Press.
Marcuse, H. (1991). *One-dimensional man: Studies in the ideology of advanced industrial society*. Boston: Beacon Press.
Merton, R. K. (1968). *Social theory and social structure*. Glencoe, IL: The Free Press.
Miliband, R. (1969). *The state in capitalist society*. New York: Basic Books, Inc.
Mills, C. W. (1959). *The sociological imagination*. New York: Oxford University Press.
Mitchell, T. (1991). The limits of the state: Beyond statist approaches and their critics. *The American Political Science Review*, *85*, 77–96.
Mitchell, T. (2002). *Rule of experts: Egypt, techno-politics, modernity*. Los Angeles: University of California Press.
National Association of Schools of Public Affairs and Administration (2008). Available at http://www.naspaa.org/accreditation/seeking/reference/standards.asp. Retrieved on January 5, 2008.
Neocleous, M. (1995). From civil society to the social. *The British Journal of Sociology*, *46*, 395–408.
Nickel, P. M. (2008). There is an unknown on campus: From normative to performative violence in academia. In: B. Agger & T. Luke (Eds), *Tragedy and terror at Virginia Tech: There is a gunman on campus* (pp. 161–186). Lanham: Rowman & Littlefield.
Nickel, P. M. (2009). Public administration and/or public sociology: Disciplinary convergence and the disciplinary dispersal of public sentimentality. *Administration and Society*, *41*(2), 185–212.
Nickel, P. M., & Eikenberry, A. M. (2007). Responding to "natural" disasters: The ethical implications of the voluntary state. *Administrative Theory & Praxis*, *29*(4), 534–545.
Offe, C. (1996). *Modernity and the state: East, West*. Cambridge, MA: The MIT Press.
Paolucci, P. (2008). Public sociology, marxism, and marx. In: H. Dahms (Ed.), *Current perspectives in social theory 25*, 353–382.
Patterson, O. (2007). About public sociology. In: D. Clawson, R. Zussman, J. Misra, N. Gerstel, R. Stokes, D. L. Anderton & M. Burawoy (Eds), *Public sociology: Fifteen eminent sociologists debate politics and the profession in the Twenty-first Century* (pp. 176–194). Berkeley: University of California Press.

Perrucci, R., Ferraro, K., Miller, J. A., & Muschert, G. (Eds) (2008). *Agenda for social justice: Solutions 2008*. Available at http://www.sssp1.org/File/Agenda_For_Social_Justice_2008.pdf

Piccone, P. (1978). The crisis of one-dimensionality. *Telos, 35*, 43–54.

Polanyi, K. (2001). *The great transformation: The political and economic origins of our time*. Boston: Beacon Press.

Poulantzas, N. (1968). *Political power and social classes*. London: New Left Books.

Poulantzas, N. (1969). The problem of the capitalist state. *New Left Review, 58*, 67–78.

Said, E. W. (1994). *Representations of the intellectual*. New York: Vintage Books.

Stacey, J. (2007). If I were the goddess of sociological things. In: D. Clawson, R. Zussman, J. Misra, N. Gerstel, R. Stokes, D. L. Anderton & M. Burawoy (Eds), *Public sociology: Fifteen eminent sociologists debate politics and the profession in the twenty-first century* (pp. 91–100). Berkeley: University of California Press.

PEIRCE, PRAGMATICISM AND PUBLIC SOCIOLOGY: TRANSLATING AN INTERPRETATION INTO PRAXIS

J. I. (Hans) Bakker

There is nothing so practical as a good theory. (Lewin, 1951, p. 169; see Kant 1999 [1793])

And in academia, even the most 'radical' seem to have no problem with the triumph of the market ideology … (Frank & Mulcahey, 2003, p. 7)

It also shows how Einstein's liberating ideas in physics derive entirely from his profound commitment to Zionist praxis. (Empiricus, 2001, in Collins, with ironic intent, in Seidman & Wagner, 1992, p. 180)[1]

ABSTRACT

This paper endeavours to translate Peirce's ideas into paradigmatic sociological theory in general, and praxis linked to that paradigmatic theory. Greater comprehension of the general usefulness of a Peircian Pragmaticist semiotic perspective (and 'fallibilism') will enhance social action and collective responsibilities. As a philosopher, C. S. Peirce interpreted the world, but he himself did not attempt to change it. How can we incorporate a unifying 'perspective' such as Peirce's theory into

Nature, Knowledge and Negation
Current Perspectives in Social Theory, Volume 26, 229–257

ISSN: 0278-1204/doi:10.1108/S0278-1204(2009)0000026012

'method' (both methodology and techniques)? How does that an emphasis on methodology improves sociology as an empirically based, rigorous discipline (a human 'science') and various forms of praxis, especially applied sociology and public sociology? If politics is going to have any long-term impact, there has to be a degree of sophistication about the theories involved (e.g. conservation of fundamental human rights and liberties, reform within liberal parliamentary democracies, transformation within neo-conservative regimes, applied sociology in professions and occupations like social work and criminology, Feminist critique and action, GLTB action, neo-Gramscian critiques and Michael Burawoy's 'public sociology' in the narrow, technical sense.) Praxis can include many forms of political activity that are not specifically informed by any well-developed and coherent theory. In terms of Peirce's Pragmaticist semiotics, we can translate theoretical awareness into praxis, to interpret and to change the world. Good theory and methodology is the most practical way to promote useful social action, applied sociology and public sociology. Anything less than Peircian Pragmaticism and semiotics tends to lead to fragmentation of 'paradigms' or postmodernist nihilism.

1. INTRODUCTION: A TRIADIC EPISTEMOLOGY AND A META META-PARADIGMATIC PERSPECTIVE

There is nothing as practical as a good theory. That idea, found in Kant (1999 [1793]), is often attributed to Kurt Lewin (1951, p. 160). If we have a bad theory, we may be climbing up the wrong ladder. That is, if the theory misleads us, we may head into the wrong kinds of policy decisions. If we regard the social problems that confront modern (and postmodernist aspects of) contemporary industrialized societies through the wrong lenses, then we will never be able to solve those problems. Indeed, we may even exacerbate them. We have seen time and again that an incomplete understanding of a situation leads to disasters. That has been true in terms of waging wars in regions and countries that are not well understood by the military planners. It has been true internationally and it has been true domestically in the United States. We can never achieve a more just society, much less a more just global civilizational order, if we do not have the correct sociological vision. We need to have excellent sociological theories. We can 'test' the heuristic value of those theories for their heuristic value in research, but ultimately, we have to do a further testing.

We have to see if the 'interpretations' of the world help us to change 'the world' (as a whole and in terms of national and other problems). When Marx scribbled down his theses on Feuerbach, he pointed out that philosophers have only interpreted the world. He did not mean to imply, I would guess, that interpretation of 'the world' is not necessary. He would not have spent so much time writing social theory (e.g. the political economy of modern capitalism) if he had felt that all we need to change things for the better is to simply forge ahead in any direction that seems intuitively attractive. The point is to change 'it', but we cannot change 'the world' (or any aspects of the world) unless we have 'good theory'.

Yet, today sociology as a discipline is – as Phillips (1991) has pointed out – a kind of Tower of Babel, a 'chaos' with very little integration (Turner, 2006). There is a cacophony of voices, as exemplified in the various 'sections' of the American Sociological Association (ASA). A member of the ASA may belong to more than one section, but no one can communicate effectively in the language of all of the sections, much less participate in all the myriad activities of so many highly divergent networks. Many of the different theoretical and methodological paradigms are seen as incommensurate.

A concrete example is the debate concerning so-called pure sociology. The term pure sociology is highly misleading. It essentially refers to the attempt to theorize about structural phenomena without any resort to ontological individuals' beliefs and even without any form of 'methodological' (or, epistemological) individualism. 'Pure sociology ... simply means sociology without psychology' (Michalski, 2008, p. 267). The 'middle range' and 'research' theory that comes out of 'pure' sociology even tries to do without any kind of 'sociological psychology' and hence is criticized by some, but not always dispassionately (e.g. Marshall, 2008). To use both social psychological and structuralist generalizations: (1) people get very passionate about their paradigmatic assumptions and (2) 'paradigmatic assumptions as structures' tend to be correlated with 'in-group solidarity' as a group-level phenomenon!

That is well illustrated by a recent analysis of 'Positivism' post-WWII (as a structure within sociology as a discipline and as a social psychological belief system). Philip Mirowski (2005) develops a persuasive argument that many European-American theorists were influenced by the challenges of WWII and were drawn to Positivistic views quite contrary to the American Pragmatism of the period between WWI and the beginning of WWII (1918–1933). The 'military-industrial complex' famously evoked in Eisenhower's speech tended to promote scientism. Mirowski (2005, p. 145, Table 1) uses John Dewey as an exemplar of Pragmatism and a defender of an approach associated with institutional economics, historicism and pluralistic democracy. The post–WWII trends are associated with a wave of recent immigrants who were forced to leave Europe. Hans Reichenbach is discussed as an exemplar of 'Logical Empiricism', decision theory (DT) and operations research (OR) (Mirowski, 2005, pp. 158–169). The term 'democracy' was endowed with an entirely different meaning than that which had originally been intended by Dewey and other Pragmatists. During and after WWII, 'democracy' became the social psychological idea of citizens as voters who as 'rational individuals' could follow epistemologically grounded, unimpeachable scientific truths and therefore avoid errors. The ordinary citizen, therefore, had to follow the paternalistic, scientific guidance of 'science', particularly the natural scientists and mathematicians. The idea of science itself became restricted to physics and nomothetic laws (cf. Michalski, 2008, p. 264).[2] Dewey was re-interpreted to be saying things he probably never intended. Within the academic discipline of philosophy, the ideas of the American Pragmatists were dropped in favour of 'analytical philosophy' as a way of emulating the Logical Empiricism and Scientism of postwar researchers (Mirowski, 2002, 2005, 2006).

There was some resistance to a Positivist agenda after the war. It is definitely not the case, for example, that all foundations and German émigrés favoured naïve Behaviouristic views. One 'black swan' in that regard is the International Programs section of the Rockefeller Foundation under Kenneth W. Thompson. Thompson was a student of Hans Morgenthau, a political scientist who sought to develop a theory of international relations (IR) that was opposed to the dominant Behavioural trend associated with the kind of 'variable analysis' advocated by thinkers such as Paul Lazarsfeld, Daniel Lerner and Arnold Wolfers. Ironically, the 'grand theorist' Talcott Parsons, so heavily criticized by George Homans for lack of scientific rigor and by C. Wright Mills for a pigeonholing approach, was also prone to accept the generally Positivistic thrust in the philosophy of science and the philosophy of social science. Behaviouralism exerted a significant pull over postwar social scientists. The Positivism that was rampant after the war involved a subtle, largely unstated, rejection of the more democratic tendencies among American Pragmatists between the two world wars. Although some of the opponents tried to use the language of Positivistic science while at the same time rejecting Positivistic epistemological assumptions, the general outcome was a rejection of approaches that were deemed to be unscientific. Among philosophers, it was felt by many that an analytical approach was far superior to the pre-war Pragmatist approach.

Of course, in the late 1960s and early 1970s, all of the social science disciplines other than economics experienced a significant shift in Meta-Paradigmatic thinking. The 'crisis' that Gouldner discussed was an epistemological crisis. Those who opposed Parsonian structural-functionalism and Lazarsfeldian empiricism were united in their opposition to a Positivistic approach. But that was pretty much all they were united by: opposition to the consensus that prevailed immediately after WWII. Humpty Dumpty fell off the wall, and the social sciences have not yet been reassembled. (Economics, again, is the exception, since 'neo-classical' marginal utility analysis still dominates completely in most academic departments.) It is a 'Tower of Babel' in sociology, psychology, social psychology, political science, geography and anthropology.[3]

At sociology conferences, people are polite to one another in a superficial manner, but subtle forms of symbolic interaction and socio-economic 'exchange' are used to exclude opinions that are regarded as not pertinent. Ideas that are not seen as relevant are not really debated so much as simply ignored.

I believe that a semiotic approach, based on Charles Sanders Peirce's Pragmatism (or, Pragmaticism), is helpful in bridging some of the gaps in discourse within the discipline of sociology. I believe that if the discipline is going to provide good interpretations of the world – theoretical and methodological insights that can lead to good empirically based research findings – then we have to find an epistemological perspective that trumps the various limited epistemologies commonly shared by different groups (e.g. sections of the ASA). Rather than Positive, Interpretive, Critical, Feminist and Post-Modern Meta-Paradigms conflicting with one another, we need to find a way for those radically different perspectives to find some common ground.

That common ground is a thorough understanding of the limitations of the Cartesian perspective, particularly Rene Descartes' subject–object (mind–body) dualism. We need to move beyond the simplistic Cartesian way of thinking about the 'subject' versus the 'objective' and 'subjectivity' versus 'objectivity'. The way to do that is to comprehend that there are no concrete material subjects or objects that can be discussed in a meaningful way using language. There is a real material world, of course. But when we use words and numbers to try to interpret that world, we quickly get lost in limited 'pragmatics' and 'semantics'. A complete understanding of any symbols requires a semiotic awareness of the way signs can only have a clear meaning in the context of an interpretive community or 'network'. The Interpretive Network (IN) decides how the words will be applied. If the IN of all Critical Theorists finds the ideas of the IN of all Neo-Functionalist Positivist sociological theorists incommensurable, then it is time to go back to the root meanings of key concepts. To move beyond the limitations of words and mathematical (statistical) signs, we need to situate every theoretical statement in its intellectual context.

We have a tendency, for example, to discuss the ideas of Marx, Weber, Durkheim, Simmel, Mead, Pareto, Nightingale, Martineau, DuBois, Park, Blumer, Goffman, Adorno, Horkheimer, Marcuse, Benjamin, Derrida, Deleuze, Dewey and hundreds of others as if they were all participating in the same conversation, in the same room.

But they did not all live in the same time and same place. They lived in widely different linguistic and political economic contexts. We cannot take a text from 1919 that was originally presented in German and then regard the English language translation as a completely relevant document for 2019. In one hundred years, across an ocean, in a completely different comparative and historical context, the words have nuances of meaning. Just as it is almost impossible to 'translate' a joke from German into English, it is incredibly difficult to fully capture the nuances of subtle theoretical insights. Time and place really matter. All 'texts' need to be 'read' in 'context'. Every statement – verbal or written – is a socio-cultural, political economic singularity. We can certainly try to generalize from a translation of a text by Derrida, but we must always remember when, where and how that text was originally delivered. Knowing, for example, that Derrida, despite his fame in literary circles in the United States (e.g. Yale), was still somewhat of an outsider in the rarefied intellectual circles of Paris helps one to think of some of Derrida's more shocking statements a bit more clearly.

The idea of a triadic semiotic epistemology in Peirce is so radical that it is very hard to make it clear to people what it means for theory and methodology. I have had long conversations with people where I found out in the end that we were still stuck in a lack of adequate conceptualization of the difference between a Cartesian and a Peircian epistemology (and subsequent ontology). I am not just saying that we can have inter-disciplinarity, although such inter-disciplinarity certainly does require going back to key concepts. What I am saying also goes far beyond the idea that we can use triangulation of quantitative and qualitative research designs, combining the two. If theorists with different Meta-Paradigms are ever going to communicate at a really meaningful level, then we need to have a Meta Meta-Paradigmatic framework for such discussions. That Meta Meta-Paradigmatic framework, I believe, is C. S. Peirce's triadic epistemology, which is the key to his semiotics and his Pragmaticism. It is precisely because he wanted to escape from a subject–object Cartesian framework that he eventually named his perspective Pragmaticism, a word so ugly that no one will steal it. He wanted to differentiate himself from William James. James should be acknowledged as a very important contributor to American Pragmatism, but even James did fully grasp the central idea of a triadic epistemology. James continued to think in terms of the individual. He was a psychologist who conceptualized the individual human being as a 'subject'. In so far as a latent Cartesianism exists in James' theories, it is Pragmatism, but not necessarily Pragmaticism. The same can be said for John Dewey. (What is not entirely clear is to what extent George Herbert Mead really escapes this dilemma.)

While it is well known that American Pragmatism (e.g. William James, John Dewey and G. H. Mead) influenced Symbolic Interactionism (SI), it is not as well known that Peirce's version of Pragmatism (i.e. Pragmaticism) cannot be separated from his mature version of Semiotics. There is no specifically 'Semiotic SI' today, although there is a 'social semiotics' that does not quite get at the same underlying epistemological questions. Many 'Interactionists' are apolitical, in part because they emphasize the study of interaction and de-emphasize the study of semiotic signs. Most research in SI is based on an acceptance of the existing social construction of reality. Many excellent descriptive, ethnographic studies based on 'fieldwork' have been done. But they have tended to simply accept the status quo. (This has also led to a post-modernist version of SI among students and colleagues of Norman Denzin.) Many Symbolic Interactionists have very little patience with the semiotic study of symbols and other kinds of signs. They do not report the process of signifying outside of the specific context of an ethnographic field study of a particular sub-culture or group. But that is highly misleading since it deviates from Herbert Blumer's stated axiomatic assumptions about SI and tends to mis-represent G. H. Mead's highly philosophical view of the 'significant symbol' as an aspect of 'Social Behaviourist' epistemology. The process of signifying is highly political. Nevertheless, even if we fully grasp what Mead and Blumer mean by 'symbol', that does not necessarily lead us directly to a Peircean framework. Peirce is not the only thinker to present a useful Pragmaticist approach to semiotics. (One could explore the relationship between Peirce and Karl Popper, for example.) But for this chapter, it will be enough to simply call attention to the value of Peirce's theory and methodology for various forms of praxis.[4]

1.1. Social Problems

The United States is a rich and prosperous country. Moreover, it is usually considered a 'modern capitalist' country. It is highly likely that after the current recession, the United States will recover and that many people will have good incomes and lead reasonably productive and happy lives. But despite over a hundred years of sociological theory, there are social problems that are severe. What is the relevance of sociological theory for the future if it has had so little effect in the past? There may have been some significant changes (like the Civil Rights laws), but many of America's problems are as severe today as they were decades ago. Some social problems have become even more significant in recent years and not just because of recent economic problems.

Let me mention an empirical fact, a fact that does not require deep sociological theorizing: there are many people in the United States who are being punished using what most people would consider cruel and unusual techniques of punishment. It is not just a matter of prisons in Cuba and elsewhere, although the issue of Guantanamo (and other, clandestine prisons) cannot be ignored. Conditions in many American prisons are not humane. That is, very few people would be able to withstand prison conditions as they are in many places. A good society should not allow such conditions to continue, even for one day. The truly Medieval punishments like drawing and quartering are not practiced any longer, as Foucault makes clear, but incarceration in U.S. prisons is not just a matter of true 'justice'. True justice would not involve putting people into a small, windowless white cell 23 hours a day for months and even years on end. The only respite is one hour in a slightly larger cell with any opening to the sky. California Prison Focus, for example, reports on such practices in the state of California. But the pattern is repeated in New York State and Missouri (Parenti, 1999). To some extent, of course, any judgment that such punishment is intrinsically 'cruel' involves a 'value judgment'. But I think it is a value judgment that most American citizens and immigrants would agree does not require deep philosophical justification. The United States would be a better country if the criminal justice system was significantly improved.

Let me draw another empirical fact to your attention. As is well known, there are more 'African-Americans' and 'Native Americans' in prison, per capita, than there are 'White-Americans'. Close to 7 out of every 1,000 African-American men are behind bars today, a rate that is seven times as high as that for White-American men. Minorities make up 65% of the prison population. Much of this is related to social class. The United States has the most people in prison per capita of any country in the world. Georgia, Illinois, Alabama, Mississippi, Louisiana, North and South Carolina, Maryland, New Jersey and Virginia have the largest Black populations in prison (as a percentage of the total African-American population in those states), but even in Tennessee, where 16% of the state population is African-American, fully 53% of the prison population is Black. (The situation in Tennessee is much like that in states like Michigan, Ohio and Pennsylvania.) There are a disproportionate number of African-Americans in prison.

Put those two facts together and what we have is a system of 'justice' that clearly discriminates against human beings who happen to be located within a particular 'space' in contemporary American society, and does so in cruel ways. American Blacks are often treated unfairly simply because they are Black.

This is not just a recent 'postmodern' phenomenon. It has been going on a long time and is well documented in hundreds of 'liberal' (i.e. not Neo-Marxist) textbooks on social problems (Coleman & Cressey, 1988, pp. 188–221, 397–436) I think that the prison system could be reformed and that the United States could become a more 'just' society? In my vision of the future, it is not merely 'utopian' to think that a greater degree of tolerance and comprehension might lead to better outcomes with regard to the prison system.

The same essential idea can be applied to all social problems and social services. When will the majority of Americans receive adequate health care? When will the unemployment rate be low enough to warrant saying that the problem of unemployment (and underemployment) has been solved? When will all Americans have an opportunity to receive adequate levels of education and training to lead productive and interesting lives? When will the credo of the American Declaration of Independence become more than just empty slogans, so that all Americans (and immigrants) can pursue and expect life, liberty and happiness?

Unfortunately, social problems abound in the United States, as well as in other nation-states. Countries that are considered to be part of the so-called 'Western industrialized world' (e.g. the G-8) (I am focusing on the United States because we are in the United States, but much the same could be said about social problems in Canada, Mexico, France, Germany and many other countries.) The idea that modern capitalism is not 'functioning' smoothly is not just called into question by the incredible ways in which the modern state has had to bolster finance capital. Even when the majority of American citizens and immigrants are doing well, there is not a reasonable distribution of wealth. (The Gini coefficient for Japan [0.24] is far better than it is for the United States [0.469], while the United States has an uneven distribution, much like Singapore [0.522]; see Tong, 2008, p. 188). National statistics on per capita income cloud the fact that many Americans are below the poverty line, a fact that becomes evident as one drives in rural areas away from the turnpikes and interstates.

We are in a situation today that some call 'modern capitalism' and that others insist is 'post-modern capitalism'. Is Critical Theory the key to solving the social problems in a modern (or postmodern) capitalist society like the United States? In his recent edited book, Harry Dahms (2008) uses the provocative title: *No Social Science Without Critical Theory*. His own chapter in that book is entitled: 'How Social Science is Impossible Without Critical Theory: The Immersion of Mainstream Approaches in Time and Space'. I would like to begin by examining that assertion, paying special attention to the notions of 'Time' and 'Space', found in comparative-historical sociological theory. The title would seem to mean that at the least we cannot have a valid approach to social science without incorporating *at least some* of the insights of Frankfurt School Critical Theory. (I would hope that he does not mean that the *only* path to good praxis will be Critical Theory in the limited sense.) But, of course, the assertion could be read in a number of different ways. A wide variety of opinions are represented. The dozen articles in this edited volume (in addition to the lead chapter by Dahms) do not all agree precisely on which aspect of the term 'Critical Theory' to emphasize.

As Dahms (2008, p. 3, emphasis added) points out: 'Any endeavour to circumscribe, *with a certain degree of precision*, the nature of the relationship between social science and critical theory would appear to be daunting'. Here, in the interest of 'precision', I will restrict myself to sociological theory as a sub-disciplinary specialization within sociology as a discipline and not attempt to discuss the issue with regard to psychology, economics or political science, much less all of social science. Even the idea of discussing all of sociology with some precision is daunting, and the other social sciences are sufficiently different to require separate treatment.

Moreover, I am mainly attempting to consider *sociological* theory in the somewhat more narrow sense. That is, today, sociological theory is the product of sociologists and only rarely incorporates insights from social theorists outside of the discipline. There is a disciplinary boundary that is difficult to cross, particularly for those who are not at elite universities in Germany, France or the United Kingdom. American citizens working in other disciplines or outside academia (e.g. as 'public intellectuals') are not likely to break into sociological theory, although they may make a contribution to the larger interdisciplinary field of social theory.

To give some idea of the range of ways in which Critical Theory is used in this edited book, it is enough to compare two excellent chapters in Dahms (2008). Mark Worrell (in Dahms, 2008) discusses the Institute for Social Research (ISR, 1944) study of Anti-Semitism and reports on the findings as well as speculating on some of the possible differences between Anti-Semitism in 'Europe' (the German nation-state) and 'America' (the U.S. nation-state). Sarah Amsler (in Dahms, 2008) examines the possibilities for use of Critical Theory and critical pedagogy (*conscientization*) for radical transformation of 'post-modern capitalism' in the United States today. Obviously, the link between the ISR (in the 1930s and 1940s) with education (and pedagogy in general) in the United States in the 2000s is a theoretical leap. One of the key links is the notion of 'critical hope'. But hope in Germany in 1939 is qualitatively different from hope in the United States in 2009. Time and space are certainly very important for understanding what we mean by Critical Theory, then and there versus here and now.

Years ago, I wrote a slim book on Mohandas Karamchand Gandhi's social theory. It has the enthusiastic title: *Toward A Just Civilization* (Bakker, 1993a, 1993b). The Gandhian perspective is usually not discussed by Critical Theorists. But there is a resonance between Gandhi's 'realistic utopia' and the notion of 'critical hope'. Gandhi was 'critically reflexive' and did not accept modern capitalism as inevitable or ideal. No one would confuse Gandhi's writings with 'mainstream social science' or 'mainstream sociological theory'. It would be interesting to compare the core of Frankfurt School [Neo-Marxian] Critical [Social] Theory with Gandhian Social Theory, but that is not my goal here. It would lead me too far astray because (1) Gandhi was not a sociological theorist, although one can reasonably argue that his ideas have relevance for social theory, and (2) the Indian sub-continent (including Pakistan and Bangladesh) in the 1930s is not even similar to Germany in the 1930s, much less the United States today. Instead of examining Gandhi's social theoretical implications, I would like to tackle a broader issue and examine another neglected theoretical orientation, which may help to sharpen some of the arguments made by Critical Theorists (and others) in social theory and in sociological theory – Pragmatism. I would particularly like to focus on C. S. Peirce's version of Pragmatism, which later in life he dubbed Pragmaticism. The main goal of this chapter is to provide a way for translating Peirce's Pragmaticist and semiotic ideas into applied sociology, public sociology and political action.

Let me make it clear that when I discuss 'Pragmatism' in this chapter, I am not referring to the common sense use of the word 'pragmatic' in everyday English. Thus, for example, the interesting article by Tong (2008) about the way in which the government and people of Singapore have gone about the quest for human happiness and 'joy' in a pragmatic manner is definitely not what I mean. He states, 'Singaporeans are also a very pragmatic people. In fact, it can be argued that the national ideology of Singapore is based on pragmatism and meritocracy ... Chan (1976) calls this the "non-ideology of pragmatism"'. The idea that 'pragmatism' (or, Pragmatism) is in no way 'ideological' is plainly incorrect. Clearly, Chan and Tong have in mind the idea that Singapore is not a Marxist–Leninist nation-state like the People's Republic of China or Vietnam. The American Pragmatist (and Pragmaticist) philosophical viewpoint I have in mind is a 'standpoint' that can be reconciled with Neo-Marxian social theory to some extent, but that clearly is a Meta-Paradigmatic (or even Meta Meta-Paradigmatic) 'ideology' or perspective. It is not, however, a political ideology or religious dogma in the simple sense of the word ideology that one finds in the mass media. I am thinking of the level of discourse that one finds in a work like Jolyon Agar's (2006) *Rethinking Marxism.*

Let me also make it clear that I would like to distance myself from much of the Postmodernist approach found in Steven Seidman's article 'Postmodern Social Theory as Narrative with a Moral Intent' (Seidman, 1992) but not from certain key insights found in Derrida (1986).[5] Norris (1990) writes cogently about Derrida and Postmodernism. For example, he points out 'How Not to Read Derrida' (Norris, 1990, pp. 134–163).

There are many social problems. There are not many good solutions. Practical solutions that require applied sociology and public sociology are not going to magically appear from atheoretical understanding. A fuller comprehension of the inherent difficulties of 'changing the world' requires a better conceptualization. That is especially true at the Meta Meta-Paradigmatic level.

2. CHARLES SANDERS PEIRCE'S PRAGMATICISM AND SEMIOTICS

As a philosopher, C. S. Peirce interpreted the world, but he did not (directly) attempt to change it. Yet, as Kurt Lewin points out, there is nothing as practical as a good theory. I would like to submit that Peirce's brilliant insights concerning Pragmaticism and Semiotics constitute an important 'Meta-Meta-Paradigmatic Framework' for sociological theory. Peirce, borrowing in part from Kant, uses the Greek words that in English are pronounced Pragmatism, Pragmaticism and Semiotics. His ideas are highly abstract. What makes it especially difficult to discuss Peirce's theory is that it is (1) not specifically directed to sociological theory and (2) interpreted in many different ways by various philosophers and social theorists. How can we take a 'grand theory' unifying 'perspective' like Peirce's theory and incorporate it into 'method' (both Methodology and techniques)?

While it is well known that American Pragmatism (e.g. William James, John Dewey and G. H. Mead) influenced SI, it is not as well known that Peirce's version of Pragmatism (i.e. Pragmaticism) cannot be separated from his mature version of Semiotics. There is no specifically 'Semiotic SI' today, although there is a 'social semiotics' that does not quite get at the same underlying epistemological questions. Many 'Interactionists' are apolitical, in part because they emphasize the study of interaction and de-emphasize the study of semiotic signs.

Peirce is not the only thinker to present a useful Pragmaticist approach to semiotics. (One could explore the relationship between Peirce and John Dewey, or Peirce and Mead, or even Peirce and Karl Popper, for example.) But for this chapter, it will be enough to simply call attention to the value of Peirce's theory and methodology for various forms of praxis.

3. SEMIOTICS AS A METHODOLOGY

One way to conceptualize Peirce's Semiotics is to think of it as an overarching Logic of Method, or 'Methodology'. It is a Methodology and not just a 'methodology' or method in the narrow sense of techniques of research, such as open-ended interviewing or ethnographic fieldwork. It is most certainly not just a technique for doing specific aspects of research, such as we often teach in our introductory (and even advanced) 'methodology' courses (e.g. ways of calculating regression equations or taking field notes). The key aspect of Peirce's Semiotic Methodology is the epistemological assumption that we do not 'observe' or 'see' the 'objective world' as it really is. Instead, whatever scientific or social scientific observations we make, we make them utilizing 'signs'. That is, the IN (or, community of scientists) always utilizes 'sign systems' to represent and assess any Operationalized Representation (or, way of designating a set of semiotic objects) through the medium of sign systems (e.g. mathematical or statistical symbols, theoretical arguments expressed in a language using words, graphs and diagrams and pictures). *Signs always mediate our observations.* We cannot have an empirical inquiry at the level of what Peirce called 'Firstness', without any conscious use of signs. In science and social science, there is no such thing as pure observation or pure, atheoretical induction devoid of signs. Every empirical statement is a statement made using signs. If there is no sign, then there cannot be any statement. Signification is all about using signs to designate a semiotic object or collection of objects (e.g. Linnaeus' *Homo sapiens*). It is a Romantic fallacy to think we can have a direct perception (or, apperception) of any phenomenon. Hence, we cannot pursue Husserl's phenomenology in the way that it is often interpreted and we cannot follow Garfinkel's ethnomethodology to the most basic roots of accomplishments. It is simply impossible to 'bracket' everything. The *eidetic reduction* will always be limited by our most basic assumptions. We take certain things for granted, a priori.

4. SIGNS AND SIGN SYSTEMS

The term 'sign' is used as a covering word for all forms of gestures, ciphers, tokens, marks, indices and symbols that convey human meaning. Some thinkers trace the beginning of human cognition by the earliest *Homo sapiens* to the use of signs. Religious thinkers emphasized some supernatural indicators of the true nature of reality; they understood 'signs' in nature as messages. This led to necromancy and other forms of divination. The Chinese *Yi Ching* was initially based on the reading of tortoise shells. In ancient times, victory in battle was often seen as a sign of the whim of the gods. That which was not understood directly had to conjectured. Greek physicians utilized somatic signs to diagnose disease. They called this process semeiosis.

The idea of signs has been extended to cover more features of reality. Modernity has made the notion that signs come from supernatural forces less acceptable. Classicist and theologian Friedrich Schleiermacher discovered that the way he carried out exegesis was no different for the pagan, secular texts than for the Christian, sacred texts. Hence, he postulated the possibility of a *general* hermeneutics. This idea was further developed by Wilhelm Dilthey. Dilthey's approach helped to provide a foundation for social sciences. His hermeneutics is based on the study of human beings as moral actors whose motivations could be understood (*Geisteswissenchaften* which used *Verstehen*).

But hermeneutics lacked a more general epistemological foundation. That semiotic foundation came with the work of Charles Sanders Peirce, founder of pragmatism. Peirce emphasized the way in which signs mediate the represented and the interpreted. Peirce's critique of Cartesian dualism makes it clear that an epistemology that focuses on the solitary individual 'subject' as an interpreter of 'objects' is severely misleading. Peirce's correction of Descartes's epistemology parallels Einstein's extension of Newton's physics. The isolated individual never exists in reality but only as a thought experiment. In reality, all scientific understanding is based on communities of scholars (Cohen, 2001).

We do not see the world precisely as it is; we only interpret stimuli with the aid of signs. Peirce had a complex typology of signs, but the most important for sociology are icons, indices and symbols. Icons are very specific images. Indices are signs that point to a more abstract level of reality. All statistics are indices. The most complex type of human sign is the symbol. George Herbert Mead's concept of the 'significant symbol' is an echo of Peirce's general theory of signs. Some philosophers (e.g. Wittgenstein) argue that the real meaning of a sign is in its use. Due to 'intertextuality', we cannot escape a certain degree of circularity in examining signs (Eco, 1999, pp. 275–9).

Theories put forward by Ferdinand de Saussure and C. S. Peirce have been further developed by other semioticians. The key ingredient is awareness of the universal function of signs. There has been considerable attention paid to signs in models of the process of semiosis (Petrilli & Ponzio, 2005).

Axioms are absolutely necessary for human communication. If we do not share any assumptions in common, then we cannot speak a language that will be understood. Wittgenstein's critique of the idea of 'private language' is relevant here. Kant attempted to specify those axioms in terms of what he considered synthetic a priori categories. But Kant did not have the empirical results of the modern discipline of anthropology as a resource for considering the wide varieties of ways human beings conceptualize historical time and socio-economic political space, as well as Universal Time and Universal Space.

5. THE ROLE OF THEORY

The word theory has many different meanings in different contexts. The concept is often used as a kind of covering label for many different features of research, from the very abstract to the relatively quite specific and concrete. The most abstract level of theory in social science can be called 'Meta-theory'. George Ritzer has elucidated three possible sub-types of Meta-theory based on the goals the thinker wishes to attain. He labels those three types of Meta-theory: deeper understanding, prelude to further development and overarching perspective. Many writers think of theory as limited to less 'philosophical' goals. Often, theory is defined as a set of interrelated hypotheses or propositions, but that definition is too Positivistic for many Interpretive, qualitative researchers, especially those who emphasize inductive approaches like grounded theory. Some even use the word theory to mean a specific hypothesis or very restricted set of tentative, abductive hypotheses. Consequently, it is difficult to generalize about the role of theory in general or even the role of theory in case study research. It is safe to say that there is a range of opinions concerning the role of theory in case study research. In the arts and humanities, there are some who wish to reject the notion of 'Theory' altogether, particularly Postmodern theory, but, the use of the term often sidesteps philosophy of social science approaches and involves more of a critique of a specific theory than theory in general. Few contemporary writers accept such pre-modern 'sciences' as astrology, alchemy, palmistry and geomancy; yet, a complete rejection of modern and Postmodern theory would require one to make no such distinctions among approaches. One's theory is intimately connected to the kinds of research that is regarded as fruitful.

5.1. Methods and Theory

Those who are experts in methods often pay less attention to theory. They are more interested in techniques for carrying out research. There is a deep ambiguity in the use of the word 'methodology'. It can refer to techniques for carrying out a specific kind of study (e.g. open-ended versus closed survey questionnaires). But it also refers to the broader 'logic of method', which is an aspect of theory. Theory and 'Methodology' (in the more abstract sense) go hand in hand. Most case study researchers emphasize qualitative techniques such as participant observation and open-ended interviewing. But there is also a more quantitative stream, as in content analysis. Some social scientists believe that all single cases should be studied inductively. Many sociologists call that approach 'grounded theory'. The word 'theory' in the expression grounded theory means tentative conclusions reached on the basis of very thorough investigation of all details of a specific case. Charles Sanders Peirce calls this 'abduction'. Once that case has been studied to the point where no new information seems to be available the case study is finished. Many grounded theorists reject the notion of the Hypothetico-Deductive Method (HDM) as a 'Methodology' or Theory of Method. There may not even be much of an attempt to compare one case to another very similar case. For example, a qualitative study of one community may lead to grounded theory concerning that specific community but the researcher may not even attempt to generalize to other communities in the same geographic area of cultural setting.

On the contrary, many social scientists do believe in the continued HDM for case study research, even though case studies are not usually associated primarily with a Positivist approach. The relevance of the HDM in an Interpretive approach relates to the way in which hypotheses can be deduced from an Interpretive set of generalizations that constitute a recognized theory. Such a theory is an inter-related set of propositions from which hypotheses can be deduced. Some theorists view theory as encompassing Meta-theoretical statements concerning epistemology and ontology. Such philosophical statements can be called Meta-Paradigmatic. The unity of science thesis is held by those who emphasize the similarities between the so-called soft and hard sciences. For example, the case study approach is often used in integrated biology and ecology in ways that are not appreciably different from case studies in sociology or anthropology.

5.2. Fuzzy-Set Social Science

Charles Ragin has argued that we need to incorporate aspects of 'fuzzy-set' theory into social science. He advocates studying cases as 'configurations' and emphasizing diversity in our interpretation of such cases, singular or plural. The contextual situation has a great deal of impact on the ways in which variables should be interpreted. Different situations alter the relationships of factors. The meaning of any statement or 'score' is shaped by the historical situation or process under investigation. Every generalization needs to be anchored in a specific time and place, or a set of times and places. Introductory textbooks, for example, often make generalizations that apply to the United States since WWII, but which do not apply to the United States in the eighteenth or nineteenth centuries and which do not summarize the situation in most countries today or historically. A comparative-historical sociology is contextual but nevertheless allows for careful generalizations. There are relatively few truly 'nomothetic' laws in social science and there are surprisingly few such laws in many of the life sciences, like ecology. In an ecological analysis of a swamp, different factors come into play than in a similar study of a field high in the mountains. Biology, like sociology and other social sciences, is largely contextual. Max Weber emphasizes this point in his theoretical work and points to the 'idealized' nature of many generalizations. Many theories are generalized statements that need to be further investigated for specific cases rather than statements of universal laws.

Even in the so-called hard sciences, there are few real laws that can be said to be true for all Time and all Space. Cosmologists discovered the vastness of the Universe only in the early twentieth century. There are very few categories that are firmly fixed and unlikely to be modified through further empirical findings and theoretical analyses.

5.3. Heuristic Postmodern Theory

Postmodernist theory is heuristic; it has led, for example, to a re-examination of classical and modern thinkers. That reanalysis of received 'texts' has led to fruitful interpretive analysis of specific cases. The claim made that it is possible to move beyond 'foundations' is really an attempt to think through implicit assumptions. This leads to interesting theoretical problems. For example, the reading of G. W. F. Hegel's works is considered by many to be quite different after seminal work by Jacques Derrida. Yet, it is possible to re-read Hegel on the basis of Derrida's exegetical hermeneutic principles. Derrida's notion of 'deconstruction' is not so much a technique as a theory of interpretation of specific cases. To 'deconstruct' is not to destroy or deny but to make the implicit axioms and contrasts explicit. By contrasting Hegel with Jean Genet, the poet, aspects of Hegel's theory come out more forcefully. Ambiguities become a basis for further comprehension rather than merely points that should be skipped entirely. The startling contrast provides for a re-examination of strongly held convictions. Some analytical philosophers at British universities feel that such playfulness is merely a trick and that Derrida is a charlatan. If, however, it is taken as serious theory in the guise of playful rhetoric, then it has many implications for case study research. When we interpret any situation or process, scene or event, we do not come to it merely as a completely neutral observer. We are not from some other galaxy. Even if we were we would still bring some kind of theoretical assumptions to any observations or participant observations we might make. The movie 'E.T.' illustrates that quite well; the Extra Terrestrial is far from neutral.

The attempt in Husserl's phenomenology to bracket all assumptions is bound to be an ideal goal. We may be able to approximate that goal, but the asymptotic curve will never be parallel to the admirable idea. We cannot easily get rid of axiomatic assumptions, our 'apriorities'. Even the most basic description of phenomena like colour is theory-laden. That is, we cannot say 'This apple is red' unless we have common agreement on a range of phenomena that we can designate as 'red' due to perceptions of 'redness'. If we follow Sapir–Whorf on this point, if we do not have names (signs) for variations in the texture of snow, then we will not fully recognize that the different kinds of snow are textured somewhat differently.

Almost every teenage boy growing up in America in the 1950s and 1960s knew how to distinguish a Chevy from a Ford or a Plymouth, but very few knew the 30 or more varieties of snow. Many meat-eating people know the difference between sirloin and T-bone, but few can distinguish among varieties of rice. What we fully grasp is that which we can 'de-sign-ate' with signs. If we do not know the names of different flowers, then all we will see is colourful flowers. If we do not know the names of birds, then we will miss 'seeing' (i.e. really observing and noticing) many varieties.

Most people do not know how to distinguish one variety of Marxist theory from another. Students who are introduced to Marx for the first time will have great difficulty understanding what he is saying. But eventually, a scholar interested in Marx and his thought will start to see connections. The pattern will become clear. Imagine trying to understand Marx without having a clue what a 'factor of production' is? Yet, many undergraduate students try to read introductory textbooks in theory with absolutely no idea what is meant by productive forces. For Peirce, we are the sum total of our signs. He did not mean that literally, of course. I take it that he meant that if I have a good mastery of the German language I am more likely to read and appreciate Kant or Husserl than if I have no knowledge whatsoever of German. Reading a translation of a work originally written in another language can only take one so far in terms of good comprehension. The better the translation the more likely the original will not be completely misunderstood. But if one has no language at all (i.e. no sign system), then translations, no matter how good, will still not work.

5.4. Theory and Practice

The relationship between theory and practice is often debated. When it comes to policy decisions and implementation, there is often a reluctance to move beyond the single case. A good case study of a specific phenomenon is frequently not considered to be generalizable to any other cases. At times, however, the resistance to transferability is more a matter of political obstructionism than scientific scepticism. A distinction is often made between practice as a kind of reformist effort at administrative change and praxis as a fundamental transformation. Theories concerning practice have been keenly debated in recent years. A Neo-Marxist and Feminist praxis is not necessarily equivalent to reformist activities such as social work, administration and public health nursing. Kurt Lewin repeated Immanuel Kant in pointing out that there is nothing as practical as a good theory.

6. METHODOLOGICAL CONSIDERATIONS AND PRACTICAL OUTCOMES: USING SEMIOTICS WISELY

I am advocating a more widespread use of Peirce's semiotics. Note that I am not dealing with other forms of semiotics or semiologie. I believe that a full comprehension of the thrust of Peirce's triadic theory of signs provides an epistemological basis for a Pragmaticist Meta Meta-Paradigmatic and Methodological approach. How does that Methodology/method help us to improve:

A. sociology as an empirically based, rigorous discipline (a human, social 'science') and
B. various forms of 'praxis', especially applied sociology and public sociology?

For sociology to be a rigorous discipline that can be considered a 'social science' (*Sozialwissenschaft*), a humanistic science (*Geisteswissenschaft*) or a true 'science' (*Wissenschaft, scientia*), it must have a general sociological theory. It is often thought that sociology has no such theory available. But I believe that we do have the makings of a reasonable sociological theory in the broad sense. My candidate for such a 'grand', general sociological theory is Peirce's Pragmaticism. The Methodology for that general theory is Semiotics.

Praxis can include many forms of political activity. Those forms of political activity include a continuum of types of collective social action, from 'conservative' and 'liberal' to 'feminist', 'radical' and 'Neo-Marxist' (e.g. conservation of fundamental human rights and liberties, reform within liberal parliamentary democracies, transformation within neo-conservative regimes, applied sociology in professions and occupations like social work and criminology, Feminist critique and action, GLTB action, neo-Gramscian critiques and Michael Buroway's 'public sociology' in the narrow, technical sense.)

7. SYMBOLIC INTERACTIONISM VERSUS SEMIOTIC INTERACTIONISM

We need to sharpen our understanding of George Herbert Mead's insights and move beyond the somewhat limited discussion of Mead's Social Behaviourism that is found in most of the literature of SI. Today, most SI research focuses on the empirical study of 'interaction'. That is well and good. But few SI researchers take the term 'symbolic' as seriously as they might. What is required as a first step towards improving contemporary modern and postmodern SI is a deeper understanding of the Semiotics of symbols. It is not just a matter of reporting the terms that a sub-group uses or reflecting on the jargon characteristic of a deviant sub-culture, as important as that is in and of itself. It is also a matter of questioning the deepest implicit axioms of our construction of the underlying general culture. Most research in SI is based on an acceptance of the existing social construction of reality.

Many excellent descriptive, ethnographic studies based on 'fieldwork' have been done. But they have tended to simply accept the status quo. (This has also led to a Post-Modernist version of SI among students and colleagues of Norman Denzin.) For example, when an American researcher does SI research in the United States, then he or she tends to accept such fundamentally 'American' assumptions as the existence of 'persons' with specific 'racial' and 'gender' characteristics. It is 'obvious' to 'everyone' that the United States consists of 'citizens' who can be classified as being 'white' or 'black' and 'male' or 'female'. Those basic axioms are hardly ever challenged. Indeed, many SI researchers wonder why anyone would challenge such deeply held axiomatic beliefs.

Many Symbolic Interactionists have very little patience with the semiotic study of symbols and other kinds of signs. They do not report the process of signifying outside of the specific context of an ethnographic field study of a particular sub-culture or group. But that is highly misleading since it deviates from Herbert Blumer's stated axiomatic assumptions about SI and tends to mis-represent G. H. Mead's highly philosophical view of the 'significant symbol' as an aspect of 'Social Behaviourist' epistemology.

The process of signifying is highly political. Nevertheless, even if we fully grasp what Mead and Blumer mean by 'symbol' specifically in terms of Peirce's semiotics, how can we translate that theoretical awareness?

We can use it to not only interpret the world but also to change it. Good theory and methodology is the most practical way to promote useful social action, applied sociology and public sociology. Anything less than Peircian Pragmaticism and semiotics tends to lead to fragmentation of 'paradigms' or Post-modernist nihilism.

8. SEMIOTICS AND CRITICAL THEORY

Just as Symbolic Interactionists do not pay attention to Peirce's semiotic insights concerning symbols as one type of sign, Critical Theory neglects the deeper questioning of symbols. Empirical work on such topics as 'Anti-Semitism', for example, does not question the social construction of the symbolic meaning of the word 'Semite'. It is taken for granted that the term is meaningful. Yet, it is blatantly obvious that if the word Semite means Jew, then we should ask what the opposite of a Semite would be. For many Jews, the opposite of Jewish is Goy. Yet, we never write or speak about Anti-Goyites. Why is that the case?

Whenever our picture of the world (*Weltbild*) is divided into two and *only two* categories, we are likely to make mistakes in our beliefs, which ultimately may lead to mistakes in our actions. There are many examples of this, but precisely because our internalized semiotic codes are so deeply ingrained it is often hard to speak of current examples. It is much easier to speak of historical examples, like the phenomenon of the Irish eventually becoming 'white' in America in the course of the nineteenth century. Today, many people use the acronym 'WASP' (Hacker, 1957), even if they are speaking about Irish Roman Catholics, like JFK. Few people pause to consider how it is possible for an Irish Catholic to become an Anglo-Saxon Protestant. Even fewer would challenge the notion that many recent immigrants from Ireland who live and work in America are still not 'White' in the original elite sense of the W.A.S.P. category (Baltzell, 1964). (Is it in some sense correct to label Barack Obama, a Harvard Law graduate, a WASP?)Terms tend to eventually lose their sharp focus and they are often used in a very simplistic way.

If we think only in terms of Jews and Goyim, then we are likely to ignore a host of other factors that should be taken into consideration. (The term Goy literally means 'nation' and is used in the Bible to mean nations other than Israel–Judah. It can also mean 'Gentiles'.) For example, it is commonly considered that American Jews are 'white'. They are certainly not considered to be 'black' or 'Latino/a' or 'Native American Indians'. But for an Ultra-Orthodox Chasidic Jew, all non-Jews are, in a sense, Goyim. While secular Jews do not use the word in that way, it is nevertheless a word that is well known to all Jews, secular or religious. So that would imply – illogically one might assume – that there is no difference among blacks, Latino/a's and Indians; they are all Goyim. None are considered to be children of Shem. This example can easily be misunderstood. But the paradox is evident to anyone who is willing to think outside of the box of the standard axioms that are implicit in American society today.

As an illustration, let me briefly summarize a recent article by a scholar from Nigeria. When he came to Canada, he did not think of himself as 'black'. In the Nigerian context, he thought of himself as 'yellow' relative to other Nigerians. But in Canada, his morphological characteristics, especially his skin colour, made many people think of him – automatically – as 'black'. He eventually chose to fit in and give up some of his sociological reflexivity to 'become black'. He let the common definition of the situation be the appropriate social construction for his self-identity. But at some level, he still retains a very real awareness that he is not 'just' a black Canadian. He is also a Nigerian!

9. SEMIOTIC SHARPENING OF PRAGMATIST INSIGHTS

The fundamental thesis is that greater comprehension of the usefulness of a Peircian semiotic perspective and 'fallibilism' will help to promote the right kinds of social action and collective responsibilities. A Pragmaticist perspective that utilizes a Methodology informed by the triadic epistemology of Peirce's semiotics provides a deeper understanding of why certain kinds of views appear to be incommensurable when in fact they are not. They require dialogue and translation, but they are not fundamentally at odds when it comes to practical kinds of issues. The real differences among Paradigms like Neo-Functionalism and SI are of not practical consequence when it comes to tackling relatively clear-cut social problems like cruel imprisonment.

10. PEIRCE'S INTERPRETATION OF THE WORLD

Peirce's worldview (*Weltanschauung*) was essentially scientific. He borrowed a great deal from Kant, particularly Kant's *Critique of Pure Reason*. In a sense, all contemporary scientists and social scientists are Post-Kantians (which should not be confused with 'Neo-Kantians').

11. PEDAGOGY

There is a different way of doing sociological theory and methodology. We need to open students' eyes to the importance of signification systems. Most students have never been introduced to this before. It is a new Meta-Meta Paradigm. We can allow students to reconceptualize one of our most basic axiom: the idea of a 'subject' viewing 'objects' (i.e. the Cartesian gaze, masculine or feminine). We can aid the students in developing a gender perspective to study the individual and society. Today, the Cartesian dualist outlook is widely accepted but that tends to lead to the reification of 'objects'. A Peircian semiotic epistemology makes it clear that all notions of objectivity are socially constructed as a result of semiotic sign systems mutually agreed upon by members of the same IN. Peirce clarifies the limitations of the 'modern' Cartesian perspective that is behind almost all of what we are taught to think of as basic to sociology and other social sciences. But the emphasis is not 'postmodern' so much as *avant garde* modernist. We can open students' eyes to a whole new perspective on the conceptualization of sociological theories and research strategies. We can aid students in developing a semiotic perspective to study social change in the context of religious, ethnic, racial, class, status, power, gender and sex differences. Examining various semiotic theories will provide a greater depth of understanding of commonly help assumptions. We can increase the depth and breadth of students' understanding in the society in which they live. It will encourage independent thinking as they are encouraged to come to terms with new information about epistemological assumptions. Since sociology is a 'multiple paradigm social science', the semiotic perspective helps students to put it all together in ways they had not previously thought of.

NOTES

1. This fictitious book review is a product of Randall Collins' imagination. In an earlier draft of this chapter, I started by quoting this book review at length. It is a wonderful parody of extreme versions of nihilistic postmodernism, but also of the other extreme found in Sokal. See Seidman and Wager (1992, pp. 180–181).

2. Joseph H. Michalski is not familiar with the work of Philip Mirowski. But when Michalski explains how he views 'pure sociology' as based on a pure conceptual framework that 'should apply across similar cases of social behavior spanning diverse cultures and across human history' (Michalski, 2008, p. 264), he is clearly invoking the Positivist, Logical Empiricist agenda that Mirowski (2005) critiques. Unfortunately, Marshall (2008) also seems to be unfamiliar with Mirowski's excellent analysis. Note that for Mirowski, there was a shift in Meta-Paradigmatic assumptions in many academic disciplines in the 1980s, and the exemplary philosopher today is Philip Kitcher.

3. Rural sociology, as a discipline separate from the discipline of general sociology, is a partial exception. However, the rural sociology of the 1950s and 1960s, unified around such topics as the diffusion of innovations, was significantly challenged in the 1970s by the newer approaches to the study of agriculture associated with theorists such as Bill Friedland, Fred Buttell, Bill Heffernan and Harriet Friedman. Another partial exception is the Society for the Study of Symbolic Interaction (SSSI). The Symbolic Interactionist approach was consistently opposed to Positivistic social science and Behaviouristic variable analysis.

4. Portions of the discussion in this chapter are drawn from my contribution 'signs' in the *Blackwell's Encyclopedia of Sociology*, edited by George Ritzer. The term 'praxis' is used here to represent the general notion of practical applications. Insofar as 'public sociology' represents a truly practical application of some fundamental aspects of Marxist, Marxian and Neo-Marxian social theory, as well as Neo-Marxian sociological theory, the term is useful.

5. See the excellent articles by Alexander, Aronowitz, Brown, Collins, Jonathan Turner and Stephen Turner. In a longer article, I would like to dialogue with that set of articles.

REFERENCES

Agar, J. (2006). *Rethinking Marxism: From Kant and Hegel to Marx and Engels.* London: Routledge.

Bakker, J. I. (Hans). (1993a). *Toward a just civilization.* Toronto: Canadian Scholars' Press.

Bakker, J. I. (Hans). (Ed.) (1993b). *Gandhi and the gita.* Toronto: Canadian Scholars' Press.

Baltzell, E. D. (1964). *The protestant establishment: Aristocracy & caste in America*. New Haven: Yale University Press.

Cohen, I. B. (2001). *Revolution in science*. Cambridge, MA: Belknap/Harvard University Press.

Coleman, J. W., & Cressey, D. R. (1988). *Social problems* (3rd ed., pp. 188–221, 397–436). New York: Harper & Row.

Dahms, H. (Ed.) (2008). No social science without critical theory. *Current Perspectives in Social Theory*.

Derrida, J. (1986). *Glas* (J. P. Leavey, Jr. & R. Rand, Trans.) (Originally published in 1974.). Lincoln, NE: University of Nebraska Press.

Eco, U. (1999). *Kant and the platypus*. San Diego: Harcourt Harvest.

Empiricus, V., Jr. (2001). *Review of multiple paradigms in modern physics: An introductory text by Juergen von Himmel*. Berkeley and Cambridge: Bootleg and Edens Fall.

Frank, T., & Mulcahey, D. (Eds). (2003). *Boob Jubilee: The cultural politics of the new economy*. New York: W. W. Norton[The term "boob" refers to H. L. Mencken's booboisie, the bourgeoisie; it is a somewhat unfortunate choice. Most articles are by public intellectuals outside academia.].

Hacker, A. (1957). Liberal democracy and social control. *American Political Science Review*, *51*(4), 1009–1026.

Kant, I. (1999 [1793]). On the common saying: That may be correct in theory, but it is of no use in practice. In: M. J. Gregor (Trans.), *Practical philosophy* (pp. 275–309). Cambridge, UK: Cambridge University Press.

Lewin, K. (1951). Problems of research in social psychology. In: D. Cartwright (Ed.), *Field theory in social psychology: Selected theoretical papers in social psychology* (pp. 155–169). Ann Arbor, MI and New York, NY: University of Michigan Research Center for Group Dynamics and Harper & Row [The original date of publication of the original working paper is 1944.].

Marshall, D. (2008). The dangers of purity: On the incompatibility of 'pure sociology' and science. *The Sociological Quarterly*, *49*(2), 209–235.

Michalski, J. H. (2008). The social life of pure sociology. *The Sociological Quarterly*, *49*, 253–274.

Mirowski, P. (2002). *Machine dreams: Economics becomes a cyborg science*. Cambridge, UK: Cambridge University Press.

Mirowski, P. (2005). How positivism made a pact with the postwar social sciences in the United States. In: G. Steinmetz (Ed.), *The politics of method in the human sciences: Positivism and its epistemological others* (pp. 142–172). Durham, NC: Duke University Press.

Mirowski, P. (2006). *The commercialization of science, and the response of the STS. New Handbook of STS*. Cambridge, MA: MIT Press.

Norris, C. (1990). *What's wrong with postmodernism: Critical theory and the ends of philosophy*. Baltimore, MD: The Johns Hopkins Press.

Parenti, C. (1999). Bring us your chained and huddled masses. In: T. Frank & M. David (Eds), *Book Jubilee: The cultural politics of the new economy* (pp. 51–60). New York: W. W. Norton. [Originally published in Baffler (a radical literary magazine)].

Petrilli, S., & Ponzio, A. (Eds). (2005). *Semiotics unbounded: Interpretive routes through the open network of signs*. Toronto: University of Toronto Press.

Phillips, B. (1991). *Beyond sociology's tower of Babel: Reconstructing the scientific method.* New York: Aldine de Gruyter.
Seidman, S. (1992). Postmodern social theory as narrative with a moral intent. In: S. Seidman & D. G. Wagner (Eds), *Postmodernism & social theory.* Cambridge, MA: Blackwell.
Seidman, S., & Wagner, D. G. (Eds). (1992). *Postmodernism & social theory.* Cambridge, MA: Blackwell.
Tong, C. K. (2008). Materialism, pragmatism and the pursuit of happiness in Singapore. In: K. Kosaka & O. Masahiro (Eds), *A quest for alternative sociology* (pp. 183–200). Melbourne, Australia: Trans-Pacific Press. Advanced Social Research Series, The Center for the Study of Social Research for the Enhancement of Human Well-being, Kwansei Gakuin University, Nishinomiya, Japan.
Turner, J. (2006). American sociology in chaos: Differentiation without integration. *The American Sociologist, 37*, 15–29.

APPENDIX A. SEMIOSIS AND BAKKER'S INSOR MODEL

This chapter is an attempt to introduce sociological theory and methodology from a Peircian Pragmatic point of view. The Pragmatist perspective utilized here is C. S. Peirce's Pragmatism, also later called by him *Pragmaticism.* The essence of the Peircian version of Pragmatism and semiotics is an emphasis on the triadic nature of all cognitive processes. Peirce's epistemological views are quite complex. Therefore, in this analysis, a somewhat simplified version of Peirce's key idea about triadic epistemology is utilized. The model proposed here is in some ways a creative misappropriation of Peirce. It is called the INSOR model and this appendix will begin to show how each of the three components of that model fits together with the other two components.

IN-S-OR

Part One: Signs (S) and Sign Systems (SS)

The basic unit of analysis for Peirce's triadic epistemology is the 'sign'. He uses the word sign to designate (de-sign-ate) any kind of icon, index or symbol. In his complex theory, the word sign can also mean any other kind of marker, indication, gesture or way of representing and signifying (sign-ifying) anything. That which is being signified by the sign need not be a concrete, physical, material 'thing'.

Part Two: Interpretive Networks

When signs are used to signify something, they signify something that members of a sign system are willing to accept. The members of a sign system are themselves designated by a sign, and so, in the abstract, they themselves are a kind of sign. For example, all those who are active in the ASA are considered to be 'sociologists'. Together, the sociologists who are members of the ASA constitute a sign system. If one person dies, then he or she is no longer a material, physical part of that sign system. But the books and articles the sociologist may have published in her or his lifetime may still be read and thus continue to influence the on-going construction and re-construction of the sign system.

Part Three: Operationalized Representations (OR)

What is accepted as properly signified by members of the IN is a set of representations of 'things' signified. The representations are 'operationalized' by members of the INs involved. There is sufficient agreement among members of the relevant INs to make it possible for the signs to be commonly accepted. If they are Operationalized Representations, then they must have some degree of acceptance by a majority of those who are member of the INs. A key part of an OR is the use of a spoken and written 'vernacular' language, such as English, French, Spanish, German, Russian, Indonesian, Arabic or Mandarin. The ASA conducts its meetings in English, and native speakers of other languages are therefore required to write and to deliver their papers in English. But use of one vernacular language is only the beginning. There is also a specialized vocabulary of signs that constitute a kind of professional, academic sign system. Since there are 44 sections of the ASA, there is even some variation among those 44 sections as to what must be known.

The term 'jargon' is often used in a derogatory way. It is certainly true that 'unnecessary jargon' may at times interfere with good communication. However, without the technical terms of the sign system that constitute the ORs of the INs, then there cannot be any 'significant' academic communication. If we only use the common sense words used in everyday conversation, we are severely limited as to what we can say, with some degree of precision. That is one reason why acronyms are very commonly used; it is an indication that the terms being used should be seen in their technical meaning as signs that are part of the professional sign system.

PART III
NEGATION

THE DIALECTIC OF SELFHOOD☆

Lauren Langman

But he who dies in despair has lived his whole life in vain.

Theodore Adorno (1951, p. 167)

1. INTRODUCTION

One of the most important and brilliant contributors to the Frankfurt School understanding of character was Theodore Adorno. For Adorno, domination was not simply due to class relations, but the totality of market society in which Reason as the logic of exchange relationships migrated into the family and was insinuated within the person to colonize subjectivity. A central moment of the critique was the internalization of the authority relations of caretakers within superego (cf. Horkheimer, 1972). The law of value, together with Instrumental Reason as a hegemonic ideology and the commodification of culture led to the formation, if not deformation of an authority seeking superego as the typical means of adaptation that sustained political economic arrangements, albeit through suffering based on the repression of desire, the suppression of self and the thwarting of human possibility. Although this critique was rooted in Marx's analysis of capitalism as alienating, dehumanizing and objectifying, the emancipatory

☆This chapter was first presented at the Remembering Adorno Session, *American Sociological Association,* Boston, July 2008.

Nature, Knowledge and Negation
Current Perspectives in Social Theory, Volume 26, 261–281

ISSN: 0278-1204/doi:10.1108/S0278-1204(2009)0000026013

quest sought the liberation of self and desire from the alienation, commodification, and objectification of bourgeois society. But so too can we find that from out of the depths of the alienation and despair of the émigré scholar, there also comes the promise of redemption and the possibility of the "good life," which requires overcoming alienation, and with that overcoming, transcendence, and emancipation from domination.[1]

Surely the influence of Wilheim Reich on Adorno's work must be noted, as it is often ignored. In addition, Erich Fromm played a seminal role in bringing Vienna to Frankfurt. Nevertheless, the most important works on character must begin with the foundational *Dialectic of Enlightenment* (Adorno & Horkheimer, 1944) and the "Authority and the Family" essay (Horkheimer, 1972), *The Authoritarian Personality* (Adorno, 1950), and *Minima Moralia* (Adorno, 1951). Adorno despaired at the suppression of self and the "damaged life." In his words Freud provides us an understanding of domination internalized as a character that leads individuals to not only accept authority and reproduce their subjugation, but also suffer in the process. Yet he also sees that Freud's promise was to free people from irrational suffering imposed by society. The dialectic between despair and hope, domination and freedom, damage and redemption, can be clearly seen by contrasting Adorno and Bloch. Although this contrast may well reflect the personalities of each, or perhaps the historical contexts of their writings, as will be argued, and although the conditions of present age might well validate Adorno's despair, so too are there forces for freedom and transcendence that, as Bloch dreamed, can provide hope. *As shall be argued, a dialectical approach to character, as articulated through identities, offers a space at the intersection between Adorno's legacy of despair, Bloch's hope and the progressive mobilizations of today.*

2. PART I – CHARACTER

2.1. A Subjectivity the Frankfurt School

2.1.1. The Western Subject

The fundamental framework for the subject in Western philosophy was Cartesian dualism that bifurcated thinking from the body. For Kant, the subject was "a fully self-transparent, self-conscious, self-consistent, and transcendent free will – what is commonly understood as an autonomous ego that depended on the self-legislation of the will (i.e., on living by rules that one gives to oneself....the Kantian subject could be described as unencumbered (not socially constructed or embedded), centered (as a single psychic entity made whole by its orientation to the universal principles of right), and unitary (internally consistent, with uniformity and conformity toward its centering elements)."[2] Hegel in turn, criticized the ahistorical notion of Kant's historically constituted subject.

In contrast for Nietzsche, the self, *motivated by will*, was an aesthetic self-creation expressed in spontaneous, creative action which was the essence of humanity, if not super humanity. And that self-creation might mean rejecting the dominant norms that took many forms. Enlightenment based Reason, like Christian religion, repressed the will. Christianity fostered a "slave mentality" in which the Apollonian suppressed the Dionysian; will thwarted from transcendence turned into self hate and loathing of the banal conformist masses with their *resentment* of the free and powerful. The quest for freedom and transcendence was repudiated by petty masses who readily submitted to authority and actively colluded in their own subjugation. Anticipating much of Freud, Nietzsche began a critique of both the separation of thought from the body and the idea that an autonomous, unitary, centered self constituted its self through reason. *Au contraire,* the self was "de-centered," not the unitary master of its destiny, but a multiplicity of subject positions driven by a "will to power" rooted in the body-facing myriad kinds of socially based repression.

Freud, as noted, most fully charted the nature of character, its development, its differentiated structure in which its parts were often at war, its multiple levels of consciousness, the dialects of desire and repression, and the ambivalences of desire, anxiety, fear, and guilt. For our purposes, *the fundamental dialectical conflict of self as a moment of character, and identity as a cultural narrative*, was framed as civilization vs. desire, and resolved through guilt based repression – as Marcuse (1955) would argue, *Freud conflated civilization with its capitalist form in the Victorian era in which surplus repression sustained subordination.* In terms of Freud's own framework, elements of the haute bourgeoisie, and many of the proletariat of this era, alienated by the demands of the market, were prone to toward anal sadistic authoritarianism that repressed freedom and self-fulfillment. But much as Nietzsche had shown, this repression, led to a willful subordination to authority (slave mentality), intense conformity and aggression to the self that was either experienced as guilt and depression, or projected or displaced as hatred to other. As Fromm put it, the "escape from freedom" was found in submission to authority, destructiveness and conformity.

More recent trends in psychoanalysis emphasize the role of attachments between people in the formation of selfhood, while identity as social, are aspects of the same developmental/interactional moment yet at different levels of analysis. Both are socially constituted, each is dependent on the other. One's self, or perhaps noting the pluralization of de-centered selfhood, one's selves, as the expression of character are expressed in various identities from familial to occupational, political or leisure, religious or secular. Thus, aspects of identities become internalized by the person, whereas the impetus to enact various identities, and the styles in which they are enacted, rest on the person's desires, consciousness, and ambivalence. As Giddens (1991) put it, the notion of "self identity" refers to the union of self and identity as expressed in reflexivity, action, feelings, goals, and meanings that link the person to his/her various communities of meaning. The important point is that identities are socially constructed, although for most of us, they are not constructed under conditions of our own choosing. That said, through social interaction, observation etc., certain aspects of collective identities become cathected and internalized as part of our self.

2.1.2. The Subject Defined

For the purposes of this chapter, we must define the subject/subjectivity, the qualities and experiences of the concrete person in terms of her character, self, and identity. Many uses of these terms are such that the same term often has several meanings, while conversely, different authors use the same term for different levels of analysis. *Character, rooted in Freudian theory, refers to the totality of and individual's personality structure, her ego/super-ego, and id.* This of course includes her personal history with caretakers now internalized as ego/super-ego and moral values, the vicissitudes of her instincts, her defenses, and interpersonal styles that shapes the way she thinks, acts, feels, and functions in the world. In Freudian theory, an essential part of that structure is that part of the ego that is reflexively experienced as the self, which while having its own developmental trajectory and unconscious roots is a representation of the totality of character to the ego. It is the primary locus of experience(s), of past, present, and anticipated. It includes its standards of self-appraisal, for example, esteem vs. shame, its [grandiose] ambitions and its idealization of parental images (cf. Kohut, 1971). But this view of the self tends to be a bit individualistic – if not solipsistic. A long tradition of social psychology, from James and Cooley to Mead and Blumer, has emphasized the more interpersonal, symbolically mediated, socially constructed aspects of self – but in the process, attenuating the drives, internal conflicts, ambivalence, and fears. Much like neo-Freudian theory, such views are more likely to foster adaptation than resistance (Jacoby, 2005).

Character, as the totality of an individual's history, structure, desires, defenses and modes of consciousness including the self as its reflexive moment is largely articulated through social identities that have been constructed through interaction with others. Social life takes place within identity granting, emotionally gratifying, communities of meaning. These of course include the family, age groups (gangs, clubs), churches, nations, etc. Identities, as aspects of communities/relationships, serve to mediate between the individual and her society. Much like the concept of role in earlier theorizing, it is through one's identity that one gratifies her needs, articulates her self, and locates her self in groups/organizations. But while roles are based on institutions, identities are more socially constructed – in some cases by large-scale institutions and often such identities are moments of hegemony; in other cases, they are constructed through interaction and, here, may assume counterhegemonic qualities.

Identity, as socially constructed, reflexive narrative of self as located in a community of meaning is a plot, a story of origins, a framework for understanding current reality, and vision for the future – values/goals for future. *As such, identities embrace shared meanings, goals and values.* But identities are the products of human interaction, negotiation, debate, resolution, and change. Some of these may have long historical roots such as religious identities whose narratives go back to the axial age. Most identities are not only socially constructed and negotiated, but when constructed by schools, media, and churches, they are often crafted to serve hegemonic functions such as national identities qua citizenship (Langman, 2006). And in the present age, as the Frankfurt School has so lamented, the culture industries that now include fashion and travel, now provide a wide range of identities that gratify constructed needs and colonize consciousness to insure both the reproduction of domination and the demise of critical thought that would critique that domination. This is the social psychological basis of "willing assent" to ruling class interests that make the historical natural, "common sense." But character structure is not the same as identity. Although the expressions of identity may be phenotypically different, an underlying genotypic social character, authoritarian or democratic, native or émigré (or even authoritarian émigré), can be expressed in many ways ranging from anti-Semitism to elite disdain of mass culture.

2.2. Authoritarian Character

Reich (1946, 1933) had noted how (sexual) repression, "bourgeois sexual morality," fostered docile acceptance of capitalist domination and disposed the embrace of Nazism – especially among certain segments of workers. With Fromm and Horkheimer, Adorno would fully examine the relationships of character, domination and the political. Adorno, as one of the most vociferous critics of traditional thought, for example, positivism/empiricism in social sciences, would ironically direct one of the largest empirical studies of attitudes, values, and character. *The Authoritarian Personality* revealed a fundament characterological pattern of subordination to superiors, domination/contempt to these below-deemed weak, inferior, and legitimate objects of scorn. Authoritarians value conformity, anti-intellectualism, are intolerant of ambiguity, and likely to project fears, weakness, and aggression to outgroups deemed inferior and/or dangerous – and such others deserve hatred, if not destruction. But this character type had an unintended, secondary consequence; by thwarting transcendence, self-fulfillment, and freedom, as Nietzsche pointed out and Freud reiterated, aggression turned upon the self as the self-hatred and loathing of banal men. Fromm (1973) reinterpreted the aggression to the self as "necrophilia," as the love of death and destruction. This was especially clear in Theweleit's (1987) analysis of the *Freikorps*, a groups of proto-fascist military men who despised women, sex, and dirt, for whom the goal of life was to die in battle, peace was the time to prepare for war, and women/sexuality were to be loathed and hated. For the FS, the authoritarian character, together with the invisible domination of Instrumental Reason (see later) and the culture industry to sustain domination as both an internalized structure of authority relations and as the colonization of desire – the production of artificial needs (Adorno & Horkheimer, 1944).

2.3. The Discontents of Capitalist Civilization

As Nietzsche argued and Freud later "discovered" (capitalist) civilization had a very high price, the repression of desire though internalized authority was sustained by guilt. As such, alienation was not simply produced when surplus value was appropriated by the bourgeoisie, but the very separation of mind and body noted by Descartes, ruled by Reason for Kant, resulted in alienation at the level of self in which desire was colonized and consciousness manipulated to reproduce the conditions of domination through conformity, consumerism, and the uncritical acceptance of authority. *But the main point was that the "normalization" of desire and colonization of consciousness was such that people actively reproduced their subjugation within a class society that rewarded this subjugation with alienation, anomie, and suffering*. If Marx suggested that religion was the opiate of the people that expressed genuine pain, 100 year later, the culture industries/consumerism offered more concrete palliatives that anesthetized the pain though the momentary gratification of "artificial needs" and embrace of mass mediated identities. But in both cases, the results were the same, individual autonomy, freedom and fulfillment were thwarted, communities were fragmented, rendered niche markets, while people were left powerless – or were they? For Adorno, between fascism and consumerism, all that was left in the "administered" society was a collection of damaged selves.

3. PART II THE DAMAGED LIFE

3.1. The Émigré in America

The fundamental theme of *Minima Moralia* as "reflections on a damaged life" was the elaboration of social critique through an introspective analysis of moments of his own life conditions and personal suffering, expressed through a number of aphorisms styled after Nietzsche's. *But while his trenchant insights were written as the musings of a "homeless" émigré, he was attempting to offer a more general social critique of damaged lives in general for whom selfhood was thwarted and despair the norm.* Otherwise said, given the rise of fascism, the domination by Instrumental Reason and the power of culture industries to dull critical thought, was human transcendence possible after Auschwitz? In face of this despair, for Adorno, critical philosophy provides redemption:

> [t]he only philosophy which can be justified in face of despair is the attempt to contemplate all things as they would present themselves from the standpoint of redemption. Knowledge has no light but that shed on the world by redemption: everything else exhausts itself in its reconstruction and remains mere technique. Perspectives must be fashioned that displace and estrange the world, reveal it to be, with its rifts and crevices, as indigent and distorted as it will appear one day in the messianic light. To gain such perspectives without velleity or violence, entirely from felt contact with its objects–this alone is the task of –this alone is what thought wants to know. (153, 247 [334]).

Although we cannot ignore the fact that he was writing in the 1940s as America was triumphant in the war and about to become the major player in the world's economy, it's geo-politics, and [mass] culture, we must also note that many intellectuals, literary or social, left or right, often disdained the banality of the American masses, their anti-intellectualism, conformity, and crass materialism, not to speak of their garish tastes and boisterous manners. Here Adorno was not much different than Burke or Ortega. But perhaps the most appalling fact for Adorno's life was not simply disdain of the "dangerous classes," but finding his life completely uprooted after those (German) classes, or masses, faced the major crisis of 20th capitalism and embraced fascism, specifically in National Socialist form, Nazism.

3.1.1. Fascism
The domination of an entire society by a single, anti-democratic political party with authoritarian political party can take a number of forms. But in Germany, given its long history of anti-Semitism, turned especially virulent by Hitler, this meant that Germany would become a very dangerous place for various intellectuals, artists, leftists, and above all Jews. For the Frankfurt School in general and Adorno in particular, this spelled danger. When the projective tests for the early studies of authority and the family revealed the degree of intensity of the hatred that significant sectors of the Germans felt toward the Jews, blamed for everything from blood libel (using the blood of Christian babies for matzoth) to "stabbing Germany in the back" in WWI, and finally, as the masters of world finance, they were blamed for the Depression. Quite prudently, they left Germany and would become exiles, eventually in Los Angeles.

3.1.2. Exile
The prototypical exile is the "stranger" one both within and without of the culture (Simmel, 1950). It is no accident that the most perceptive critiques of American society have included such outsiders, consider only the tradition of Crevecoeur, de Tocqueville, or more recently Baudrillard. But for Adorno, despite the material comforts, and quite often a rich social/cultural life among other émigrés, his experiences were twofold. On the one hand, he offered some highly perceptive observations about the United States and its mass culture – even if in some cases, jazz, he got it wrong. On the other hand, his own cultural elitism made it impossible to ever fit into the egalitarianism of the United States where elites were given little respect, where anti-intellectualism was celebrated and its mass culture lowered the bar to ground level. As he put it, every émigré is mutilated.

3.1.3. Instrumental Reason – Identity Thinking
One of the central themes of Frankfurt School cultural critique, rooted in Hegelian Marxism, was the work of Lukacs on the reification of

consciousness embedded in the categories of bourgeois thought that made it difficult for the proletariat to gain a standpoint for self reflection. For the FS, the reified thought of bureaucratic logic (Weber) informed the understanding of the Enlightenment that promised emancipation from superstition and ignorance and would let humanity take charge of its own destiny. But as Nietzsche had noted, the logic of Reason had instead sustained old forms of domination and control.

For Adorno, the fundamental flaw of the Enlightenment was that Reason promoted identity thought, "dialectics seeks to say what something is, while identarian thinking says what something comes under, what it exemplifies or represents, and what, accordingly, it is not itself" (Adorno, 1987, p. 149). Identity thinking, as seen as the bureaucratic logic of classification, serves the interest of power and control. The understanding of self and world has become synonymous with its representations within the logic of Instrumental Reason. Foucault, also influenced by Nietzsche, would make similar arguments. Thus the logic of the Enlightenment, what Marcuse (1964) called "one dimensional" thought, became insinuated within individual consciousness as "normal, natural and common sense," served as a means of reproducing domination through "willing assent" to the status quo in Gramsci's terms. But furthermore, that one dimensional thought was valorized in the mass-mediated understandings of the world in which class power was cloaked. Discontent was assuaged by the culture industries, and acceptance of the system was rewarded by consumer goods and identities. Adorno went so far as to call such thought totalitarian – and only the critique of that thought might provide redemption.

3.1.4. The Culture Industries

Informed by their concerns with Nazi propaganda, the Frankfurt School was especially sensitive to means by which mass media in general served ideological functions. Fascist propaganda spoke to the fears and anxieties of the masses and promised amelioration through support of Hitler and the Nazi party. After the war, the Frankfurt School remained sensitive to how the mass culture systematically produced by the entertainment industry had had become the major source of domination by insinuating ideology within its many products, that however seemingly diverse, produced a mindless uniformity, subverted freedom and fulfillment, and eroded critical thought that might challenge domination.

Mass media, as one of the primary sites of identity thinking, tends to foster a valorization of the individual as he/she consumes culture, tastes, leisure activities, fashions, and home décor, while at the same time, massifies the individual by locating him/her within a marketing niche, audience or category in which his/her unique qualities are ignored. Thus individual psychological development, qua freedom and fulfillment are subverted, while his/her capacities for critique are limited to the evaluations of Betty Boop or Brittney Spears. Perhaps the more sophisticated might compare Madonna to Jenna Jameson. But with the demise of the family as a socialization agency, the mass media assumed ever greater role in fostering artificial needs, colonizing consciousness, providing models with whom people could identify, values they might internalize, and branded identities that could be purchased. The great leveler and equalizer of media was also the means by which hegemony was sustained through consumerism while the capacity for critical thought was erased.

3.2. Ambivalence and Resistance

For Adorno, the consequence of exchange relationships, the domination of Instrumental Reason and the culture industry was a character structure in which genuine human freedom and fulfillment were thwarted. But central to a dialectical understanding of character, following upon Nietzsche's notion of will and Freud's concepts of desire, *repression was problematic and ambivalence inherent to personal life*. As Wrong (1961) so elegantly put it, *the self can never be fully socialized and controlled – save as the ideal of bourgeois sociology and psychology*. (Read: advertising agencies.) It is at this point, noting the ambivalence within one's character, and the ambivalent relation of the self to the social that I would like to return to a central premise: *character, as used in this essay, is largely a quality of the individual, regardless of the extent to which that character has been constituted, if not fully, by the social; character is not the same as identity*.

Domination in general, and of Instrumental Reason, was historically instantiated. There are times when social changes in political economy are such hegemonic ideologies fail to adapt to new circumstance, and the relationships between character, self, and identity become problematic, vulnerable to change. At such times, there are spaces for counter hegemonic alternatives. As Marx put it, there are moments when certain patterns of social relations are no longer useful; they may act as a hindrance on social progress must be discarded. Here the roles of crises, organic intellectuals, and mobilizations are crucial in ushering change if not transformation through mobilizations. As Adorno notes, there is a contradiction in Freud between the liberation of desire and its repression to preserve domination.

Freud's "unenlightened enlightenment plays into the hands of bourgeois disillusion" (37, 60 [72] "As a late opponent of hypocrisy," [Freud] "stands ambivalently between desire for the open emancipation of the oppressed, and apology for open oppression" (37, 60–61 [72]). This contradiction or ambivalence in psychoanalysis "gives us the means for discovering" the "common root" of "mind [or Spirit] and pleasure" but "unintentionally" reproduces a "hostility towards both" (37, 61 [72]). Yet it would be the fate of psychoanalysis in America to move from the critique of repression to America's popular culture where "adaptation" became "desirable" without critical interrogation as to one might adapt to. Dr. Phil consoles Betty Boop.

Social life is always historically [materially] constituted and as such, constantly in a state of flux. Therefore, simply on the basis of cohort flow, every generation, and subgroups within that generation, exposed to different economic/political and cultural realities at the times of adolescence and early adulthood, needs to refashion existing identities to forge era appropriate identities (Mannheim, 1952). *These processes of cohort flow which require reconstituting identities, inject themselves into the spaces between character, selves and identities*, and thus create conditions that make constructions and/or renegotiations of identity especially likely. This takes place within social movements that seek to maintain the status quo, return to an earlier era, of or not of fashion new, progressive projects that would in turn lead to a better world.

4. PART III – FROM DESPAIR TO HOPE

4.1. Contesting Domination

From what has been argued, identity, as a central site of hegemonic process, as the point of mediation between individual character and his/her communities of meaning must be constantly renegotiated in face of changing conditions to serve hegemonic functions.

That said, the interstitial spaces between character and identity provide openings for challenge, contestation and change. The "cunning of Reason" often creates opportunities for contestation and transformation; there are moments for what Marcuse called a "great refusal" in which mass mobilizations can challenge received wisdom, logic, and identities to subvert the reproduction of hegemony. At times when things seem most pessimistic and despairing, there are often things going on in the liminal zones of the society quite below the radar – even of insightful philosophers. Minerva's owl depends on hindsight to notice what was missed at the time.

In the mid/late 1940s, despite the defeat of Nazism, for Adorno, the ascendance of the United States and explosion of its consumerism, its mass media, had ominous consequences portending dominating conformity at best or resurgent fascism at worst. But at the same time, there were growing tendencies that would eventually critique the alienation, conformism, materialism and massification of US society. Furthermore, not acknowledged by most thinkers at the time, but a vast number of returning soldiers went to college. Among them were a number of African Americans who, years later, would become major players in struggles for human rights.

Meanwhile, by the 1950s, a number of writers began to note the growing alienation of youth – and indeed, the very concept of alienation gained currency outside of Marxist circles. *The anti-heroes of the age, Holden Caulfield, Marlon Brando or James Dean, were not only products of the culture industries, but at the same time, personifications of the alienation and ennui of youth.* Second, a small number of intellectuals began to read existentialists such as Kierkegaard, Heidegger, Sartre, Camus, and I would include de Bouvoir. Although much of this literature stood apart from critiques of political economy, it did raise questions about mass society, blind conformity, inauthenticity and the vapidity of mass consumption. This literature was a part of the same cultural moment as the Beat generation of the 1950s. Although the Beats tended to be apolitical, they did give further voice to the alienation and dehumanization of (capitalist) mass society and they did challenge the repression and hypocrisy of the dominant society – at least on cultural grounds. The seeds were planted. And little acknowledged by the elite critiques of the culture industry, "Negro music" was changing into R & B, and eventually became vastly popular when marketed as rock'n roll.

It was in this context that Adorno's pessimism and/or perhaps his fears of resurgent fascism eventually led him to repudiate the student movement of the 1960s. At this time, however, the work of Bloch, extolling hope as the essence of humanity became a counterfoil to Adorno's despair. Like Adorno, Bloch had been influenced by Freud. But while Adorno was largely concerned with the internalization of authority as a moment of the reproduction of domination that was paid for in the suffering of repression, for Bloch, the day dream with its utopian hopes for a better life was rooted in the Freud's theory of the dream as wish fulfillment and drive gratification. And the 1960s became a time of hope. And it also became a time of a great deal of then transgressive drive gratification. The celebration of sex and drugs and rock and role was a critique and repudiation of repression, mass culture, and inauthenticity as much as an expression of yearning for a more "natural" life without the constraints of either necessity or repression.

4.2. Social Movements and Social Transformation – Here Comes 1968

When Habermas (1973) attempted to reframe Marx's theory of revolution as "legitimation crises," he noted how economic and/or political crises could migrate to the realms of identity and motivation. Here crises emerge within the cultural framework, for example, anomie exists when systems of norms and meanings become dysfunctional; they neither provide people with systems of meaning nor values to guide action. Between the individual logic of capitalism and the fragmentation of the social, especially when consumerism fosters a pseudo-individualism of niche marketed branded identities, there is a privatization of consciousness and a withdrawal of subjectivity from the political whether as the privatized hedonism fostered by the culture industries, the "civic privatisms" (Habermas, 1973) or not of political abstention through retreat to career/home, or the intellectual withdrawals to apolitical existentialisms, Eastern religions or beat literature. But as was earlier noted, the very conditions that led to the alienation of mass society decried by Adorno also contained seeds for what might become progressive mobilizations and transformations. In the postwar years the shadows of Nazism, the cacophony of Tin Pan Alley, the ersatz world of Hollywood dreams and syncopation of jazz were never far from Adorno the haute bourgeois émigré. Unbeknownst to him, as he and his Frankfurt School colleagues, especially Max Horkheimer, offered trenchant critiques of postwar alienation, reification, and commodification rooted in Instrumental Reason and identity thinking, these very conditions would fester, grow, and seek the negation of their negation.

As the war/post-war cohorts matured, little recognized at the time, but certain structural strains and contradictions were growing, namely the contradictions that would give rise to civil rights struggles, feminism, and sexuality. Large numbers of women held jobs – from Rosy the Riveter to Patty the Pilot who flew bombers to the war zones. For growing numbers of educated African Americans, the status quo of racial subordination was ever more intolerable. Much the same can be said as more and more women pursued higher education. Similarly, and in part due to WWII, when GIs found themselves in more liberal contexts, while at home many women had to compete for same man, there were growing tendencies for people to become more sexual.

By the 1960s, between the struggles for civil rights, and then anti-war and feminist mobilizations, the typical conformist identities of the dominant society faced challenge and contestations ranging from the counter culture with its culture of drugs and sex and rock and roll, and its be-ins of naked, yet painted, dancers high on pot, to the growing anti-war movement. It is important to note that in 1962, a group of UM grad students in sociology and philosophy, gathered together as SDS in Port Huron Michigan and developed a set of statements that decried the alienation and dehumanization of society, while rejecting the classical model of proletariat revolution. Accordingly, this marked the birth of what came to be known as the New Left. Two years later, when "*One Dimensional Man*" appeared, it would inspire a number of young people. It would offer a diagnosis of the ills of late capital – primarily through a critique of "one dimensional thought". But even more important, besides its critique of capitalist culture, as an essential aspect of Frankfurt School critique, it also offered an emancipatory vision of what the "good life" might be when domination is overcome. Moreover, as part of the Frankfurt School tradition Marcuse suggested that given the extent to which the proletariat had been incorporated into the society – especially through consumerism that erased class differences through niche marketing – a new revolutionary subject was needed.

It is at this point where critical theory intersects with social movement theory, especially the New Social Movement theory associated with Touraine, Castells, Melucci, Klandermans, etc., which has itself been influence by the FS. These frameworks were in part inspired by the events of the 1960s that required rethinking both the theories of the "irrational mobs" (Le Bon/Freud) or the traditional Marxist frameworks in which economic crises engendered a revolutionary proletariat that would be led by a vanguard party. For NSM theory, *au contraire,* the precipitants of mobilization may be economic, cultural, psycho-social, or indeed, a number of interdependent factors may coalesce to precipitate "legitimation crises." A culture that valorizes technology and its logic, Instrumental Reason, typically fosters dehumanization, meaninglessness, and denies people recognition of their humanity, in turn prompting anger and/or self destruction.

At certain historical moments, whether for individuals or the society, the relations between character and identity become problematic, often shattered and subject to reconfigurations. Individuals may join together and collectively negotiate new identity formations through which they may express themselves. One consistent theme of NSM theory argues that *its goals are not simply a reform, a new department or new program, but fundamental changes in the nature of identity, community, and frameworks of meaning that would in turn change the direction of the society toward a more participatory, inclusive democracy with more humanistic goals and the freedom for self fulfillment.* As Touraine (1985) and Melucci (1989) argued, in a post-industrial information society, social movements were more typically concerned with questions of culture (meaning/morality) and identity. From what has been said, I would like to suggest that among the most important questions raised by the 68 movements concerned identity, community, and culture [meaning]. Moreover, many of the actors were from the middle classes rather than peasants or workers who had traditionally been at the forefront of interest based social movements. Such groups, then more economically secure, were more likely to press for transforming identities. Although certain earlier movements ranging from the Romantics and Transcendentalists to the *Vandervogel* may have also been concerned with such issues, these were limited to small circles of the elites whose family fortunes enabled the cultural capital to pursue aesthetic/literary pursuits and life styles.

As Melucci (1995) pointed out: "Collective identity is an interactive, shared definition produced by several individuals (or groups at a more complex level)…that must be conceived as a process because it is constructed and negotiated by repeated activation of the relationships that link individuals (or groups) [to the movement]."[3] Although activists may realize their identities and values in their political/cultural activities, they are at the same time, involved in the submerged networks of everyday life, networks of meanings they produce and reproduce (Melucci, 1989, p. 71). The everyday life networks and communities are the spaces where identities are fashioned and negotiated. It is important to note that identity is not so much given as *the product of interaction and negotiations that are often moments of struggles.* The movements of 1968 were thus marked by three important features that were quite different from earlier kinds of movements. They were bottom up, or perhaps marginal to center movements may be more apt. Although there were wide scale uprisings that often did become violent, these were not "irrational mobs" seeking anarchy. Most violence came from the State.

Crises of legitimacy are often very sudden and unexpected. Although there may have been a long gestation period in which they slowly grow, such, there come times in a society, when mobilizations suddenly erupt. For various reasons, the typical identities become problematic. At various "moments of madness," all things seem possible (Zolberg, 1972). At these moments, social movements seek to articulate new kinds of collective identities and cultural understandings to foster political economic and/or cultural transformations that would result in a different kind of society. Thus, a crucial aspect for social movements is rejection or refashioning of identities/values. From what has been said, collective identity can be seen as a contested terrain in struggles for hegemony. For Melucci (1992, p. 133), a key element in the ideology of a movement is the "negation of the gap between expectations and reality. Ideology thus overcomes the inadequacy of action." It is at this point where fundamental insights of the Frankfurt School compliment New Social Movement theory. *Furthermore, unless and until there are emotional consequences to legitimacy crises and/or official policies, most people do not interrogate their identities or values.* Crises of legitimacy can trigger the emotions that lead some folks to question their identity, whereas others more radically affirm the long-standing identities.

Since identity is a central moment of hegemony and the terrain of struggles over historicity, how and why are identities interrogated and transformed? We might first note that one's identity is determined by a multiple of factors and even hegemonic identities are never quite fixed; rather, they are provisionally open (Laclau & Mouffe, 1985).[4] Furthermore, an individual maintains his/her collective identity insofar as it provides him/her with certain emotional gratifications, often compensations, namely attachments to a group, recognition of one's self, realms of agency, and a connection to frameworks of meaning. These gratifications of group membership that in turn sustain collective identities serve to enhance loyalty to the group, positive feelings toward the group, and positive feelings about one's self because of membership.

To paraphrase Marx, at certain stages of development, the perpetuation of "typical" identities comes into conflict with new and changed social circumstances, such identities act as a hindrance, a social psychological barrier to adaptation; people must then renounce certain identities and embrace new ones more suitable for the new social circumstances.[5] At this point, the newly emergent social conditions initially foster new configurations and constructions of identity among activist cadres who might be considered the harbingers of transformation. These movements aim at both changing the structures and ideologies that shape historicity, for example, selfhood in the future, as well as persuading others, larger audiences to reconfigure themselves to if not embrace, at least support these transformations.

From what has been said, crises of legitimacy typically evoke various emotions from anxiety, fear, and anger to shame and despair. But for the present purposes, we must understand these emotions within a group context which provides people with frames for understanding the basis of their emotions and feelings, and these frames are a part of an identity. A social movement will attempt to use such feelings to impel action beginning with enlisting others, often within "submerged networks to join in concerted action. As Melucci (1989) has pointed out, these networks allow people to interact and communicate with each other, and as they frame reality, debate interpretations, plan actions, so too do they create and negotiate identities within this context. Thus having places, spaces, and proximity is essential for identity based movements dependent on interactions.[6]

For many people, the very existence of others organizing or demonstrating leads them to shift their own notions of self and they may attend meetings, join demonstrations, and themselves become activists. Finally, in the process of negotiating collective identities, actors articulate visions of what a democratic society might look like. In face of alienation people seek agency, community, and meaning. In face of domination by authoritarian governments unresponsive to popular sentiments, people seek democratization, not just as a right or reform, but as a different kind of governance.

Castells has suggested that three are three kinds of identity formations, (1) *legitimizing identities*, attempt to maintain the status quo, but are often under assault by social/cultural changes. (2) *Resistance identities* of exclusion, opposition, and subordination are defensive attempt to forestall social change and demise, they often valorize traditions. Consider that racism, patriarchy and virginity have faced major assualts. Despite the reactionary mobilizations, they are now on the losing side of history. Finally (3) *project identities* aritulate new identity formations that are typically far more progressive than the society as a whole. The convergence of the Frankfurt School with NSM theory can be found in the common quest to foster project identities, expressions of subjectivity that valorize freedom, democracy, community, and includion. Such movements that envision a common humananity regardless of race, ethnicitiy, gender, or gender orientation, in turn, portend a better world. Instead of the bleak world of Adorno's damaged life, we can see that the Global justice movements tell us "a different world is possible."

5. CONCLUSIONS

Social theory, itself historically contingent, easily lends itself to fads, fashions and fissures. In the 40 or so years since the death of Adorno, we have seen the rise and demise of structural functionalism, structuralism, existentialism, and the late great post modernism. Critical Theory, rooted in Hegelian philosophy, Marxist political economy, Weber's interpretive sociology and Freudian depth psychology has never lent itself to being the fad theory of the month, or at least the year. It is far too complex, dialectical, and nuanced to be quickly absorbed and/or reduced to a few pages of a social theory text.

It has itself critiqued the positivism of "traditional" theory that has been embraced Instrumental Reason; it has sustained emancipatory critique and theory that would explain its marginalization in the academy. Moreover, besides its philosophical roots, its complexity and its interdisciplinarity, *it dares commit the ultimate sin of social science – it offers a value based critique of domination, and within that critique there is a utopian hope with a vision, if not prescription, for the "good life" in which fulfillment does not depend on power, the exploitation of others, domination over nature, and the accumulation of consumer goods*. But much like the horror movies produced by the culture industries, after the stakes are driven into the hearts of Freud and Marx, as the caskets are lowered into the dustbins of history, aka theory textbooks, a few fingers shoot up from the ground to signal that the news of the death of the Frankfurt School has been a bit premature.

Notwithstanding the pessimism and despair of the emigrant stranger in an anti-intellectual land, Adorno's many writings, inspired perhaps by a latent hope that such writings may yet negate the conditions of which he wrote, have shed profound insights about modernity and its contradictions. But at the same time, his very despair has led others to find optimism and hope. As has been argued, despite his disdain of the social activism of the 1960s, the fundamental basis of modern domination rests not so much on the power of the elites, "but the civic privatism" of the masses that are typically indifferent to the conditions of their subjugation, indeed as the Frankfurt School in general has shown, and Adorno in particular, the very nature of character as socialized and mediated through mass produced and mediated identities leads people to actively collude in their own subjugation.

That said, the Frankfurt School understanding of how domination becomes internalized, remains a major, and enduring contribution to the understanding of domination, and a far more nuanced understanding than similar arguments of Althusser or Foucault whose frameworks, devoid of affect, agency or resistance, reproduce the very domination they would critique. The Frankfurt School understanding, as rooted in the dialectical relationship of character and identity, that was not fully theorized, offers a space, both theoretically and empirically, for resistance and challenge.

Here, the legacy of Freud, the unconscious, and ambivalence offer both a way to understand domination and despair, but at the same time, despite the despair of Freud, as well as Adorno, their pessimism was tied to both their own characters and the historical context of their lives and writing.

As was shown, the very analysis of the "damaged life," Adorno's, the victims of Nazism and those alienated by postwar consumerism/mass culture, led not to "joyous consciousness," but either civic privatism or a politics of despair. As Adorno might note, thinking of Beethoven, must it be? I have suggested the very alienation that Adorno critiqued, also generated resistance and contestation – as was evident in the various mobilizations that have come to be known as the 1960s. While the 1960s did bring about some cultural changes, especially regarding the place of minorities in the United States, the role of women and sexual agency, absent a progressive political party, there was reaction and indeed political stagnation if not retrogression. But again, must it be?

While in some ways the 1960s was a brief moment of hope, which was crushed by tanks in Prague, machine guns in Mexico City, a failure of nerve by the Communist Party of France (PCF), and the hedonism of the American counter culture. But again, we can ask, must that be? And I would argue NO. Although the conditions of alienation, domination, and reification have joined with reactionary politics that would preserve resistance identities facing demise, these very conditions offer new possibilities for mobilization.

And as could be argued, and has been, one of the legacies of the 1960s has been the Global Justice Movements. Although these movements are not political parties per se, we have now seen how various mobilizations in Latin America have moved most of that that continent sharply to the left and less subject to US domination or intimidation. From Chile to Venezuela, the social movements, reactions to domination, and alienation have forged new identities and paths to more participatory democracies.

At the same time, we are today witnessing a vast mobilization of youth in support of Barack Obama. Although it may be easy to decry the general apathy of youth, and question the extent that any mainstream politician might be an agent of progressive change, I would argue that in the very process of mobilization, new identities are being crafted and negotiated. Notwithstanding the fact that Obama is hardly the progressive some would imagine, and he will surely disappoint those who think so, he nevertheless has captured the *Geist* of a new generation that would seek a new political imaginary. The important question then asks whether this mobilization can fashion new identities with progressive political imaginaries. Let us meet again in 40 years to judge this question, and again give thanks to Theodore Adorno whose despair can only inspire our hope. Without Adorno, we cannot embrace Bloch, appreciate the importance of dreams, we cannot recall the past, understand the present, nor imagine that the good life is possible.[7]

ACKNOWLEDGMENT

The author expresses his deep gratitude for the many insightful comments of Professor Steve Bronner.

NOTES

1. For an extended critique of Minima Moralia, see Israel (1997).
2. http://science.jrank.org/pages/9748/Multiple-Identity-Critique-Subject.html
3. "The Process of Collective Identity," in Johnston and Klandermans, eds., 1995. *Social Movements and Culture*.
4. For example, consider a person at the intersection of say gender, race, and class, which will be most salient.
5. Parenthetically, it might be noted that for many, not only are changes in subjectivity and identity unlikely, but in face of challenge, such people are likely to become angry, and some may even turn to violence to retard social change. Thus, for example, we see various expressions of irrational hatred toward President Obama who represents an emergent multi-cultural society. As the hegemonic status of white, male conservatives wanes, especially at a time of economic retrenchments, we can see various expressions of anger from the barely disguised racists birther movement that he was born in Kenya.
6. McAdam (1988) showed the importance of churches as places where during "Freedom Summer" activists could gather, live, work, and plan for social justice. As argued, identity movements are quite dependent on place, in many cases, more so than they are dependent on leadership.
7. An excellent introduction to Bloch can be found in Kellner, http://www.uta.edu/huma/illuminations/kell1.htm

REFERENCES

Adorno, T. (1950). *The authoritarian personality*. New York: Harper.
Adorno, T. (1951). *Minima Moralia: Reflections on a damaged life*. London: New Left Books.
Adorno, T. (1987). *Negative dialects*. New York: Continuum.
Adorno, T., & Horkheimer, M. (1944). *The dialectic of enlightenment*. London: Verso.
Fromm, E. (1973). *Anatomy of human destructiveness*. New York: Holt, Rinehart and Winston.

Giddens, A. (1991). *Modernity and self-identity: Self and society in the late modern age*. Stanford: Stanford University Press.
Habermas, J. (1973). *Legitimation crisis*. Boston, MA: Beacon Press.
Horkheimer, M. (1972) Authority and the Family. In: *Critical theory: Selected essays* (pp. 47–129). New York: Herder and Herder.
Israel, N. (1997). Damage control: Adorno, Los Angeles, and the dislocation of culture. *The Yale Journal of Criticism, 10*, 85–113.
Jacoby, R. (2005). *Past imperfect*. New York: Colomiba University Press.
Kellner, D. Ernst Bloch, Utopia and ideology critique. Available at http://www.uta.edu/english/dab/illuminations/kell1.html
Kohut, H. (1971). *The analysis of the self*. New York: International Universities Press.
Laclau, E., & Mouffe, C. (1985). *Hegemony and socialist strategy: Towards a radical democratic politics*. London: Verso.
Langman, L. (2006). The social psychology of nationalism: To die and kill for the sake of strangers. In: G. Delanty (Ed.), *Handbook of nations and nationalism*. London: Sage.
Mannheim, K. (1952). The problem of generations. In: P. Kecskemeti (Ed.), *Essays on the sociology of knowledge by Karl Mannheim*. New York: Routledge & Kegan Paul.
Marcuse, H. (1955). *Eros and civilization*. Boston: Beacon Press.
Marcuse, H. (1964). *One dimensional man*. Boston: Beacon Press.
McAdam, D. (1988). *Freedom summer*. Oxford: Oxford University Press.
Melucci, A. (1989). *Nomads of the present*. Philadelphia, PA: Temple University Press.
Melucci, A. (1992). Frontier land: Collective action between actor and systems. In: M. Diani & R. Eyerman (Eds), *Studying collective action*. London: Sage.
Melucci, A. (1995). in The process of collective identity. In: H. Johnston & B. Klandermans (Eds), *Social movements and culture*. London: Routledge.
Reich, W. (1946, 1933). In: T. P. Wolfe (Trans.), *The mass psychology of fascism*. New York: Orgone Institute Press.
Simmel, G. (1950). *The sociology of Georg Simmel (K. Wolff, Ed. & Trans.)*. Glencoe, IL: Free Press.
Theweleit, K. (1987). *Male fantasies*. Minneapolis: University of Minnesota Press.
Touraine, A. (1985). *Social movements and social change*. London: Sage.
Wrong, D. (1961). The oversocialized conception of man in modern sociology. *American Sociological Review, 26*, 183–193.
Zolberg, A. (1972). Moments of madness. *Politics and Society, 2*, 183–207.

UNDER SURVEILLANCE: HERBERT MARCUSE AND THE FBI

Stephen Gennaro and Douglas Kellner

ABSTRACT

This article is the first in a series that seeks to examine the Federal Bureau of Investigation' (FBI) surveillance of social philosopher and activist Herbert Marcuse between 1943 and 1976. We intend to map in parallel lines local, national, and international media representations of Marcuse, scholarly analysis of Marcuse's writings, Marcuse's own correspondence, speeches, and texts in comparison with the presentations of Herbert Marcuse in the collected FBI documents. Our goal is to assess what the Marcuse's FBI files tell us about the FBI, Marcuse, the New Left, and U.S. society in the 1960s. In particular, close attention is paid to examining events described inside the FBI documents occurring in the mid-1960s when Herbert Marcuse was emerging as a self-proclaimed Marxist radical, a father figure to New Left and countercultural activists, an influential author, public speaker, and teacher, and was beginning to be perceived as a threat by the FBI to U.S. national security. We seek to clarify if FBI documents can provide information and insight to help illuminate and understand U.S. social and cultural history, in this particular case, to assess how FBI documents measure up against scholarship and perceived views of Marcuse and the 1960s. We are thus interested both in what we can learn about Herbert Marcuse's life and

Nature, Knowledge and Negation
Current Perspectives in Social Theory, Volume 26, 283–313

ISSN: 0278-1204/doi:10.1108/S0278-1204(2009)0000026014

times from these documents and what FBI surveillance and documents tell us about the FBI and U.S. intelligence services.

In the 1960s, Herbert Marcuse was emerging as one of the most important intellectuals in the United States. His 1941 book *Reason and Revolution* provided an introduction to Hegel, Marx, and social theory for English-speaking audiences. Marcuse's *Eros and Civilization* (1955) provided an audacious synthesis of Marx and Freud and powerful critique of contemporary U.S. society that also anticipated the 1960s counterculture with its celebration of Eros, emancipation, play, and the aesthetic-erotic dimension of experience. Marcuse's *One-Dimensional Man* (1964) radicalized his critique of "advanced industrial society," providing powerful critical perspectives on both contemporary capitalist and communist societies.[1]

Marcuse, a German-Jewish emigrant from Nazi Germany, was a member of a remarkable group of radical intellectuals who were initially rooted in Frankfurt, Germany, and moved their Institute for Social Research to Columbia University where they continued their research projects into fascism, the authoritarian personality, and contemporary forms of power and social control. Marcuse was a key figure in this group and continued developing the "critical theory of society" on his own in his American context after Horkheimer, Adorno, and other members of the group returned to Germany after the war.

Marcuse was not only of increasing interest to intellectual and the emerging radicals of the New Left and counterculture but also elicited the attention of the Federal Bureau of Investigation (FBI) and its Director J. Edgar Hoover.[2] In this study, we examine the FBI's surveillance and records on Marcuse and are interested both in what we can learn about Herbert Marcuse's life and times from these documents and what FBI surveillance and documents tell us about the FBI, the New Left, and U.S. society in the 1960s. This preliminary study focuses on documents that portray Marcuse's initial scrutiny from the FBI during his government service in the 1940s and then his political activism in the early through middle 1960s, before he had become an icon of the New Left and media celebrity. Hence, we begin with the FBI documents on Marcuse through an examination of how these documents positioned Marcuse's place in U.S. society and politics in the 1940s and then the 1960s in relation to the social and political contexts of their time.[3] We start in the next section by noting how Marcuse came to the attention of the FBI in the mid-1960s after having passed FBI security clearances for his work with the U.S. Government during the period between 1943 and 1951.

MARCUSE, THE FBI, AND U.S. INTELLIGENCE SERVICES

On May 13, 1966, the FBI officially began systematic surveillance on Herbert Marcuse, then a Professor at the University of California, San Diego, for his perceived role in the rising student movement and countercultural protest against U.S. military involvement in Vietnam and Indochina. However, this was not the first formal investigation of Hebert Marcuse conducted by the FBI.[4] In fact, it was at the very least, the third formal investigation of Marcuse who had been the subject of constant surveillance of the FBI over the previous two-and-a-half decades, largely as standard security checks in relation to his work with the U.S. Government.

The scope of the official surveillance that began in May 1966 sheds light on earlier routine FBI investigations of Marcuse between 1942 and 1952. Ironically, between 1942 and 1952, at the same time that the first two FBI investigations of Marcuse were underway, his main form of employment came from the U.S. Government itself. Although he still had a brief connection to Columbia University (e.g., through a series of four lectures he gave at the Russian Institute in 1948), Marcuse held positions for U.S. intelligence services during World War II and then the State Department.[5]

Previously, Marcuse had worked as a researcher, translator, and informant at the Office of War Information (OWI), Bureau of Intelligence, beginning in December 1942 and then with the Office of Strategic Services (OSS) from April 1943 until September 1945, before becoming the Chief of the Central European Branch of the Division of European Research for the State Department.[6] What began as a special FBI inquiry into Marcuse's activities in the spring of 1943 as a prelude to his entering the OSS escalated over the next decade to include a full-scale investigation under Executive Order 9835 in 1950, according to which the FBI in conjunction with the Attorney General and the House Committee on Un-American Activities investigated government employees on suspicion of disloyalty and/or violation of Federal Law.[7] As a federal government employee in the period following the World War II – a period that became known as the "Red Scare" – Marcuse had been under investigation for his supposed connection to Communist countries, organizations, ideologies, and sympathizers. And even though Marcuse was officially cleared of all charges and no longer under formal investigation by the end of 1952, the stigma of suspicion had been attached to his name and would remain there for the rest of his life.

In retrospect, it is easy to see how Marcuse's first official FBI investigation came into being. As a German-Jewish radical intellectual fleeing Nazi Germany, Marcuse came to the United States in 1934 after several of his colleagues had been invited by the President of Columbia University to teach and conduct research in New York.[8] It was during this time that the Institute for Social Research, now known as the Frankfurt School (and home to thinkers such as T. W. Adorno, Max Horkheimer, Leo Löwenthal, Erich Fromm, Franz Neumann, Marcuse, and others), relocated from its original home in Frankfurt, Germany, to Columbia University in New York.[9] Along with Marcuse, who worked at the Institute branch at Columbia from 1934 to 1939, his colleagues who conducted work with the Institute at Columbia from Horkheimer and Adorno to Franz Neumann and Leo Löwenthal found themselves either as the subject of FBI surveillance or as providing testimony concerning their colleagues who sought U.S. Government positions during World War II.[10]

In fact, even beyond the Frankfurt theorists, the FBI became immediately suspicious of Jewish immigrants from Germany, especially members of the intelligentsia, and kept their activities under close surveillance.[11] While the work at the Institute in New York had by the post-war period appeared to have gained the respect of the academic community in the United States, and the Institute was viewed as a "credible institution," within the FBI the work conducted at the Institute in Frankfurt and then at Columbia was seen as pro-socialist, Marxist, and a house of sympathizers for communism.[12] No doubt then, Marcuse's record as an academic – in particular his having worked (although briefly) at the Institute in Frankfurt and Geneva, combined with his authorship of two books on Hegel and many articles on Marx, critical philosophy, fascism, and the totalitarian state – only enhanced the extent to which the FBI took an interest in what he had to say.[13] (Indeed, this is true of many who had conducted research and published writings on such subject matters). Ironically, it was neither of these scenarios that first brought Marcuse to the attention of the FBI during the World War II. Rather, it was his position working *for* the U.S. Government that made him appear as a possible security threat and ideological danger.

The first FBI report about Marcuse that is present in his Freedom of Information FBI dossier, inside file #77-27880, was dated April 20, 1943.[14] Marcuse at the time was employed within the OWI, Bureau of Intelligence, which was apparently under the supervision of the Office of Emergency Management (OEM) and he had applied to work at the OSS.[15] Upon applying, internal investigations on Marcuse's character, background, and even credit history were conducted, and all were part of the regular routine for vetting U.S. Government employees. As outlined in the report, four confidential sources were interviewed as character references, and all spoke positively of Marcuse as an academic, as a person, and as a supporter of American ideals. On May 21, 1943, J. Edgar Hoover sent an internal memo to the FBI SAC (Special Agent Center) office in New York City with the following text: "Herbert Marcuse, OEM, Discontinue Investigation."[16] For a brief moment, the FBI considered Herbert Marcuse a loyal American and not to be a security risk.

However, this was not the end of the first FBI investigation of Herbert Marcuse. Although he had been cleared by the FBI internal review, his freedom from FBI suspicion was short-lived. In fact, in the August 9, 1950 report that started the second major FBI investigation of Marcuse mention is made of Marcuse inside FBI investigations from fall 1948 through to spring 1949.[17] Therefore, it appears that in contradiction to Hoover's 1943 memo, the surveillance of the activities of Marcuse was in fact not discontinued. This additional surveillance is corroborated in a September 14, 1950 letter to J. Edgar Hoover from a SAC agent in New York that makes reference to "an extensive investigation conducted by the State Department concerning MARCUSE at which time some 25 to 30 persons were interviewed."[18] The same letter also mentions "'an extensive investigation conducted by the agency [The Civil Service Commission]' in late 1944 and early 1945 'concerning HERBERT MARCUSE' in connection with his employment as a research analyst with O.S. Services ... [where] some persons were interviewed."[19]

In both cases, the agency concluded that "none of whom [the interviewees] questioned MARCUSE's loyalty," but immediately contended that many of those interviewed made reference to Marcuse's connection to the "Institute of Social Research," which according to the FBI had brought Marcuse into direct connection with known Communists and therefore, as one respondent (later identified as Dr. Eric Franzen of New York) noted in 1950, "any information which came into MARCUSE'S hands would ultimately reach Drs. HORKHEIMER and POLLOCK and, that there was absolutely no question that MARCUSES primary loyalty was to the International Institute of Social Research."[20] Over the next three decades, the information that had been used to clear Marcuse's name in 1943 would come under this type of continual re-interpretation by the FBI as the information was gathered and files maintained, and the first investigations would be reopened and become an important part of the multiple FBI investigations that would follow. So, although formally closed on

May 21, 1943, Bureau surveillance of Marcuse remained active between 1943 and 1950.

This is a theme repeated throughout FBI documentation of Marcuse, where the 1943 investigation, although proving Marcuse to be loyal, appears in later Bureau reports as reason to keep Marcuse under suspicion. The treatment and re-interpretation of documents clearing Marcuse in formal FBI investigations in 1953 and 1966 was the same.[21] The individuals listed as references on Marcuse's initial application for State employment in 1942: Franz Neumann, Max Horkheimer, and Friedrich Pollock – whose names and reputation helped earn him employment at the OWI and OSS, would all be mentioned in later reports as evidence of Marcuse's connection to radical un-American thought and as reason for suspicion. Furthermore, the "confidential sources," who had professed to FBI agents that Marcuse was a loyal American and good worker, were replaced by "confidential sources" suspicious of his actions and associations. By the end of the 1940s, the Marcuse FBI file had been officially re-opened and Marcuse became the subject of a formal loyalty investigation of government employees based mostly on his connection to the Institute for Social Research, the work of his intellectual colleagues, and his previous 1943 investigation by the FBI.

On August 9, 1950, at the request of J. Edgar Hoover, the FBI SAC Washington Office was asked to conduct a preliminary report on "loyalty" of Herbert Marcuse, then Chief of the Central European Branch of the Division of Research for Europe within the State Department.[22] Marcuse's connection to radical thinkers, pro-Communist and pro-Socialist intellectuals, had garnered some interest within the Bureau. During his tenure with the OWI and then with the OSS, Marcuse had spent his time conducting research for the government and as such was to refrain from publishing personal and academic works.[23] However, many of his Frankfurt School counterparts used this period to publish articles, essays, and books widely critical of capitalism, politics, culture, and modes of thought, such as Adorno and Horkheimers' highly controversial *Dialectic of Enlightenment*.[24]

Throughout this period, Marcuse kept regular correspondence with his colleagues from the Institute of Social Research. It appears that it was Marcuse's relationship to the Institute and to other radical refugee intellectuals that resulted in his 1950 loyalty inquiry. For example, one of the testimonies used as the basis for questioning Marcuse's loyalty in 1950 (as it had in the 1943 investigation) was that a librarian at UCLA testified having witnessed Marcuse and other German scholars (believed to have been Adorno and Horkheimer) coming into the UCLA library. According to this "confidential source" in the April 20, 1943 report, the gentlemen looked at books about Marxism and "appeared suspicious."[25]

Marcuse's connection to social critics and public figures also under investigation for possible un-American activity continually raised eyebrows within the Bureau. This was another theme repeated throughout the FBI documents on Herbert Marcuse, where more was made of his connection to other people than of his own work and character. At first, in 1943, it was Marcuse's connection to the influx of European (and more specifically German-Jewish) immigrants fleeing Nazi occupation, and the public notion that immigrants from Germany were either un-American or unappreciative of the American way of life that raised suspicion. In the mid- to late 1940s, it was Marcuse's connection to the Institute for Social Research and to intellectual Marxists that raised suspicions. In the 1950s, it was his connection to known or suspected communists and Marxists and his own espousal of critical Marxist philosophy that incited attention inside the Bureau, while in the 1960s, it was his connection to the "student movement", the "New Left", the counterculture, Angela Davis, and European radicals, such as Rudi Dutschke. In each and every scenario, Marcuse was treated within the Bureau as guilty by association and more often than not, it was the actions of those people around Marcuse and the media, and the public perception of Marcuse's connection to these people and/or movements that became the framework that defined Herbert Marcuse in FBI documents as a Marxist, a revolutionary, an anarchist, and a dangerous supporter of the New Left and world revolution.

MARCUSE'S 1960s ACTIVISM AND THE FBI

After leaving the State Department in 1951, Herbert Marcuse spent more than two decades teaching at well-established academic centers such as Columbia, Harvard, Brandeis, and the University of California, San Diego. It was Marcuse's teaching and his writing, lectures, and speeches, including *One-Dimensional Man*, "The Obsolescence of Marxism," "The Problem of Violence and the Radical Opposition," and other texts dealing with the New Left, socialism, and revolution that not only caught the attention of the FBI but also caught the creative imagination of a generation of student protest in the 1960s.[26]

FBI documents contain a wide and impressive record of Marcuse's activism throughout the 1960s. One FBI document, labeled SD100-13701, dated November 8, 1967, which covered the investigative period of October 9 to November 1, 1967, under the rubric "SECURITY MATTER –C," features reports by five Bureau informants (labeled REGISTERED) and three San Diego informants who compiled a wide-ranging report on Marcuse's politics, writings, and activities of the early through mid-1960s. This document indicates that "On January 18, 1967, SD T-2 furnished the following information concerning the previous activities of subject:"

> On May 10, 1961, subject signed an open letter to President KENNEDY regarding Cuban policy.
>
> On Marcuse 12, 1963, subject reportedly signed a petition for clemency regarding CARL BRADEN and FRANK WILKINSON.
>
> On November 1, 1967 SD T-2 advised that CARL BRADEN and FRANK WILKINSON, both known Communist Party, USA, members, were both convicted of contempt of Congress in 1961 for refusing to testify before the Committee on Un-American Activities ...
>
> On December 5, 1963, subject signed a petition on behalf of three students indicted for going to Cuba, under the auspices of the "Emergency Civil Liberties Committee" ...
>
> On February 16, 1965, subject reported to be signer of open letter to President JOHNSON urging negotiation in Vietnam.
>
> In March 1965, subject contributed an article to the issue of "Monthly Review."
>
> On September 17, 1965 subject's signature appeared on an advertisement in the "Washington Evening Star", entitled "Eternal Light Vigil", under the auspices of the American-Jewish Conference on Soviet Jewry.
>
> On October 31, 1965, the "New York Times" carried an advertisement which was signed by the subject and which was entitled "Ad Hoc Committee on Vietnam War."
>
> On March 25, 1966, the Los Angeles Free Press' indicated that subject was a scheduled speaker on March 25, 1966, at a teach-in held under auspices of the University of California at Los Angeles' Vietnam Day Committee (a protest against U.S. policy in Vietnam).
>
> [The next entry is blacked out by the FBI and unreadable, but the succeeding FBI description indicates that Marcuse was involved in an effort to obtain "clemency for Hugo Blanco, a Peruvian revolutionary freedom fighter"] ...
>
> On May 14, 1967, SD T-5 reported that Dr. MARCUSE had recently lectured in connection with a course being offered by the University of California Extension Service entitled "The New Radicals." In this lecture Dr. MARCUSE stated that the "vested interests" exist and seek to preserve the status quo and that the conservatives may bring more evil than the radicals. He said that the conservatives advocate the intensification of

> the war in Vietnam, the repression of civil rights, anti-intellectualism and censorship in education. He stated that the New Left is not socialistic, not tied to labor movement [sic] and not committed to one class of society, whereas the Old Left was primarily socialist and labor oriented. He stated that the New Radicals and the New Left are the youth intellegentsia [sic], the "hippies", civil rights leaders, and radical "squares" like professors, writers, and artists. He said that the Youth of the New Left have no established leadership or effective communication but they represent incongruous social classes and show spontaneity and flexibility. He said that the New Left is suspicious of any new ideology or action of the "establishment". MARCUSE said that the immediate goals of the New Radicals are the termination of war and "participatory democracy." Source stated that Dr. MARCUSE appeared sympathetic toward the New Left and in his lecture advocated the teaching of non-conformis [sic] ideology, independent of government, and free universities. He pointed out that participatory democracy would entail the dissolution of the military-industrial complex, a vast extension of welfare, the breaking up of information monopolies and the abolition of foreign policy.[27]

The same memo indicates that "On February 15, 1967, SD T-2 made available information concerning Dr. MARCUSE's writings and statements" and the next several pages of the memo provide summaries of a large number of publications by Marcuse, some critiques of his work, and summaries of reports of Marcuse's activities in newspapers such as the *New York Times* or *San Diego Union*.[28] Other files in Marcuse's FBI dossier indicate that the FBI hired out individuals to provide summaries of Marcuse's books and writings under the rubric "Research Satellite Division, Domestic Intelligence Division."[29]

In re-examining the FBI documents from the Bureau's surveillance of Marcuse, in addition to writings by and about Marcuse, what becomes evident is that there is a clear battle ensuing around the social construction of the public image of Herbert Marcuse, starting in 1966 and continuing through the 1960s and into the 1970s. At this time, the FBI was engaged in an ideological battle with the youth of the country in the midst of a counter cultural revolution. And at the same time as young people and others were actively creating a public image of Herbert Marcuse as idol and icon of the emerging New Left and antiwar movement and activist for social change, the Bureau was equally hard at working for a collective construction of Herbert Marcuse as a threat to National Security. To effectively construct Marcuse, then a 70-year-old Professor as a clear and present danger to American society, the Bureau worked to paint him as an individual who stood at odds with the core values of U.S. democracy, and this battle was waged through the emphasis on representing Marcuse as the "other," that is, as "a self-proclaimed Marxist," a Jewish-German intellectual and immigrant, as an anarchist, and as a revolutionary. In addition to highlighting the differences in Marcuse's life with the social construction of what it meant to be an American, the Bureau also worked to demonstrate Marcuse's connection to student rebellion, counter-culture, and people such as Angela Davis and Rudi Dutschke and filed reports under rubrics like "Security Index C" and categories such as "anarchist," New Left revolutionary, and "threat to national security."[30]

Beginning in the spring of 1966, Herbert Marcuse's connection to the growing student and antiwar movements in both Bureau circles and the mass media intensified. His perceived connection to student and New Left uprisings at home and abroad and a backlash against Marcuse witnessed through the public protest to have him removed from UCSD as well as a series of death threats from the Ku Klux Klan all attested to his growing radicalism, public influence, and interest for the FBI. The surveillance and construction of Marcuse as a dangerous radical by the FBI in the period around 1966 will be explored in the following sections through a close scrutiny of three speeches given by Herbert Marcuse in the spring of 1966. The talks took place across the country (one in the Midwest and one on each coast), at three well-established academic centers, and as part of officially sanctioned University symposiums or teach-ins. Coverage of each of the three speeches given by Marcuse can be found inside his FBI dossier and appeared to be instrumental in establishing his identity as a self-proclaimed Marxist, a father figure to the New Left and countercultural activists, and as a threat to national security in the spring of 1966.

In particular, Marcuse's April 25, 1966, talk at the University of Notre Dame, at a symposium on "Marxism in Academia" entitled "The Obsolescence of Marxism," garnered national attention and was used by the FBI and the national media to portray Marcuse as "the self-proclaimed Marxist." A month prior, at a teach-in at the University of California, Los Angeles (UCLA), Marcuse openly attacked U.S. foreign policy and called for student activism, actively participating in the role of "countercultural father figure" and "threat to national security." Only days after his talk in Notre Dame, and already in the national spotlight, Marcuse delivered the keynote address at the Yale Socialist Symposium, in a talk that appeared to mix the "Marxism" of his Notre Dame talk with "the anti-war sentiment" of his UCLA speech, in a talk that was closely scrutinized by the FBI.

THE OBSOLESCENCE OF MARXISM: MARCUSE SPEAKS IN SOUTH BEND

April 25, 1966, may not have been Herbert Marcuse's high point as an educator, philosopher, or social critic, but in retrospect, it holds significant importance to the public representation of Herbert Marcuse and the FBI's interest in him. In fact, evidence suggests that it was a rather typical and ordinary moment in his career. As one of many speakers at an international symposium on Marxism, academia, and intellectual thought at the University of Notre Dame, in South Bend, Indiana, he spoke on re-thinking the use value of Marxist thought in relation to contemporary American society and politics and some of the challenges inherent in doing so. Marcuse was not even the keynote speaker of the event. And yet, it was this speech at Notre Dame that captured not only the attention of the FBI (and then became the basis for re-opening his Bureau file that had been closed for more than a decade) but also the attention of major sources of mass media such as the *New York Times*.

Marcuse's speech in South Bend, Indiana, made the headlines of the April 26 edition of the *New York Times*, where on page two, the article "Marxist views Vietnam in the Context of Capitalism," immediately became the topic of national discussion. For example, in a personal letter from Arnold Tovell Marcuse's editor, in Boston, to Marcuse's home in La Jolla, CA on May 3, 1966, he writes as an afterthought to the work-related letter: "My wife says she's in hearty agreement with Mrs. Marcuse's letter to the New York Review of Books! So am I, but I was much more intrigued by the way the *New York Times* treated your appearance in South Bend. Did you have a good time?"[31]

The national discussion sparked by Marcuse's speech extended beyond friendly correspondence to include even the FBI. For example, FBI Director, J. Edgar Hoover, noted in his May 13, 1966 letter from his Virginia office to the San Diego office, "[e]nclosed herewith is a Xerox copy of an article which appeared on page two of the April 29, 1966 'The New York Times' ... you are instructed to discreetly identify subject, review indices and contact established sources and pertinent confidential informants concerning subject; and submit results in a report-form suitable for dissemination."[32] Perhaps, the April 1966 speech at South Bend, which at the time was an ordinary event in his life as a public intellectual, has found a space in the construction of his role in media representations as one of the most important public talks Marcuse ever gave.

Although Hebert Marcuse was always a scholar with a clear interest in Marxist theory and Hegelian philosophy, it was only in the aftermath of the South Bend talk that the construction of Marcuse as a "self-proclaimed Marxist" began to crystallize in the public media representations of Herbert Marcuse. And while Marcuse never actually says the words "I am a Marxist" or "as a Marxist" anywhere in this talk, it appears that a great deal of the focus springs forth from one of the first lines of his talk, where in explaining how his title "The Obsolescence of Marxism" is more of a question than a statement, he notes "[t]he title of my paper is not supposed to suggest that Marx's analysis of the capitalist system is outdated; on the contrary I think that the most fundamental notions of this analysis have been validated."[33] The *New York Times*' coverage of the event immediately framed Marcuse as "the Marxist Professor," claiming "Professor Marcuse of the University of California, San Diego, noting that he was a Marxist, contended that Marxian analysis was relevant to the internal structure and development of affluent capitalistic societies such as the United States."[34] Ensuing Bureau documents, such as the FBI Directors' letter dated May 13, 1966, frame the need for continued surveillance of Marcuse based on the fact that the "subject [Marcuse] identified himself as a Marxist."[35] It appears that missed in the *New York Times* and FBI synopsis of the South Bend lecture was Marcuse's larger point that "the factors which have led to the passing and obsolescence of some decisive concepts of Marx are anticipated in Marxian theory itself," concluding, "the affluent society corroborates rather than refutes the internal contradictions which Marx attributed to capitalist development."[36]

Thus, Marcuse's talk at South Bend spoke more to the inherent Marxist nature of U.S. society where corporate powers dominate social, political, and cultural life than it did to a call for proletarian revolution or socialism, both of which Marcuse argued were unlikely in the contemporary U.S. context.

Although the talk at South Bend was the first public lecture of Marcuse to garner significant press and public response, it was not the first time Marcuse had been openly political in challenging U.S. foreign policy. For example, on May 3, 1961, at Brandeis University, Herbert Marcuse, then a Professor in the History of Ideas program at Brandeis, participated in an organized protest against America's current stance on the Cuban Revolution.[37] In one of his first public criticisms of U.S. foreign policy in the post-war era, Marcuse laid the framework for many of his future arguments against U.S. intervention and occupation in developing nations and its practice of violent intervention in the affairs of foreign countries. Many of these themes were repeated in the April 1966 "Obsolescence of Marxism" talk at South Bend as well as in his April speech "The Inner Logic of Vietnam" at a UCLA teach-in and in his April 1966 talk at the Yale Socialist Symposium, which we engage below.

In his 1961 speech at Brandeis, Marcuse argued that the Cuban crisis, in which the U.S. Government found itself, was a self-created problem.

According to Marcuse, the decision of Castro to rely on the foreign aid of Communist nations like the Soviet Union was simply the most basic response (in the best interest of his people) to the U.S. embargo, based on perceived U.S. interests but wrapped up and sold to the public under the banner of Cold War ideological battles Marcuse argues that the U.S. rhetoric of using military force in protecting Cubans from violations of civil rights under the Castro regime was hypocritical since the United States was happy to support violent and hostile takeovers in other countries from militant rightwing or Fascist groups not associated with Communism, which systematically violated human rights (he lists Spain, Formosa, Guatemala, and Haiti as just a few examples).

Marcuse then argues, "[i]t seems, in other words, that we are only against repression of civil rights and liberties if it comes from the left, but certainly not if it comes from the right."[38] Marcuse furthers his critique of U.S. foreign policy to suggest that the "hypocrisy" of the policy lies in the fact that the U.S. Government is moving further and further away from a society that protects the civil liberties of its own citizens. He argues, "what we see is the rapid transformation of our own society into a *un*free society which already shows the tendencies which we so valiantly deplore in other countries."[39] Marcuse then lists the change in how we exchange ideas and communicate in society as the key indicator of this loss of freedoms: "the restriction of the freedom of the press ... a self-imposed censorship, ... a moratorium on criticism, ... a *mis*information of the public, and [the emergence of the importance and value invested in] the cult of personality."[40]

What is particularly interesting about Marcuse's talk on Cuba is how he ends the talk by challenging his audience at Brandeis to take up the cause to make their voices heard. "I do not overestimate what can be done, but I think we have a duty to make use of the democratic means and instruments still available to us and to let the President know – not the CIA: American policy is made by the President, and we should not make the CIA a scapegoat – let the President know how you feel."[41] Marcuse does not call for violent opposition or radical social change but instead promotes change through "the long march through the institutions."[42]

And this is precisely what Marcuse did. One week after his talk at Brandeis, on May 10, 1961, Marcuse joined his students in signing an open letter to President Kennedy regarding U.S. foreign policy in Cuba.[43] According to an FBI report filed on November 8, 1967, between the period of 1961 and 1965, as we reported earlier, Marcuse was actively involved in protesting U.S. domestic and foreign policy through the existing channels of power. On February 16, 1965, Marcuse signed an open letter to President Johnson urging for negotiation and not escalation in Vietnam, and again on October 31, 1965, where Marcuse was one of many signatures on an open letter that was printed as a large advertisement in the *New York Times* produced by the "Ad-Hoc Committee on the Vietnam War."[44] Although he was not under direct FBI surveillance at this time, it appears that his actions of protest and activism remained visible and were duly noted.

Furthermore, an article from the spring 1965 edition of the Brandeis student newspaper, *The Justice*, indicates that Marcuse spoke at a large rally during this period in Boston Commons called to protest United States involvement in the war in Vietnam.[45] Marcuse noted that the U.S. intervention in Vietnam violated all the commandments of international law, "especially," in the words of the journalist summarizing Marcuse's speech: "since it is a war not in defense of our own country but of a faraway land and one in which we are slaughtering thousands of helpless and miserable people in the name of the American way of life." The author noted that Marcuse drew the analogy between the people of Nazi Germany and the American people since the Germans surrendered their freedom to Hitler and the Nazis, while people in the United States were surrendering their freedom to the Johnson Administration. Moreover, Marcuse predicted that the "best" outcome in Vietnam could lead to "the task of policing a huge hostile population," while nuclear war was a "negative consequence."

In retrospect, "The Obsolescence of Marxism" presented at South Bend was not an isolated talk, but instead was part of a larger dialogue that Marcuse had been participating in for several years. As noted above, Marcuse's 1961 talk on Cuba first outlined his views on the important role that developing countries play in the sustained maintenance and expansion of capitalism. This would later become one of his major arguments against the U.S. occupation of Vietnam; mainly that Vietnam should not be looked at in isolation as an event happening "over there" but instead to recognize the global elements of capitalism and the interconnections between policy "at home" and "abroad." As Marcuse so eloquently states in "The Obsolescence of Marxism," "[w]hat happens in Asia or Africa is not external to the system but has become an integral part of the system itself."[46]

As already noted above, the ideas found in Marcuse's South Bend talk had earlier philosophical roots. In 1963, his lecture "The Obsolescence of the Freudian Concept of Man" Marcuse posited how psychoanalysis, although at the time out of fashion in academic circles, "sheds new light on the politics of advanced industrial society."[47] Although the lecture focused primarily on psychoanalysis, it allowed Marcuse insight into two key areas that later became the backdrop for his 1966 lecture on "The Obsolescence of Marxism." First, in examining the connection between psychoanalysis and the affluent society, Marcuse concluded "that which is obsolete is not, by this token, false." Where for Marcuse in 1963, claims about the obsolescence of psychoanalysis provided the basis for revisiting psychoanalytical thought in the current social and political context, in 1966, he used the same logic to examine the importance of Marxism as a critique of contemporary capitalism and in relation to America's role in Vietnam.

A second important point expressed by Marcuse in his 1963 lecture on psychoanalysis is his concern for the removal of individual freedoms in the United States through the controlling of information. According to Marcuse, "[t]he danger signs are there. The relationship between the government and the governed between administration and its subjects is changing significantly – without a visible change in the well-functioning democratic institutions."[48] Where Marcuse lays out the transformation of democratic control away from the hands of the people in his talk on "The Obsolescence of the Freudian Concept of Man," he outlines the process for regaining this democratic control at the end of his talk on the "The Obsolescence of Marxism" with a discussion of the role of the intelligentsia in promoting truth in a society where information is deliberately suppressed and spun. "And as repression is flattened out and extended to the entire underlying population, the intellectual task, the task of education and discussion, the task of tearing, not only the technological veil but also the other veils behind which domination and repression operate, – all these 'ideological' factors become very material factors of radical transformation."[49]

On April 27' 1965, in his last public talk at Brandeis University, Marcuse's speech, "The Obsolescence of Socialism" asked if the concept of social revolution had become a misnomer in an affluent society.[50] Once again, Marcuse notes that the title of the talk is meant only half ironically, since on one hand a Marxist analysis in his view captures dynamics of capitalist societies and, on the other, there is an "acute need of the present, of moving past this society." The May 4, 1965 edition of *The Justice*, in summarizing the talk suggested that "in attempting to unite, in this address, the delineation of his latest modes of thought with a call to social action, Dr. Marcuse both exemplified and crowned the approach to student action and to the world at large that he has consistently held at Brandeis."[51] Here, Marcuse's argument for the need for student protest in social revolution gets an early articulation, which is taken up again in "The Obsolescence of Marxism" where Marcuse argues that the syndrome of revolutionary potential exists in uniting the underprivileged locally and abroad with radicalized youth and members of the intelligentsia.

Hence, Marcuse's April 25, 1966 talk at the University of Notre Dame in South Bend, Indiana, was more than just a conference paper delivered by a university professor at an international symposium. Instead, it was snapshot that captured in a single moment, more than five years of public discourse by Herbert Marcuse on the negative aspects of U.S. foreign policy, the interconnectedness of global capitalism, the suspension of civil rights, and the decline of democracy and freedom inherent in the affluent society.

During this time, in addition to his public lectures on Marxism, and his teaching at both Brandeis and UCSD, Marcuse also penned two key critical works: *One-Dimensional Man* (1964) and *Repressive Tolerance* (1965), in this same vein of thought.[52] Therefore, the obsolescence that Marcuse appears to be questioning in his speeches between 1961 and 1965 is not Marxism, psychoanalysis, or socialism. In each and every case, it is the intensifying *obsolescence of U.S. democracy* that Marcuse is calling attention to. The democratic values that were believed to be self-evident and which underpinned the formation of the United States, namely that all men were held equal under the law and therefore had the freedom to life, liberty, and the pursuit of happiness, are for Marcuse disappearing alongside the suspension of freedom of speech, attacks on political criticism and dissent, and repression of oppositional groups, all of which are fundamental to the maintenance of a democratic public sphere which Marcuse saw on the decline in the United States in the 1960s (although New Left radicalism later in the decade would provide him with renewed hope).

What appears special in retrospect concerning this particular moment in April 1966 in South Bend, Indiana, and what may have contributed to the notoriety that it received in the public press, private correspondence, and FBI memos was the vigor with which Marcuse suggests that not only was the United States in the *wrong* for its participation in Vietnam but that people had both a moral duty and a civil right to oppose government policy. This was also the focus of Marcuse's keynote speech at an anti-Vietnam teach-in at UCLA one month prior, entitled "The Inner Logic of American Policy in Vietnam."

THE INNER LOGIC OF AMERICAN POLICY IN VIETNAM: MARCUSE SPEAKS IN WESTWOOD

Even though his April talk at South Bend garnered significant public notice, only one month earlier, Marcuse gave a much more politically charged talk at UCLA. Unlike at South Bend where Marcuse examined the academic importance of Marxism and used the America–Vietnam conflict as a case study for the value of Marxist analysis in the face of advanced industrial capitalism, his talk at UCLA was a pointed and direct attack at the way in which the Kennedy and Johnson Administrations had dealt with Vietnam.[53] The title of Marcuse's lecture, "The Inner Logic of American Policy in Vietnam" while drawing on Marcuse's Hegelian notions of logic, is in fact meant to be a question even though it is poised as a statement. Marcuse uses this same style of prose, that is, he couches the irony of asking a question into the subtext of a direct statement; in creating the title for his talk at South Bend, "The Obsolescence of Marxism," where he noted, "A most important thing was omitted [from the title]-the question mark. For me the question mark is the most condensed symbol of the dialectic in Marxian theory."[54] Herein lays the root of Marcuse's discussion at UCLA – that the basis of American foreign policy in Vietnam is rooted in an *inner logic* that is hidden from U.S. citizens through propaganda and misinformation and, of even greater importance, flawed at its core.

The student union building at UCLA housed the "teach-in" on March 25, 1966. The antiwar protest at UCLA was part of a larger national protest "National Teach-In Week" organized by the Students for a Democratic Society (SDS). The SDS were well known to the FBI and often referred to as a housing ground for communists inside Bureau documents. In fact, the Yale-New Haven local chapter of SDS would sponsor the symposium at Yale that Marcuse would speak at in the spring of 1966. We have noted earlier in this article the affinities between SDS and Marcuse and how Marcuse thoroughly embraced the SDS goals of an end to the Vietnam War, addressing poverty, and promoting participatory democracy. Marcuse spoke out at SDS rallies, supported their goals, and agreed to participate in a 1965 SDS Radical Education Project, which his own teaching was already embodying.[55]

Teach-ins as a form of protest, according to Louis Menashe and Ronald Radosh in *Teach-Ins: U.S.A. Reports, Opinions, Documents*, started in March of 1965 at The University of Michigan, Ann Arbor, and "sprang from the feeling of large sections of the academic community that Government leaders were no longer listening to criticism of U.S. policy on Vietnam and the American public, subjected to the massive persuasive powers of the information media, would soon stop offering it. Accordingly then, the academic community turned to their most reliable constituency, their students ... and together, teachers and students built a movement."[56] The UCLA teach-in, as a form of antiwar protest brought together local teachers, students, and political activists.[57] By the spring of 1966, the teach-in had become a strong and viable option of democratic protest and was used quite widely across the country.

Likewise, by 1966, Southern California had become a politically charged environment ripe with opportunity for protest. However, this was not always the case. In September of 1964, in seeking the advice of close friends regarding his decision to either stay at Brandeis or accept the offer to relocate to La Jolla and UCSD, fellow Frankfurt School theorist, Leo Löwenthal, in a letter dated September 29, 1964, tells Marcuse "[i]t is true that Southern California is politically lousy, but one is always in a minority and there are quite a few people who are decent enough. And don't forget that I'm in the neighborhood."[58] And yet, less than one year later, almost immediately upon arriving at the University of California, San Diego, Marcuse found himself front and center at UCSD in politicized discussions with students and colleagues on topics such as U.S. foreign policy, academic freedom, and the value of civil protest.

An FBI report on Marcuse dated, July 13, 1966, notes that Marcuse spoke at an anti-Vietnam rally at UCSD on October 15, 1965.[59] The rally or teach-in, as it was described by the FBI, was in connection with the International Days of Protest and was attended by over 400 people. Marcuse was advertised in leaflets and flyers to be one of four keynote speakers at the event, where he is said to have spoken openly about protesting U.S. foreign policy in Vietnam. History is made by the marriage between individuals and circumstance; Marcuse may have entered into a Southern California that was devoid of radical politics according to Löwenthal in 1964, but certainly by fall of 1965 that was no longer the case. Marcuse would play a role in the politicization of this region over the course of the next decade.

However, Marcuse's political voice was not simply reserved for students but was shared with faculty and university administration when necessary. For example, on October 11, 1965, only days before the scheduled protest, Seymour Harris and Harold Urey, Professors at UCSD, issued a memo to UCSD faculty asking for their support with regard to a public statement on academic freedom to be distributed and released to the press in the aftermath of the October 15 protest depending on its outcome.[60] And while the statement itself appears to support academic freedom by stating that faculty members "will be protected against attempts of the Administration or others e.g., newspapers and other media of communication or special groups of citizens to silence him," it continues to argue for a level of distance between the academic and the individual arguing: "above all, no faculty member or group of faculty members have the right to speak for the University or give the impression though well organized demonstrations that they in fact reflect the position of the university."[61]

Marcuse's response to the memo was swift and bold. In a letter to Harris and Urey dated October 14, 1965, Marcuse argued that the memo was a violation of the very protections of academic freedom that University claims to protect. According to Marcuse, the Chancellor of UCSD had already publicly addressed this issue in responding to public outcry toward the event being held at UCSD published in an editorial in *The San Diego Union* on October 8, 1965, when he stated that "his administration will sustain the freedom and responsibility of a great university against any intimidation from whatever source."[62] Marcuse pointedly explained to Harris and Urey how circulating the memo "before the scheduled demonstration has taken place, but to be released dependent on the manner in which the demonstrations are conducted, is ... such an intimidation." Marcuse concluded his letter to Harris and Urey with a statement of intimidation of his own arguing that "The Chancellor has unequivocally stated where his administration stands. It would be a catastrophe to this university if his policy would be counteracted from within his own faculty."[63]

The political climate in Southern California had become actively charged and it would not be long before campuses nationwide would follow suit. In fact, six months later, on April 23, 1966, only two days before Marcuse would speak at South Bend, the notion of academic freedom (especially in light of criticism directed to U.S. foreign policy) again became a front-page news when Rutgers Professor of History, Eugene Genovese, at a teach-in at his home institution stated, "I am a Marxist and a socialist. Therefore, unlike most of my distinguished colleagues here this morning, I do not fear or regret the impending Vietcong victory in Vietnam. I welcome it."[64] The backlash against Genovese, teach-ins, and left of center academics was almost immediate. For example, Genovese was removed from his role at the state-funded Rutgers University, and only two days later, another "self-proclaimed Marxist" who was also critical of American foreign policy now found himself the subject of an FBI investigation. And while the words from Marcuse's UCLA or South Bend talks never met with such attacks, he did use similar rhetoric at Westwood when he ended his "Inner Logic" lecture by boldly stating that "[t]he nation that was once the hope of all liberating forces the world over has become the hope of all counter revolutionary forces the world over. The United States has become the advance guard of repression and reaction."[65]

The speech itself, "The Inner Logic of American Policy in Vietnam," shares more similarities with Marcuse's 1961 arguments against America's role in Cuba than with his arguments on Vietnam from his "Obsolescence of Marxism" talk at South Bend. The thematic lens of the lecture explores the disappearance of rational, critical discussion on U.S. foreign policy and provides a sharp and in retrospect accurate critique of U.S. foreign policy that can be directly traced to his earlier talks on Cuba. However, in 1966, the ideas first raised as questions by Marcuse re-appear here with a more direct purpose. Whereas, in 1961, Marcuse had argued that the United States appeared headed down a path of mis-information; by 1966, Marcuse was arguing that the U.S. Government was directly engaged in a large-scale propaganda campaign, similar to the advertisements for soap or candy, which aimed to sell the American public on foreign policy issues without any direct participation of the people.[66] For Marcuse, such propaganda was undemocratic and clouded in a logic that was deliberately hidden from public discourse. And while Marcuse did explore the role of the United States in Vietnam in his talk at South Bend, he spoke more of the potential for revolutionary action against such foreign policy instead of directly critiquing the policy as he did at Brandeis in 1961 or again at UCLA in 1966.

The speech, "Inner Logic," begins with a very frank statement, "the official justification for the American policy in Vietnam is couched in Orwellian language; as such it defies rational discussion."[67] Marcuse's critique that the U.S. role in Vietnam lacked rational logic, spoken at UCLA, was twofold. Firstly, Marcuse argued that U.S. participation in Vietnam was unethical since it was rooted not in the advancement of freedom and democracy as it was sold to the public but instead that it was rooted in the economic needs of monopoly capitalism on a global scale that required the affluent society to exploit underdeveloped nations to provide the necessary conditions for advanced industrial capitalism. As Marcuse argued, "the existence of a gigantic military establishment is an integral, stimulating factor of the U.S. economy … [and that] the affluent society is in need of an enemy, against whom is people can be kept in a state of constant psycho-social mobilization."[68]

Secondly, Marcuse's critique centered on the notion that the real motives for American foreign policy were hidden from the people, couched in terms of "the fight for freedom" that directly manipulated the American people into participating in the exploitation and abuse of fellow global citizens. He argued using the rhetorical device of irony so prevalent in his work at the time, "'we are fighting for freedom' – that is to say, on behalf of a military dictatorship which wouldn't last twenty-four hours without American bombs. 'We are fighting for freedom' by protecting the social groups and interests whose power is based on exploitation and slavery. 'We are fighting for freedom,' in short, by supporting a military junta which fights *against* the economic and social changes which might create the very preconditions of freedom."[69] For Marcuse, what was missing from the foreign policy and the propaganda used to sell war to the American public was the freedom of a rational, critical, national discussion as those who chose to speak out were immediately branded as un-American and ostracized by local, state, and national governments and media outlets.

Marcuse concludes his critique of the "The Inner Logic of American Policy in Vietnam" with a powerful analysis of how the war unleashes aggressive and brutal instincts that are not satisfied in the technologically mediated war involving bombing and high-tech weapons system and assuaging guilt as soldiers did not see individuals who they bombed or shot at log distance. The result is "brutalization on a massive scale, a quality which is expressed also in our daily life at home in the form of violent language, images, and mass behavior."[70] The conclusion is that the "war against Communism, waged on this basis of brutalization, becomes – by the logic of prevailing conditions – a war for reactionary military dictatorship." This leads to counterrevolutionary putsches and reactionary dictatorships all over the world, and the United States, once a beacon of liberation and democracy, becomes a bulwark of repression and reaction.

SOCIALIST PERSPECTIVE OF MODERN MAN: MARCUSE SPEAKS IN NEW HAVEN

Hence, Marcuse's critique of U.S. foreign policy was radicalizing as the New Left and the antiwar movement were expanding. Interestingly, Marcuse's talk at the April 1966 Yale Socialist Symposium provides academics and historians with a speech of Herbert Marcuse's previously unknown. And while the words to the speech or Marcuse's participation at the Symposium are nowhere to be found in any previous edited collections or biography pieces on Marcuse, or even in his personal notes or correspondence, a significant amount of discussion surrounding the speech is recorded inside the FBI dossier. For example, where the UCLA or Notre Dame speech is mentioned and referred to only in small paragraph sections, a September 30, 1966 FBI report includes almost 5 typed pages of quotes delivered by

Marcuse in his symposium speech and close to 10 pages of discussion on the event.[71] The FBI coverage of Marcuse's participation at Yale appears to come from three sources: the April 1966 "Yale Weekly Calendar" that promoted the event in advance of it happening; the April 30, 1966 issue of *New Haven Register Journal Courier* (NHRJC) that published the article "Socialist Attacks Affluent Society" summarizing the event; and the testimony of one to two confidential FBI informants.[72] According to the FBI coverage of the event, the Yale-New Haven chapter of SDS and The Yale Socialist Union (described by the FBI as "the focal point for radical student activity on the Yale campus") co-sponsored the "Third Annual Symposium" entitled "Socialist Perspective of Modern Man" and paid the expenses of Herbert Marcuse to attend and speak at the event.[73]

According to the *NHRJC*, Herbert Marcuse's speech in New Haven lasted for more than two hours and was "well received" by the 500 people in attendance. Coming only days after he received national attention in the *New York Times* for his remarks on American participation in Vietnam while speaking in South Bend, and only a week after Historian Eugene Genovese found himself in trouble with New Jersey state officials for outwardly criticizing U.S. involvement in Vietnam, Marcuse continued at Yale his radical critique of U.S. foreign policy. As the *NHRJC* reported, "A leading American socialist told an audience ... that the affluent society is irrational and dominated by blind contradictory forces."[74] Marcuse's speech itself appears consistent with themes raised as early as his 1961 speech at Brandeis on Cuba and with ideas that he would continue to hone in speeches such as "The Problem of Violence and the Radical Opposition" in 1967.[75]

According to an FBI confidential informant, Marcuse's speech dealt with four themes.[76] First, Marcuse argued that "the capitalist system is incompatible with peace and the struggle for peace is the most important goal of Socialism." Next, Marcuse suggested that in 1966 three types of countries existed: "Industrialized Advanced, (such as the United States), Less Industrialized (such as France, Italy, and Germany), and Underdeveloped" and that the survival of the first kind comes only through the exploitation of the third. Marcuse also argued that "the radicalized intelligentsia is the vehicle of change" and that it would take an energized youth united with the intelligentsia to bring about real change. Lastly, Marcuse spoke about the role of the U.S. mass media in intellectually and politically dulling the individual and related this back to his initial argument that the capitalist system requires a vicious and threatening "enemy" to channel its libidinal aggression toward. When examined in relation to the earlier talks at UCLA and Notre Dame, it appears that the talk at Yale was not only a combination of the two earlier speeches but also a culmination of the growing radicalism of the New Left.

The *NHRJCI* account of Marcuse's speech also stresses the theme laid out in *One-Dimensional Man* and articulated in many talks of the time that today's capitalist society is increasingly shaping thoughts, behavior, needs, and aspirations and that liberation requires new needs and consciousness. Hence, for radical social change, there "must be new social institutions and a new way of life."[77] The threat of Communism was responsible, Marcuse argued, for the unification of capitalist countries, but national liberation movements and an alliance of radical youth and the intelligentsia might pave the way for radical social change, a theme Marcuse would continue to develop throughout the 1960s and into the 1970s.

CONCLUSION

On September 30, 1966, the San Diego Office of the FBI published a full report for the FBI director, which outlined in detail all of Marcuse's activities known to the FBI either through direct surveillance and interviews or through information gathered in other non-related investigations over the past two decades. And while Marcuse had been busy openly protesting U.S. foreign policy over the past five years, with heightened activity at home and abroad in the spring and summer of 1966, at the conclusion of the Bureau's 21-page report, the San Diego Office stated "[i]n as much as previous investigation of the subject and current informant coverage of his activities have failed to reflect membership or affiliation with the Communist Party, Socialist Workers Party or other subversive organization, no additional investigation of his activities is being conducted and the case is being closed."[78]

FBI Director J. Edgar Hoover's response to the September report, in a letter dated October 18, 1966, was to agree with the synopsis presented by the San Diego office but rather than closing the Marcuse file, to move it from the category of "Security Matter –C" to "Reserve Index A" for future reference in case things were to escalate.[79] As Hoover noted, "Although subject is not a member of a basic revolutionary organization, he is a self-admitted Marxist who travels extensively making frequent speeches in which he espouses Marxism. He is also currently participating actively in protest demonstrations against the United States intervention in Vietnam ... [and] is an author and a philosophy professor ... [and therefore] in a position to influence others against the national interest in a time of national emergency."[80] Marcuse's time on "Reserve Index A" would be short-lived, and when he would emerge in Bureau investigations under heightened surveillance in 1968, his file would no longer be under the classification of "Security Matter –C," but instead under the classifications of "revolutionary" and "anarchist" as the FBI and media construction of Herbert Marcuse passed in the period up to 1966 from "self proclaimed Marxist" and "grandfather of the New Left" to the more explosive "anarchist" and "threat to national security."

In retrospect, it appears that the publicity that the Marcuse's talk at South Bend in 1966 garnered was more the result of media publicity through the *New York Times*, resulting in the public image of Marcuse as a revolutionary in the 1960s, in conjunction of corporate media and FBI reports on Marcuse. As Marcuse's speeches increasingly came to be described in the mainstream media, the FBI intensified its surveillance of Marcuse. As noted above, in instances such as his protest of foreign policy in Cuba or letters to the President, Marcuse had presented on numerous occasions oppositional speeches and actions of protest in the half-decade before his lecture at South Bend that failed to register with either the FBI or the American public.

After all, "The Inner Logic" talk given a month earlier at UCLA was much more direct and poignant in its attacks against U.S. foreign policy than the more philosophical and theoretical "Obsolescence of Marxism." However, after the April 23rd speech of Eugene Genovese, supporting a Viet Cong victory in Vietnam, the U.S. Government, the media, to a larger extent, and public appeared more attentive to the rising voices of dissent within academia, and in some quarters appeared ready for the repression of radical professors and students. With Marcuse's discussion of the role of underdeveloped nations in the maintenance of advanced industrial capitalism and the affluent society, it now appears that for the third time since his naturalization in 1940, the German-Jewish immigrant Professor Herbert Marcuse was at the center of a formal Federal Bureau of Intelligence investigation into his participation in allegedly subversive or "un-American" activities.

When one revisits the public talks at UCLA in March and at Notre Dame and Yale in April 1966, it is not surprising to find Marcuse under Bureau surveillance. What is surprising is the relative speed at which the case emerged, the role that the mass media played in the framing of Marcuse, and the re-emergence of earlier investigations in the construction of Marcuse in 1966. FBI scrutiny and media attention would increase in the later 1960s, especially in 1968, the year that ultimately came to define Marcuse evermore in public consciousness as a person whose ideas and actions were oppositional to the U.S. society and foreign policy. Hence, FBI surveillance played a hitherto insufficiently recognized role in the public image of Herbert Marcuse as a radical in the 1960s, and FBI scrutiny would continue to intensify, a development that we shall pursue in further studies.

NOTES

1. See Marcuse (1941, 1955, 1964).

2. A file titled "FBI" found among Marcuse's personal papers contains a letter typed on University of California, Santa Cruz, Adlai E. Stevenson College stationery by a typewriter similar to that of Barry Katz with handwritten edits on the text also appearing to be in Katz's script, leading our intelligence analysts to assume that Katz induced Marcuse to write to the FBI asking for the "full contents of my FBI file under the provisions of the Freedom of Information Act." The letter contained no date, but Marcuse received a response by the FBI in a letter signed by Michael E. Shaheen, Jr. Counsel, indicating that "A review of the FBI files ordered by the Attorney General indicates that you have been affected by an FBI counter-intelligence program in 1969." The file in Marcuse's folder contains evidence of a program labeled "Cointelpro-New Left" against San Diego radical groups, but does not list any of Marcuse's activities. Another dossier in Marcuse's personal files contained evidence of a FBI memo dated 2/14/69 labeled "Cointelpro-New Left" that authorized authorities to circulate a leaflet captioned "The Gigantic 'Pick the Fag Contest' is Here." The memo indicates that the leaflet "has been prepared by the New York Office and contains the pictures of Dave Dellinger, Che Guervera, Mark Rudd, and Herbert Marcuse." According to the memo, "New York feels that this leaflet will ridicule Dellinger and embarrass him and create further dissension between the Coalition for an anti-Imperialist Movement consciousness-AIM and NWC." The memo evidences the puerile gay-baiting employed by the FBI against the New Left.

The FBI lists its current dossier of Freedom of information documents on Marcuse as containing 518 pages (see http://foia.fbi.gov/foiaindex/foiaindex_m.htm), but an earlier dossier obtained by Douglas Kellner contains less documents, while a more recent request by Stephen Gennaro elicited a larger file, so Marcuse's FBI files released appear to be variable in terms of what and how much is released to individual requests.

3. Our contextualization draws on Kellner (1984), Marcuse (1998, 2004). These books draw on a vast literature on Marcuse and his life and socio-historical context and we will use these and many other sources as well as we proceed with this study.

4. Marcuse had been subject to standardized FBI checks in the 1940s when he worked with the Office of War information (OWI), the Office of Strategic Services (OSS), and the State Department. For an overview of Marcuse's activities in U.S. intelligence services in World War Two and post-war work with the State Department, see Katz (1989) and Marcuse (1998).

5. In February 1948, October 1949, and November 1949, Marcuse gave public lectures at the Russian Institute of Columbia University on the topics: "What is Dialectics?," "The Marxian Theory of History and Society," and "The Development of Marxian Theory after Marx," cited in FBI documents beginning with a letter from FBI Director to SAC Washington Field, dated November 14, 1950, File 121-24128. Obviously, Marcuse's lectures on Marxism evoked interest in the FBI.

6. See Marcuse (1998) Marcuse's U.S. Government employment is cited repeatedly in FBI documents beginning with a letter from FBI Director to SAC Washington Field, dated August 9, 1950, page 3, File 121-24128. The contents of the letter provide the Washington Office with all of the current information on file for Herbert Marcuse (then Chief of the Central European Branch, Division of Research for Europe, State Department) including an outline of all his previous employment and positions within the U.S. Government.

7. See FBI documents: Letter from FBI Director to SAC Washington Field, dated August 9, 1950, page 6, File 121-24128, and Letter from J. Edgar Hoover to Samuel P. Boykin, Director, Office of Consular Affairs, Department of State, dated September 22, 1950, where Hoover states "A loyalty investigation ... has been instituted by the Bureau based upon allegations that he is a Marxist and possibly possesses a pro-Communist sympathies."

8. On the Frankfurt School, see Jay (1973), Wiggershaus (1995), and Wheatland (2009). On Marcuse's relation to the Frankfurt School, see Kellner (1984, 1989) and Marcuse and Kellner (2001).

9. The Institute retrospectively became known as the Frankfurt School after key members such as Horkheimer and Adorno returned to Germany in the late 1940s. Their work is more accurately described at this point as working toward developing a critical theory of society within the Institute for Social Research; see Kellner (1989).

10. Files on Marcuse, Horkheimer, and other German refugee intellectuals are found in the FBI FOIA Reading Room (http://foia.fbi.gov/). On Adorno and the FBI, see Jenemann (2007) and Rubin (2002). For a more detailed account of FBI surveillance of German refugees during World War II, see Stephan (2000).

11. For example, Albert Einstein was the subject of extensive FBI surveillance as highlighted in Jerome's (2002) book.

12. Marxist sympathies of the Institute are cited repeated in FBI documents beginning with a teletype to Washington Office from New York Office dated August 29, 1950, File 121-24128.

13. For an overview of Marcuse's writings at this time and their relation to Marxism, see Kellner (1984).

14. This is the first official document inside of the FBI dossier of the Marcuse files released by the FBI. However, it is certainly not the first document about Marcuse. The document itself makes references to a letter on April 11 of the same year. Also, in Marcuse's own personal correspondence, such as his November 11, 1942 letter to Max Horkheimer, Marcuse mentions how his current hiring for U.S. intelligence work still has to go through FBI approval but appears only to be a formality at the moment. See Marcuse to Max Horkheimer in Marcuse (1998, pp. 234–235).

15. The Office for Emergency Management (OEM) was established by administrative order May 25, 1940, to serve as an executive of the President of the United States to clear information on defence measures. It was terminated on

November 3, 1943, and replaced by the OWI and OSS agencies in which Marcuse served.

16. FBI documents, internal memo, J. Edgar Hoover to SAC New York, dated May 21, 1943, File 77-27880.

17. FBI documents, letter from FBI Director to SAC Washington Field, dated August 9, 1950, page 6, File 121-24128.

18. FBI documents, letter from SAC New York to FBI Director, dated September 14, 1950, File 121-24128.

19. *Ibid.*

20. FBI documents: letter from SAC Washington Agent Guy Hottel to FBI Director, dated September 14, 1950, page 2, File 121-24128. We found a copy of this report in Marcuse's FBI dossier, which uncharacteristically names the sources on the report that are usually blacked out, as is material that suggest FBI surveillance strategies and sources.

21. This re-use of these documents will be examined again in later articles dealing with the activities of Marcuse in 1968 and after.

22. See FBI documents: letter from FBI Director to SAC Washington Field, dated August 9, 1950, page 6, File 121-24128.

23. See Marcuse (1998).

24. Adorno and Horkheimer (1972) and Horkheimer and Adorno (2002).

25. Cited repeatedly in FBI documents, beginning with FBI Report, SAC Los Angeles, dated April 20, 1943, named in report to be part of File 77-1327, but currently archived in File 77-27880.

26. See Marcuse and the New Left.

27. FBI Documents: FBI Report, SAC San Diego, dated November 8, 1967, pages 4-1, File 100-445771. In particular, the speech of Marcuse that is paraphrased above is very similar in text and in theme to his June 1967 speech at the Free University Of West Berlin titled "The Problem of Violence and the Radical Opposition" available online at http://www.marxists.org/reference/archive/marcuse/works/1967/violence.htm and published in *Five Lectures* (Boston: Beacon Press, 1970, pp. 83–108) and collected in *Herbert Marcuse: The New Left and the 1960s* (pp. 57–75).It is also striking here that Marcuse takes the concept of "participatory democracy," laid out in the 1962 Port Huron Statement, a founding document of SDS, as a basis for a New Left, thus showing Marcuse's philosophical and political affinity with the New Left; the Port Huron Statement can be found online, http://coursesa.matrix.msu.edu/~hst306/documents/huron.html

28. FBI Documents: FBI Report, SAC San Diego, dated November 8, 1967, pages 4-1, File 100-445771.

29. This is a practice that appears continually throughout the later FBI documents but first appears in FBI Documents: letter from FBI Director to SAC, dated April 10, 1968, File 100-445771.

30. The FBI applied varying categories of classification to Marcuse that are noted on each and every FBI document underneath Herbert Marcuse's name in the top left-hand corner of the document under the heading of SUBJECT. The categories changed but seemed to indicate the varying degree of subversive threats and level of surveillance for each subject; as we are indicating, Marcuse received a number of classifications over his long career as a government worker and then as an academic

critic and activist. For more information on the categories and their significance, see Swearingen (1999). Swearingen reports that the Bureau had two indexes: the Security Index, which consisted of known Communists (or internal government workers) whose actions were viewed to be hostile or as potentially threatening to National Security, and a Secret index, unknown to the President or Attorney General referred to inside the Bureau as the Communist Index or Reserve Index. The Reserve Index housed a significantly larger number of people and was a catchall for those individuals in a position to influence others (such as doctors, professors, teachers, labor union leaders, writers, and media) that could not be directly linked to subversive organizations. For example, on p. 41, Swearingen states "Together, the Security Index and the Communist Index [later referred to as the Reserve Index] totaled approximately 50,000 names in Chicago [in 1955] … the Security Index totaled approximately 5,000 names … the Reserve Index … 45,000." For more information on the secretive nature of FBI surveillance of individuals found on the security index, the history of the index, previously known as the Custodial Detention Index, and its connection to the illegal detainment of German, Italian, and Japanese Americans during World War II, see Theoharis (1998).

31. Letter from Arnold Tovell, Marcuse's *Beacon Press* editor, to Marcuse, May 3, 1966, found in Herbert Marcuse personal collection.

32. Cited in FBI Documents: letter from FBI Director to SAC San Diego, dated May 13, 1966, File 100-445771.

33. Marcuse (1967).

34. Handler (1966).

35. FBI Documents: letter from FBI Director to SAC San Diego, dated May 13, 1966, File 100-445771.

36. Marcuse (1967, pp. 409, 416).

37. A text found in the Herbert Marcuse archives (#231.00) has the long title "PROF. HERBERT MARCUSE speaking at Cuba protest meeting, Brandeis University, 3 May 1961: *NOT FOR PUBLICATION*." The six-page double-spaced typewritten text contains Marcuse's first significant known critique of U.S. foreign policy in the early 1960s and sketches out the lines of a strong anti-imperialist critique that he would follow throughout the decade and into his final years in the 1970s; we will refer to it below as "Remarks on Cuba."

38. Marcuse, "Remarks on Cuba," p. 3.

39. Marcuse, "Remarks on Cuba," p. 4. (Emphasis by Marcuse).

40. *Ibid.* (Emphasis by Marcuse).

41. Marcuse, "Remarks on Cuba," p. 6.

42. See Douglas Kellner, "Introduction," *Herbert Marcuse and the New Left.*

43. This letter is cited in the FBI Report, dated November 8, 1967, that we had presented in detail above.

44. *Ibid.*

45. Hirschfield (1965). A date for the rally was not mentioned in the article, but since the article ends with a summons for people to come to the big march on April 17, 1965, to "End the War in Vietnam," we assume the speech was given in the spring of 1965 before the April Washington march.

46. Marcuse (1967, p. 416) (citations to this text will refer to the published version referenced in note 31).

47. Marcuse (1970). On Marcuse's long engagement with Freud and psychoanalytic theory, see Kellner, *Herbert Marcuse*, Chapter Six, and the Introduction by Douglas Kellner, Tyson Lewis, and Clayton Pierce to *Philosophy, Psychoanalysis and Emancipation,* Volume Five, *Collected Papers of Herbert Marcuse*, forthcoming.
48. Marcuse (1970, pp. 46ff).
49. Marcuse (1967, p. 417). On Marcuse's concept of radical education, which fleshes out the task of intellectuals and education, see the collection edited by Kellner, Lewis, Pierce, and Cho (2009).
50. Marcuse's personal office file contains an article by Acker (1965). We are relying on what appears to be a highly detailed and seemingly reliable report on the lecture by Robert Acker in explicating the talk.
51. *Ibid.*
52. See the discussion of these texts and their context and influence in Kellner, Marcuse, Chapter 8.
53. Marcuse's lecture, "The Inner Logic of American Policy in Vietnam" was first published as an edited text of his March 25, 1966 lecture at UCLA in Menashe and Radosh (1967, pp. 64–67) and is republished in *Herbert Marcuse and the New Left* (pp. 38–40, page references below will be to this edition and hereafter referred to as "The Inner Logic").
54. Marcuse (1967, p. 409).
55. Thus, Phillip Wheatland's attempt in a recent book to engage in alleged myth-busting by claiming that study of early New Left documents and interviews with key figures in early SDS groups suggests that Marcuse's impact on the New Left was exaggerated is shown to be highly problematical. While Wheatland is correct (pp. 439ff) that Marcuse was not widely known by sectors of the American New Left until around 1968, he exaggerates the claim by interviewing people in SDS who indeed were not influenced by Marcuse, like Todd Gitlin, who remains hostile to Marcuse's work to this day. As we have shown, Marcuse was involved in many SDS rallies and had a philosophical and political affinity with the organization and the New Left. On Marcuse's critical relation to the SDS radical education project, see Kellner, Lewis, and Pierce (2009).
56. Menashe and Radosh (1967, Preface v).
57. In fact, according to an FBI report on Herbert Marcuse from September 30, 1966, notable participants at the event included Dorothy Healy, the Chairman of the Southern California District of the Communist Party. And yet, even with the FBI present at this event, there is no mention of the content of Marcuse's speech at the event inside the FBI files. The only mention of Marcuse's connection to the teach-in is a one-paragraph note that simply says he was there, first appearing in a July 13, 1966 report summarizing Marcuse's political activity and then again in the September 30, 1966 report, but this time with the addition that mentions the participation of Dorothy Healy.
58. Leo Löwenthal Archive, Stadt- und Universitätsbibliothek Frankfurt am Main.
59. FBI Documents: FBI Report, SAC San Diego, dated July 13, 1966, File 100-445771.
60. The October 11, 1965, memo from Seymour Harris and Harold Urey, Professors at UCSD, to UCSD faculty and Herbert Marcuse's typed response were found in Herbert Marcuse personal collection.

61. *Ibid.*
62. Herbert Marcuse, cited in memo.
63. *Ibid.*
64. For a full text version of Genovese's speech and commentary on the social and political reaction to his comments see Menashe and Radosh (1967).
65. Marcuse, "The Inner Logic" p. 40.
66. Marcuse, "The Inner Logic" pp. 38ff.
67. Marcuse, "The Inner Logic" p. 38. On Marcuse's use of Orwellian language where, following Orwell's novel *1984*, peace is war, freedom is slavery, and so on and on the connections and differences between Orwell and Marcuse's thought, see Kellner (1990).
68. Marcuse, "The Inner Logic," p. 39.
69. Marcuse, "The Inner Logic," p. 38.
70. Marcuse, "The Inner Logic," p. 40.
71. FBI Documents: FBI Report, SAC San Diego, dated September 30, 1966, File 100-445771.
72. *Ibid.*, pp. 12–16.
73. *Ibid.*, p. 12.
74. *Ibid.*, p. 14.
75. See notes 25 and 35.
76. FBI Documents: FBI Report, SAC San Diego, dated September 30, 1966, pp 12–16, File 100-445771.
77. From FBI Documents: FBI Report, SAC San Diego, dated September 30, 1966, pp. 12–16, File 100-445771, quoting from an article "Socialist Attacks Affluent Society" (no author) in New Haven Register Journal Courier, April 30, 1966, no page number.
78. FBI Documents: letter from SAC San Diego to FBI Director, dated September 30, 1966, File 100-445771.
79. FBI Documents: letter from FBI Director to SAC San Diego, dated October 18, 1966, File 100-445771.
80. *Ibid.*

ACKNOWLEDGMENT

For expert editing and editorial advice, we thank Harry Dahms.

REFERENCES

Acker, R. (1965). Marcuse farewell. 'The obsolescence of socialism'. *The Justice* (May 4), 4.

Adorno, T. W., & Horkheimer, M. (1972). *Dialectic of enlightenment* (J. Cumming, Trans.). New York: Seabury Press.

Handler, M. S. (1966). Marxist views Vietnam in context of capitalism. *New York Times* (April 27), p. 2.

Hirschfield, P. (1965). Dr. Herbert Marcuse warns Against U.S. surrender of freedom to LBJ. *The Justice* (Spring), 3.

Horkheimer, M. & Adorno , T. W. (2002). In: G. S. Noerr (Ed.), *Dialectic of enlightenment. Philosophical fragaments* (E. Jephcott, Trans., 2nd ed.). Stanford, CA: Stanford University Press.

Jay, M. (1973). *The dialectical imagination* (second edition, University of California Press, 1996). Boston: Little, Brown and Company.

Jenemann, D. (2007). *Adorno in America*. Minneapolis: University of Minnesota Press.

Jerome, F. (2002). *The Einstein files: J. Edgar Hoover's secret war against the world's most famous scientist*. New York: St. Martin's Press.

Katz, B. (1989). *Foreign intelligence research and analysis in the office of strategic services 1942–1945*. Cambridge, MA: Harvard University Press).

Kellner, D. (1984). *Herbert Marcuse and the crisis of Marxism*. Berkeley and London: University of California Press and Macmillan Press.

Kellner, D. (1989). *Critical theory, Marxism, and modernity*. Cambridge, UK, and Baltimore, MD: Polity Press and John Hopkins University Press.

Kellner, D. (1990). From 1984 to one-dimensional man: Reflections on Orwell and Marcuse. *Current Perspectives in Social Theory*, *10*, 223–252.

Kellner, D., Lewis, T. E., & Pierce, C. (2009). *On Marcuse: Critique, liberation, and reschooling in the radical pedagogy of Herbert Marcuse*. Rotterdam, the Netherlands: Sense Publishers.

Kellner, D., Lewis, T. E., Pierce, C., & Cho, K. D. (2009). *Marcuse's challenge to education*. Lanham, NJ: Rowman and Littlefield Publishers.

Marcuse, H. (1941). *Reason and revolution: Hegel and the rise of social theory*. New York: Oxford University Press.

Marcuse, H. (1955). *Eros and civilization*. Boston: Beacon Press.

Marcuse, H. (1964). *One-dimensional man*. Boston: Beacon Press.

Marcuse, H. (1967). The obsolescence of Marxism. In: N. Lobkowlcz (Ed.), *Marx and the western world* (pp. 409–417). Notre Dame-London: University of Notre Dame Press.

Marcuse, H. (1970). The obsolescence of the Freudian concept of man. *Five Lectures*.

Marcuse, H. (1998). Technology, war, and fascism. In: D. Kellner (Ed.), *Collected papers of Herbert Marcuse* (Vol. 1). London and New York: Routledge.

Marcuse, H. (2004). The new left and the 1960s. In: D. Kellner (Ed.), *Collected papers of Herbert Marcuse* (Vol. 3). London and New York: Routledge.

Marcuse, H., & Kellner, D. (2001). Toward a critical theory of society. In: D. Kellner (Ed.), *Collected papers of Herbert Marcuse* (Vol. 2). London and New York: Routledge.

Menashe, L., & Radosh, R. (Eds). (1967). *Teach-Ins, USA: Reports, opinions, documents*. New York: Praeger.

Rubin, A. (2002). The Adorno files. In: N. Gibson & A. Rubin (Eds), *Adorno: A critical reader* (pp. 172–190). Malden, MA and Oxford, UK: Blackwell.

Stephan, A. (2000). *Communazias: FBI surveillance of German émigré writers*. New Haven, CT: Yale university Press.

Theoharis, A. G. (Ed.) (1998). *The FBI: A comprehensive reference guide* (pp. 18–24). Phoenix: Oryx Press.

Swearingen, M. W. (1999). *FBI secrets* (pp. 41ff). Boston: South End Press.

Wheatland, T. P. (2009). *The Frankfurt school in America: A transatlantic Odyssey from exile to acclaim*. Minneapolis, MN: University of Minnesota Press.

Wiggershaus, R. (1995). *The Frankfurt school*. Cambridge, UK, and Cambridge, MA: Polity Press and MIT Press.

THE ACTUALITY OF CRITICAL THEORY: A REPLY TO DAHMS' LATE PROLEGOMENON

Steven P. Dandaneau

The widely held but typically unstated claim made by adherents of Frankfurt School critical theory for whom Jürgen Habermas' towering second generation oeuvre inspires more disappointment than inspiration is that Max Horkheimer and his principal Institute for Social Research partners – Theodor W. Adorno and Herbert Marcuse and, secondarily, Walter Benjamin – articulated versions of critical self-consciousness that have yet to be surpassed. We resist acknowledging this claim, no doubt, because of its prima facie arrogance and because it seems impossible, contra Habermas, to rationally deduce the relative value of forms of self-consciousness by any established means of interpretation, scientific inquiry, or rational reconstruction. Instead, the contention remains a latent albeit animating commitment akin to what C. Wright Mills once observed with respect to his mid-century scholarly peers, that he and they lived off the ideas of previous generations, Karl Marx's and Max Weber's especially. Today's non-Habermasian Critical Theorists implicitly acknowledge the same relationship to Horkheimer, Adorno, Marcuse, and Benjamin, superior minds who benefited, crucially, from learning and lived experience largely (and in some respects, thankfully) unavailable to thinkers today. Thus, we acknowledge, if we are honest with ourselves, that they articulated

Nature, Knowledge and Negation
Current Perspectives in Social Theory, Volume 26, 315–325

ISSN: 0278-1204/doi:10.1108/S0278-1204(2009)0000026015

the highest or most compelling form of Western-style enlightenment as yet achieved, and, through sustained criticism of notable competing Western philosophical and methodological alternatives, advanced human self-understanding – thus far, at least – to its limit.

What at first blush appears as so much hubris, if not rubbish, is less easily dismissed when one notes comparable claims that are taken for granted. To return again to the figure of Mills (1959), his book, *The Sociological Imagination*, asserts that the so-called classic tradition of social theory, when properly distilled, provides *the* form of self-consciousness appropriate to the demands of "post-modern society." And is it not the case that Habermas (1968, 1981) himself, in *Knowledge and Human Interests* and later in his magnum opus, *Theory of Communicative Action*, advances what he regards as *the* necessary and unavoidable approach to knowing-in-the-world, not only *this* world but *any*? To believe that Adorno, for one, represents the high-water mark of critical intellectual achievement is no different than attributing the same achievement to John Stuart Mill or Thorstein Veblen. And other than with respect to the significance of historical specificity, upon which much more will be said momentarily, neither are such assertions inconsistent even with the thrust of conservative elitist thought from such diverse sources as Alasdair McIntyre and Leo Strauss, who situate the modern within the *longue durée* and valorize the possibility of enduringly rare and normally quite impenetrable insight. To underscore that instances of rare and probably unparalleled intellectual achievement are the result of unique constellations of historical and societal as well as biographical factors is nothing more, really, than sociology of knowledge.

The above may be considered a terse and unsubtle restatement of the central thesis advanced over nearly 60 pages in Harry F. Dahms' (2008) "How Social Science is Impossible Without Critical Theory: The Immersion of Mainstream Approaches in Time and Space", which I read as a prolegomenon to any future Critical Theory. Evidently, Dahms is of the view that the original Frankfurt School did more than any other alternative to advance the cause of social and cultural theory. Moreover, Dahms stresses Critical Theory's penchant for the reflexive application of such learning in self-assessment as intellectuals, social analysts, and historically specific human beings. On the basis of this reading of the original Frankfurt mandate, Dahms questions the viability – both doubtful and much-needed – of social science under contemporary conditions. In what could be described as his late contribution to *The Positivist Dispute in German Sociology* (1969), Dahms effectively demonstrates – one hopes against hope that it is once and for all – that "social science" can be neither truly social nor truly scientific without full and complete incorporation of the historically specific reflexivity that is the defining characteristic of the work of Horkheimer, Adorno, Marcuse, and Benjamin. Given the pathological trajectory of societal development and the systematic cannibalization of available cultural resources worldwide, Dahms also implies that social science needs the theoretical chutzpah that Benjamin insisted was required by any attempt to live fully awake in the time-of-the-now.

Theoretically obstinate, to be sure, Dahms' prolegomenon is odd or decidedly non-mainstream in other ways as well. To wit, its novella length, which immediately reduces its likely audience to a handful of probably already sympathetic readers, such as myself; its insistent and uncompromising title, which is matched in its assertiveness only by Dahms' title for the volume in which the prolegomenon resides as introduction and for which he also serves as editor: "No Social Science Without Critical Theory": its insistent and uncompromising thesis, which states, in effect, that social science as we know it is the greatest impediment to social science as we need it. And as I have intimated, the statement is profoundly peculiar in that this prelude to all future Critical Theory (think chutzpah) arrives at the end of a century's worth of development of Critical Theory (think untimeliness, lateness). Recall that Georg Lukács' *Soul and Form* appeared in 1910, the same year that Karl Korsch earned his doctorate in Jena. "How Social Science is Impossible Without Critical Theory," a prolegomenon, appears at the end of a straight line leading from the likes of Lukács and Korsch. As postscript, it is an elegy for a tradition begun in Germany but articulated fully only in America by émigré Germans like Dahms. As prologue, it is a starting point for social science, which aims to produce practically useful knowledge – knowledge necessary and suitable for history-making – from within the hyper-mediated, mirror-of-mirrors world of the postmodern, a "context of delusion" if there ever was one.

The disjuncture between these temporal/spatial moments provides the basis for my – let me be clear – wholly sympathetic reply. My intention is to discuss the two implications of Dahms' treatise that matter most to me and which I believe follow most directly from the approach therein elaborated. In reply to Dahms, then, I wish to state what I regard as the political as well as the personal implications of Critical Theory's simultaneous lateness and urgent orientation to the future. Radical hypotheses are needed, for the actuality of critical theory, its emphatic spatio-historical self-fulfillment, is an impossible project of unsurpassed significance and urgency.

1. THE POLITICS OF THE MAINSTREAM

Dahms (2008, pp. 43–45) distinguishes Critical Theory from "Critical Liberalism, Cultural Pessimism, and Public Sociology". He means, first and foremost, to underscore the political gulf that has opened between the original Frankfurt School thinkers and their most celebrated second and third generation heirs, Habermas and Alex Honneth. The latter's affirmative theoretical embrace of progress and cautious optimism render them constitutionally incapable of understanding their putative subject matter, modernity, and thus also of articulating a radical politics sufficient to provide orientation to the specific – the specifically totalizing and lethal – spatio-temporal challenges that confront humanity at the end of modern society. Having barred themselves from fully considering their own participation in a contradictory and deadly system, Habermas' and Honneth's a priori ideological commitments render them unable to "face facts," as Dahms (2008, p. 44) stresses, and thus unable to execute a discerning, or even a useful, critical social science. Unwilling to practice Marx's dictum that critique must be followed to its logical conclusion and without regard to opposition from the powers that be, Habermas and Honneth's otherwise very considerable erudition thus fails Critical Theory's original and still most essential litmus test.

Critical Theory eschews cheerleading for, á la Hegel, *bourgeois right*, however significant it is for life in John Maynard Keynes' omnipresent short-run; the economic growth and political rights associated with Western liberal democracies are understood as manifestations of a society that manages and displaces barbarism while at the same time it mass manufactures alienation and cultural banality. Critical Theory sets its sights on the built-in and therefore unavoidable contradictions and irrationalities as well as the resulting human degradation necessitated by the routine functioning of a pathological modernity. Who doubts that Horkheimer, Adorno, Marcuse, and Benjamin understood war and holocaust as essential features of twentieth-century modernity, not as the unfortunate mistakes of an unfinished modernity? And just as E. P. Thompson and C. Wright Mills defined the New Left some 50 years ago against the defuse doyens of the "Age of Apathy" and the self-serving ideologues of the "End of Ideology," Dahms argues that a rigorously reflexive contemporary Critical Theory parts company with today's spokespersons for cynical complacency, including those whose call for a "public sociology" denotes little more than a fantasy wish for media exposure for a fading discipline's shiny happy talking heads.

Dahms' exposition is more developed and more subtle than my rendering, but the political upshot of his position deserves, I believe, unsentimental formulation: Critical Theory is as radical and uncompromising as the social theory of Marx and Weber. It is therefore not surprising that the political implications that flow from Critical Theory's spatio-historical specific studies of ideology and domination are also radical, also go to the roots of what it is that is subjected to critique, also bite the many hands which offer to feed it crumbs. What are these implications?

First, the modern system has incorporated mainstream social science into its workings. Mainstream social science has therefore forfeited its theoretical autonomy and with this its capacity to project representations of the totality. Mainstream social science thus bears no genuine relation to classical social theory or to the original Enlightenment project that first motivated it. It is, rather, a mere object of study among others with which Critical Theory must cope. Without the capacity for critical self-reflection and the capacity to provide culturally or politically useful orientation, mainstream social science is, politically speaking, part of the problem, not unlike what Marx derided as "bourgeois economics" or the mainstream social science Mills decried in *The Sociological Imagination.*

Second, since the modern system is violently self-destructive, it is untenable to await an Owl of Minerva for insight into its ultimate nature. In the seconds of awareness before a worldwide nuclear holocaust, more might judge favorably the most radical of critiques, but then it would be too late to learn from them. And likewise, as Earth spirals into a post-fossil fuel greenhouse madhouse, we might recall that Marx and Weber more or less explicitly projected this exact future, that is, short of revolutionary or at least extraordinary countermeasures. Instead of viewing the dying system at its end, the alternative theoretical tact is, of course, recovery of insights born at its origin, which is the sociological explanation for the vital importance of the classic tradition of social theory – Marx and Weber, most of all – and, of course, the reason why their greatest students – Horkheimer, Adorno, Marcuse, Benjamin, Mills, and others too – have yet to be surpassed. Today's politics of liberation must redeem what Ernst Bloch (1986, p. 316) called "the memory of future times," not to mention the live insights of dead theorists.

Third, and as Adorno ([1969]1976) emphasized more than any other, it is not only theory that must be recovered, it is also the diverse history with respect to which the theory addressed itself that demands redemption, however impossible redemption at this level may be. It may be, for example, as Jean Baudrillard (1993) feared, that the history of the modern system is no longer intelligible by means of social science inasmuch as social science lacks the autonomy to discern the system as a totality and as a result of the fact that the system is dedicated to fabricating its own history, fabrications that a naïve social sciences accepts as its raw data.[1] But it may be also that the moral depravity and concomitant loss of humanity, which characterizes modernity's tumultuous short existence, deprives historical reflection of its power to redeem. As daunting as history and historical social science are under such conditions, complacency before the systematic falsification of the historical record, the systematic confusion of collective memory, and the systematic suppression of the irredeemably senseless violence wrought on millions in factories and concentration camps and every other specific location of domination, is tantamount to a politics of surrender. Dahms and his Critical Theory comrades insist that scholarship, including especially historical and anthropological scholarship, must continue to inform today's politics, despite the barriers posed by mainstream social science's pseudo-scholarship. Inasmuch as they can muster alternative empirical descriptions and superior interpretations of the given, Critical Theorists resist the temptation to lead private lives in quiet despair. The form as well as the content of Dahms' novella-length prolegomenon is a testament to his commitment to genuine scholarship as an essential moment in the critique of systematically distorted consciousness.

Fourth, as the total system speeds largely unabated toward its self-destruction, seemingly progressive actions that remain construed and enacted below the historical level of history-making become, in fact, increasingly regressive just as the urgency to grapple with the totality increases. Paper or plastic, Obama versus McCain, bio-fuel versus electric cars: pseudo-choices that bespeak a systematically distorted understanding of the nature of the system and the nature of the crisis, both of which demand qualitative transformation. In contrast, a politics able to "face facts" is radical on an anthropological scale: the question that must be posed, that has been posed for decades but without any practical recognition of such, is *What way of life is needed to begin to repair humanity's relationship with its planet and to sustain the prospect for an enlightened and perhaps even genuinely happy earthly human existence?* A subsidiary question: *What sort of worldwide revolution will permit the actuality of such a way of life?* And subsidiary to that: *How – how specifically – to effect this revolution in the least amount of time possible, and with the least human suffering possible?* These questions point to a politics that is, as it were, relevant. By all means, choose paper, Obama, and electric cars, but do so without self-delusion concerning the relationship between such acts and the demands of history-making in our epoch.[2]

Fifth, societal/cultural endings are dangerous. In the case of the modern system, its ending is predictably apocalyptic in a manner that is unprecedented and planetary. So-called soft landings are the self-delusions of good-natured and mild-mannered bureaucrats. In this specific spatio-historical context, the politics of human survival is wedded as never before to the immediate and world-embracing pacification of existence. Regressive tendencies – fascist exploitation of fear and prejudice, liberal embrace of fictive market beneficence, social democratic higgledy-piggledy, and so on – are intolerable inasmuch as they facilitate further distraction from the radical action that is urgently necessary and long-overdue, and which would actually address – address in actuality – the causes and consequences of the system's self-destruction. It may be too late – better, it is very clearly too late – but the task of Critical Theory today remains unchanged from its original formulation: to supplant our collective willingness to ignore, suppress, and deny self-knowledge, which has been the principal motive force of our recent history, with the highest form of self-consciousness as yet embodied. In other words, Critical Theory's political work is to democratize itself.

2. LIVING WITH AND WITHOUT CRITICAL THEORY

Readers of Marcuse's (1964) *One-Dimensional Man* will recall that his empirical method included channel-surfing AM radio. What listeners today hear broadcast on such stations includes agitation and propaganda that is only dissimilar to that analyzed by the original Frankfurt School by virtue of its linkage to even more pervasive and omnipresent media, especially, of course, to television programs and Internet sites. Glenn Beck is on the radio, and Glenn Beck is on FOX. Lou Dobbs is on the radio, and Lou Dobbs is on CNN. Rush Limbaugh is on the radio, and Rush Limbaugh is streaming live at RushLimbaugh.com. But the delusional content is essentially the same and includes appeals to the so-called little man beleaguered by forces out of his or her control to scapegoat convenient Others, on the one hand, and promulgation of related fantastic conspiracy theories, which purport to identify the true source of the masses' fear and anxiety on the other. Thus, the estimated one-million listeners of *The Michael Reagan Talk Show*, for example, are introduced to the idea that President Obama is not, despite overwhelming evidence to the contrary, a natural born U.S. citizen and is therefore not eligible to serve as President of the United States. This thesis – that the election of the 44th President of the United States was, in effect, as illegitimate as the 2009 re-election of Mahmoud Ahmadinejad as the President of the Islamic Republic of Iran – is then repeated by each of the above-named agitators until there emerges a "viral" feeding frenzy among their audience, who are, to be blunt, predominately uneducated white males overwhelmingly affiliated with the Republican Party, who admire former Alaska governor, Sarah Palin, and who hail from her "real America," and who are thus ideologically predisposed to accept racist scapegoating as legitimate analysis. In 2009, fully one-quarter of self-identified Republicans nationally and 47% of those in the South did not accept that President Obama was a natural born U.S. citizen, while other national polls revealed that 40% or more of voting age U.S. citizens regarded Sarah Palin favorably, so much so that the former Vice Presidential candidate was widely deemed a viable candidate for President of the United States in 2012.[3]

The vitriol of the so-called birther movement is best viewed on-line, in snippets of outspokenness before befuddled officialdom and in mass protests, which when so viewed leads one to understand the consistency this virtual "movement" has with the racially tinged class resentment of the 1930s. Of course, in the 1930s, fringe parties were more apt to adopt and serve as the ideological home of such violence-laced political rhetoric and were deemed "fringe" precisely for this reason up until, in Germany and Italy, violence-laced political rhetoric achieved dominance. As yet, Mr. Reagan, the adopted son of Ronald Reagan and Jane Wyman, has not achieved dominance but nor is he unlike Father Charles Coughlin in his influential marginality. Nor are the members of Congress who sponsored the Presidential Birth Certificate Bill or former Governor Palin, whose authentic prattle makes her perfectly popular and parodied alike, as yet as dangerous as the ridiculous officialdom in Germany and the United States who pandered to similarly regressive mass sentiment 80 years ago, but how close we are to closing this gap is anyone's guess, that is, without social science as critical theory there to address the question.

And so it goes in the lifeworlds without Critical Theory, where, for the most part, mainstream social science fears to tread. Were it to do so, it would risk its hard-won if also still-tenuous legitimacy in the eyes of the system (e.g., university administrators, university students, government officials and administrators, the wealthy and their corporate managers, charitable foundations and town fathers, paying clients, etc.). It might also be distracted from its pursuit of overt and in-kind underwriting for research and publication from these same sources of power and prestige. Moreover, mainstream social science would have to part company with its most cherished self-delusions, that irreversible progress has occurred since the 1930s such to render comparisons with current events overdrawn if not absurd, and that cautious optimism, say, about President Obama's progressive leadership, warrants leaving studies of what I would call the 40% lunatic fringe to whatever underfunded and overwhelmed researchers at the Southern Poverty Law Center can muster. Living without a social science informed by – indeed, *formed by* – Critical Theory means living with irrational and quite often pathological social-psychological tendencies manifest in nearly everything and everyone one meets, despite how otherwise pleasant things can be and how friendly folks are, and quite apart from any shared interests in sports, food, babies, and rock-and-roll. Does anyone imagine there were no light moments or playful distractions in Nazi Germany?

Living *with* Critical Theory is, however, no walk in the park either. The personal implications are legion but include living a rigorously mediated life in which what Habermas distinguishes theoretically as a systems and a lifeworld perspectives are in practice fused within a single consciousness who seeks stability and solace from an albeit dispersed and tenuous community of the like-minded, a Generalized Critical Other; vacillation between performance anxieties associated with living out objective contradictions, on the one hand, and the paranoid delusions that are risked when internalizing generalized fear and anxiety associated with the pursuit of truth, alas, under conditions of absolute untruth, on the other; recognition that the old "personal is political" slogan refers also to the sicknesses, stresses, and other somatic burdens that afflict us as a result of living life in a Fast Food Nation, where, as I overheard one new mother in my town recently note, children are likely to grow up "obeast"; and constant investment of cognitive and emotional energy in an effort to daily and repeatedly "face facts," as Dahms stresses, which is to say, to eschew the false comfort of convenient self-delusion, which means sundering many if not most opportunities, rare though they may be in any case, to live free of alienation from others and institutions.

Others, however, have it worse, including much worse. Others act in the public square and, as a result, are shot dead on the spot or dragged off to shadowy prisons where they are tortured to death, or are tortured and then paraded through show trials before being executed. And many, many more have not enough to eat, or simple medicines that might prevent entirely preventable blindness, say, or an infant's death. Lamentation upon the personal struggles of the Critical Theorist, whether with respect to Karl Marx's carbuncles or Walter Benjamin's suicide, must never overshadow the struggles with respect to which these intellectuals and their kin orient themselves and which will still be with us, it seems likely, long after even contemporary theorists like Dahms have achieved relief from that which ails us all: the struggle for a life worth living.

But, in reply and for what it is worth, I note that Dahms' prolegomenon, always simultaneously on point and on edge, nowhere explains from where or whence the Critical Theorist is to find the wherewithal needed to enact the knife-edge project of Critical Theory, enact it politically nor carry it along with them in their lives. Clearly, considerable formal education – including in logic, history, politics, and political economy – stand back of Dahms' approach, but there are many astute and highly learned scholars for whom Dahms' theoretical obstinacy and political commitments are anathema. Stated differently, my concern is with what makes for a Theodor W. Adorno versus a Martin Heidegger? Why C. Wright Mills and not Robert K. Merton? Simply, where do people like Dahms come from? And how should a student, say, who were to read and declare intellectual fidelity to Dahms' classical treatise respond when they awake the next day?

Critical Theory has never often ventured to answer such questions, at least not explicitly, although no doubt much of the appeal of reading Horkheimer, Adorno, Marcuse, and Benjamin is, as I began this chapter, to be found in the identification, the often secret identification, which serious readers often make with the extraordinary forms of self-consciousness there given expression. Sheer mimicry is not what is at issue. More it is the affirmation that one's own thoughts have a history, that we, as individuals, are not alone. The potential for narcissism and elitism of the sort usually associated with conservative thought bars most Critical Theorists and would-be Critical Theorists from dwelling in such waters, but I suggest that we should permit ourselves appropriate spatio-historical reflexivity in this regard as well, for we are in so many ways irretrievably alone. Not only our fate – death, the biological universal, is the supreme act of individuation – but also the historically specific fact, as Fredric Jameson has rightly noted, that any reader awake in the time-of-the-now has to consider that "the question about poetry after Auschwitz has been replaced with that of whether you could bear to read Adorno and Horkheimer next to the pool" (1990, p. 248), which is to say, whether and how we are to muddle on toward a future when so much in our past militates against our so doing and as this realization then bears down upon us and makes it difficult to breathe. Perhaps additional thought on the reproduction of Critical Theory, its background sensibilities and emotional prerequisites, and not only late reminders of its eclipse, also requires our attention. My hope is that it is here encouragement enough to simply suggest this as a possibility.

NOTES

1. Baudrillard (1993, p. 160)af0 once noted: "We tend to forget that our reality, including the tragic events of the past, has been swallowed up by the media. That means that it is too late to verify events and understand them historically, for what characterizes our era and our de siècle is precisely the disappearance of the instruments of this intelligibility. It was necessary to understand history when there was still history".

2. Liberal critics of genuinely Left politics often point to the Left's inability to distinguish between Albert Arnold Gore, Jr., and George W. Bush or, likewise, to grasp that the election of Barack Obama as President of the United States would be more apt to promote social justice at home and abroad than would election of John McCain. Liberal critics cannot seem to imagine Left analysts who maintain the integrity of their analysis while also actively pursuing the election, say, of Senator Obama over and against Senator McCain, this, through the usual means of political mobilization such as making financial contributions to campaigns, volunteering, and acting as informal opinion leaders. I mean to say that such liberal criticism is hogwash inasmuch as it implies a false choice between sound, if also radical, political analysis and simple participation in mainstream electoral politics. The two are in no way necessarily exclusive. Dialectical thinkers are accustomed to living (with) such contradictions. Moreover, they understand the unavoidable need to learn to do so.

3. Polling data was viewed on August 4, 2009, at *United Press International*, *New York Times*, and *Gallup Poll* web sites (see http://www.upi.com/Top_News/2009/08/01/Poll-Birthers-Republicans-Southerners/UPI-70871249143676/, http://opinionator.blogs.nytimes.com/2009/07/22/birther-boom/, http://www.gallup.com/poll/121514/Americans-Political-Future-Palin.aspx)

REFERENCES

Adorno, T. W. ([1969]1976). *The positivist dispute in German sociology* (G. Adey & D. Frisby, Trans.). London: Heinemann.

Baudrillard, J. (1993). In: M. Gane (Ed.), *Baudrillard live: Selected interviews*. London: Routledge.

Bloch, E. (1986). *Natural law and human dignity*. Cambridge: MIT Press.

Dahms, H. F. (2008). How social science is impossible without critical theory. In: H. Dahms (Ed.), *No social science without critical theory. Current perspectives in social theory*, (Vol. 25, pp. 3–61).

Habermas, J. ([1968]1972). *Knowledge and human interests*. Boston: Beacon Press.

Habermas, J. ([1981]1984). *Theory of communicative action* (Vol. 1). Boston: Beacon Press.

Habermas, J. ([1981]1987). *Theory of communicative action* (Vol. 2). Boston: Beacon Press.

Jameson, F. (1990). *Late Marxism*. London: Verso.

Marcuse, H. (1964). *One-dimensional man*. Boston: Beacon Press.

Mills, C. W. (1959). *The sociological imagination*. Oxford: Oxford University Press.

BEYOND 'FEMINISMS': REFOCUSING THE WOMEN'S MOVEMENT THROUGH THE LENS OF LIBERATION

Vanessa Walilko

In 1974, in a lecture at Stanford University, entitled, "Marxism and Feminism" (published in the same year in the journal, *Women's Studies*), Herbert Marcuse (1974b, p. 279) stated that the Women's Liberation Movement was "perhaps the most important and potentially the most radical political movement that we have, even if the consciousness of this fact has not yet penetrated the Movement as a whole". Marcuse saw the liberating potential of the movement in its critical consciousness – the light it shone on the invisible structures of oppression inherent in a patriarchal, capitalist system and the knowledge that the dismantling of this system was necessary to have a truly free world.

My experience in a contemporary undergraduate Women's Studies department, however, showed me that this consciousness had been blunted. The penetrating light that the Women's Liberation Movement had once shone on normalized and assumed social constructions of domination was now diffused. The writings I encountered seemed less about dismantling and re-creating and more about understanding the ways in which social constructions of race, gender, class, and sexual orientation affected one's daily life and experience of domination and how one could alleviate concrete

Nature, Knowledge and Negation
Current Perspectives in Social Theory, Volume 26, 327–338

ISSN: 0278-1204/doi:10.1108/S0278-1204(2009)0000026016

problems like domestic violence within the system. As a movement, feminism had been fractured into many "feminisms," to the point that the word itself carried no political or philosophical meaning. Furthermore, there was no mention of Marcuse or his work in any of women's studies classes or any of the literature. It seemed to me that feminism's general avoidance and silence regarding Marcuse was ill-fated: the strength of the movement depended on refocusing feminism through the lens of liberation.

I worked in the Women's Studies department at my university for two years, took classes with a number of the department's professors, and did extensive work on domestic violence as a research assistant. One experience keeps coming back as pivotal in my understanding of contemporary feminism and the state of the women's liberation movement. It was the first week of a feminist frameworks class, and the professor asked each of us what we were most interested in studying. One of the students mentioned that she was interested in the positioning of the letters "LGBTQA" and how that positioning revealed a certain hierarchy among non-heterosexual groups of people. It was a seemingly innocuous statement, but for me, it marked a point of cognitive dissonance. I was stunned that she was focused on the hierarchical organization of letters instead of the hierarchies in society that informed this ordering. It was then I realized that somewhere along the way, feminism had been led astray. It was only after reading Marcuse that I discovered what had gone wrong.

My initial interest in feminist scholarship was essentially intuitive. Exposed to feminist theory in various classes, it was an intense feeling of relief to read the works of Adrienne Rich, Andrea Dworkin, Robin Morgan, and others and find words for the unspeakable sense of being suffocated in society. Rich (1976, pp. 56–57) discussed the "sexual understructure of social and political forms ... in which the female is everywhere subsumed under the male". Dworkin (1988, p. 134) stated that "the stance of the women's movement is that [gender inequality] is not 'just life.' It is politics; it is history; it is power; it is economics; it is institutional modes of organization: it is not 'just life'", illuminating how institutional support of gender inequality exists everywhere. Morgan (1977, p. 310) rejected "the patriarchal dichotomizing of intellect and emotion", and she and other radical feminists advocated a holistic approach toward thinking, toward life.

While reading these authors, it was as though a veil had been lifted, and someone was explaining in clear language how and why women the world over were repressed, oppressed, belittled, despised, and feared, simply for being women. It described the barriers I was facing, the interactions I was having. It illustrated the ways in which my life was being defined by my gender, which itself was a social construction crafted to achieve political ends.

Feminist theorists went further than other theorists I encountered, describing systems of domination inherent in patriarchy – which they identified as the source of all oppressive systems – and illustrated how all systems of domination were interrelated. Furthermore, it located women in multiple marginalities – it showed how race, class, and sexuality changed the pitch of sexism and misogyny experienced. While conscious of systems of oppression operating in our society, feminism first offered me a language for my oppression and a tool for analysis of a repressive society. Furthermore, radical feminist theorists illustrated how the creation of a free and equal society depended on challenging all institutions that perpetuated inequality and dominance. Dworkin (1988, p. 131) warned against supporting "male ideology, male interpretation, or male intellection". The creation of a free society would depend not only on rethinking our assumptions but also on qualitatively changing how we thought.

And yet the more I began to work in women's studies, the more dissatisfied I became with the direction of feminism. Perhaps, I read too much about the beginning of the women's liberation movement and was expecting too much. I had thought that being in the women's studies department, I would find myself among a group of young radicals, bra-burners, and lesbian separatists living at the edge of society and trying to free everyone from a repressive social apparatus. What I found there was what I seemed to find in every department: complacency – a contentment in endless theorizing and a lack of radicalism.

Contemporary feminist theorists I encountered seemed more interested in giving everyone a voice and understanding how every potential cross-section of life was understood and given its own feminism. Authors argued for "post-colonial feminism" to represent interests of women living in the Third World (Mohanty, 2003; Narayan, 1997). Feminist post-structuralism illustrated how subjective realities shape the use of language and how language is used to maintain unequal power distributions in society by making women responsible for their own victimization (Whedon, 1997). While silence serves no group of people, it seemed to me that this theoretical tactic of "understanding" reinforced the system as it was. Theorists and practitioners were looking for representation of marginalized groups within a repressive system. Everyone was talking about equality, but no one was asking about how this could be achieved in a social system built on and sustained by domination. Furthermore, no one seemed to be talking about the ways in which we should and could begin to re-imagine a society in which everyone could be free. The endless theorizing reminded me of a fractal, spinning out infinitely and looking exactly the same.

My senior thesis was the culmination of two years of research about domestic violence. In all the literature I read about changing violent men, domestic violence programs seemed oriented on teaching men non-violent ways of interacting with their partners and reinforcing anti-sexist messages. The specific program on which I based my research taught violent men that violence was a learned choice rooted in power and control and perpetuated in social and cultural attitudes. Researchers and practitioners located domestic violence as a system of men's control and dominance (Campbell, 1993), identified gender as a routine accomplishment (West & Zimmerman, 1987), and viewed interpersonal violence as a mark of masculine gender production (Anderson, 1997). However, the research did not offer any reasonable explanations how interpersonal violence could end *when society is contingent upon the present repressive apparatus*. Not one said that since interpersonal violence was a mirror of social violence, men's domination of women was a component of the capitalist system's domination over all people, that the only way to end violence was to create a new society. No one seemed willing to go far enough. One article even indicated that "theoretical and empirical work is severely handicapped by failure to attend to the distinctions" of various types interpersonal violence (Johnson & Ferraro, 2000, p. 959), as if the failure to distinguish types of violence was even more troubling as the lack of attention paid to a repressive system, which fostered and encouraged violence. That being said, the feminist understanding of violence as a social system of control of men over women was still more desirable than researchers who claimed that reports of gender symmetry in violent relationships were evidence of women's increased standing in society (Archer, 2000).

I suppose this is why my encounter with Marcuse's works was so fulfilling and transformative. While feminist theorists seemed to have retreated from their initial radical positions, Marcuse offered an uncompromising radical vision in line with the feminists I admired. *Eros and Civilization* offered a vision of a world based not on repression but on liberation. Marcuse (1974a, p. 3) directly challenged Freud's proposition that "civilization is based on the permanent subjugation of the human instincts" by reinterpreting the nature of instincts themselves. This is Marcuse's visionary step. Marcuse argued that instinctual composition was not static but adaptable and shaped by historical and social forces. The rationalization for repression based on the dangerous quality of the Death instinct was recast: Thanatos was "an expression of the eternal struggle against suffering and repression" (*ibid.*, p. 29). The instinct's objective was "not the termination of life, but of pain" (*ibid.*, p. 235). As society was reshaped to free Eros from surplus-repression and promote instinctual gratification, "[d]eath would cease to be an instinctual goal" (*ibid.*) for "the objective of the death instinct is not destruction *per se*, but the elimination of the need for destruction" (*ibid.*, p. 271).

Marcuse offered a vision of humanity I hadn't encountered before, one that negated the constraining ideologies I had encountered everywhere else. People from every walk of life had the same idea: if people were left to their own devices, they would destroy society. Therefore, the instincts must "be deflected from their goal, inhibited in their aim" (Marcuse, 1974a, p. 11). But Marcuse had faith in the power of Eros: the creative, libidinal energies, the unifying source of all life. The rationality of civilization had indoctrinated us into believing that the human instincts needed to be repressed in the name of progress, but Marcuse (1974a, p. 109) argued that it was "the *failure* of Eros, lack of fulfillment in life, which enhances the instinctual value of death". The instincts, destructive as they may appear, "testify to the destructiveness of what they strive to destroy: repression" (*ibid.*). In a free world, the elimination of surplus-repression would allow the gratification of the instincts that would "absorb" or "neutralize" destructiveness (*ibid.*, p. 131).

While freedom seemed possible in a world of such abundance, under advanced capitalism, surplus-repression was maintained to further the interests of those in power, and the aggressiveness of the Death instinct was channeled into aggressiveness that maintained a repressive system. Surplus-repression was a "specific *organization* of scarcity" (Marcuse, 1974a, p. 36), not a fact of life, and a re-organization of society was possible, focused on "making life an end in itself … the enjoyment of being, for the emancipation of the senses and of the intellect from the rationality of domination: creative receptivity versus repressive productivity" (Marcuse, 1974b, p. 286). Marcuse (1974b, p. 280) sought the liberation of Eros: "libidinal energy, in the struggle with aggressive energy, for the intensification, gratification, and unification of life and of the life environment".

Marcuse even challenged the primacy of reason – thus anticipating one of Dworkin's central points – by contending that "[r]eason is the rationality of the performance principle … an instrument of constraint, of instinctual suppression" (Marcuse, 1974a, p. 159). A hierarchical ordering of consciousness had led to the idea that "whatever belongs to the sphere of sensuousness, pleasure, impulse has the connotation of being antagonistic to reason" (*ibid.*). Marcuse argued that we are all engaged in voluntary servitude, in "collaboration in reproducing a society which made servitude increasingly rewarding and palatable" (Marcuse, 1974a, pp. xiii–xiv). In its current form, the performance principle had convinced an increasing number of people that this reality is the best we can do, no matter how much we suffer in it. If you talk to people about making "the human body an instrument of pleasure rather than labor" (*ibid.*, p. xv), you'd likely be laughed at. Marcuse, however, took seriously concepts of happiness, play, love, and liberation. I still continue to consider his work a triumph of philosophical and intellectual courage.

Marcuse's philosophy also seemed in line with radical feminism: abolition of a repressive social apparatus, integration of a personality split by repression, into a fully formed human being – a "sensuous rationality," the vision of a free world based on human fulfillment, not the curbing of human desires into socially desirable ends. Marcuse and feminism seemed a natural fit. Both seemed to be linked by a Great Refusal of the status quo: feminism in its rejection of "natural" institutions of servitude; Marcuse in his rejection of accepting a "natural" society of repression. My first encounter with Marcuse, however, took place in a political theory class. He seemed conspicuously absent from feminist writings.

One author, Haunani Kay-Trask, seemed willing to take on the challenge of integrating Marcuse with feminism. Trask (1986, p. 166) defines the "feminist Eros" as "a vision of a future feminist world where love and power, redefined, are the basis of a healthful, nurturant society". The new Eros "speaks to a release and refashioning of the life instincts from the vantage point of women's essential experiences—physical and emotional gratifications ... intended to free society from divisive competition and relentless repression" (*ibid.*, p. xi). Women's historic position as a non-person in society ironically offered them unique privilege: "greater access to the memory of instinctual gratification than men" (*ibid.*, p. 89). This proximity to instinctual gratification explains feminist theory's desire "to unify and embody ... to interpret the world as an interdependent whole" (*ibid.*, p. 86). The critical consciousness that Marcuse (1974b) praised in the women's movement was a result of social construction: women's exclusive mothering allowed them to develop deeper relational capacities than men. Women are, as Trask (1986, p. 91) indicates, "less subjugated by the performance principle, closer to the sources of the pleasure principle". This "access to the memory of instinctual gratification" fosters a consciousness "critical ... of the patriarchy, of a dominating, repressive system" (*ibid.*, p. 90). The patriarchal equation of "woman" with "nature" kept women outside of the institutions of reason, which meant that they could still retain a memory and experience of instinctual gratification. This instinctual access made them critical of patriarchal, repressive systems, allowed them to expose the injustice inherent in the repressive social apparatus.

Trask also provides gendered qualifications to Marcuse's philosophy. Marcuse does not discuss the ways in which women "make their lives within a praxis of pain, limiting the expression of their own humanity as they cushion man's inhumanity to himself and others" (Trask, 1986, p. 82). Women have access to instinctual gratification, the experience of close interpersonal relationships and the libidinal satisfaction those bring, but they pay a price. In a world divided into "work" and "home," alienated workers seek refuge in a safe, nurturing home. As such, "[w]omen's *work* becomes the sacrifice of their *life* through an infusion of *life* into others" (*ibid.*, p. 90; emphasis added). A woman must pay for her relational capacities by losing her own life and expressions through the mothering of everyone else: women must mother men as well as children. It is this surplus-repression of the sexual understructure of human relations that needs to be eliminated. A Marcusean feminist would argue that not only do physical resources need to be redistributed but also emotional resources. A society founded on gender equality and freedom necessitates the mothering of children by men and women. A consciousness of nurturing, support, and care must also follow with Marcuse's consciousness of play, love, and happiness, since due to distinct gendered socialization, "love is expressed unequally by women and by men" (Trask, 1986, p. 82).

Considering how well Trask integrated *Eros* with feminism, I was surprised that few other feminist writers were discussing Marcuse. Since I view the two theoretical frames as compatible, I can only offer conjectures as to why feminists have come to refrain from appropriating Marcuse. Perhaps, they rejected Marcuse's discussion of "polymorphous sexuality" or the repressive nature of "monogamic genital relations." Or maybe it was his kind words about "perversions" that kept feminist theorists at bay (Trask makes no mention of these in her work). At the moment, there are many "perversions" occurring in society that feminists would be happy to do without, such as rape and incest. Marcuse, however, never advocates practices that reflect and reproduce a system of domination. Because definitions of sensuality have been confined to sex, one envisions the release of libidinal energies as a cataclysmic orgy. But Marcuse and feminists seek to reconcile nature and human existence, eros and civilization, by advocating the possibility of overcoming the split of mind and body resulting from the Performance Principle. Trask calls it a "return to the body" and a "return to the mother": a redemption of what Western culture has historically decried. Both Marcuse and radical feminism seek free expression of all that it means to be human, and the end to systems of domination that hinder us from experiencing true liberation.

As I see it, feminism in its present form needs reinvention and could use the single-minded focus of Marcusean liberation. In the years following the women's movement, the definition of feminism has become muddled. There no longer seems to be a consensus as to what feminism is, and seemingly everyone can call herself or himself a feminist. Gerda Lerner (1986, p. 236) stated that feminism is a term "commonly and quite indiscriminately used", either referring to the belief that women should have rights equal to those of men or the movement devoted to attaining those rights or the belief that large-scale social change is needed to change the social position of women. Adrienne Rich (1986) argued that feminists needed to recognize heterosexuality as a political construction to achieve political ends. Dorothy Allison (1994, p. 14) argued that a feminist analysis needed to take into account the matrix of systems of domination and privilege that shape the lives of women, particularly class inequalities as we live in a "world that despises the poor". As Trask (1986, p. 174) would put it, personal preference "masquerades as feminist theory without any material basis except the assertion of individual likes and dislikes". Ariel Levy (2005, p. 56) indicates that Hugh Hefner considered himself a feminist before feminism and the creator of the *Girls Gone Wild* videos thinks his videos are "liberating for women" (*ibid.*, p. 12). Strippers "have come to symbolize sexual liberation despite the fact that it is their job to fake arousal" (*ibid.*, p. 20) and pro-sex feminists declare that pornography "breaks cultural and political stereotypes", and therefore "benefits women, both personally and politically" (McElroy, 2002).

The women's movement fought for equality within the system, but the only equality that exists seems to be that women can now exploit themselves and other women in the same ways men have historically done. In Marcusean terms, women have now been included in the repressive capitalist society to such an extent that they are losing the critical consciousness that challenged the performance principle and are now becoming cogs of repression. If women's increased use of violence is an indication of equality in our society, it is because "as equals in the economy and politics of capitalism ... women must share with men the competitive, aggressive characteristics required to keep a job and to get ahead in the job" (Marcuse, 1974b, p. 285).

In this veritable Tower of Babel of "feminisms," the word itself seems to have lost all meaning. The Combahee River Collective (2003, p. 166) once argued that "liberation is a necessity ... because of our need as human persons for autonomy". Now multiracial feminists such as Zinn and Dill (2003, p. 359) try to understand how "the existences and experiences of all people—women and men, different racial-ethnic groups, and different classes—shape the experiences of each other". I would argue that a

Marcusean-feminist synthesis could help refocus feminism through the lens of liberation. Instead of endless theorizing about how race, class, sexuality, country of origin, and myriad other factors affect one's experience within a patriarchal system, liberation shifts theory toward ending domination and repression and envisioning new worlds based on freedom. Marcuse insisted that the goals of the women's liberation movement could be attained "only by a change in the entire social system [and that it required] a change of consciousness, a change in the instinctual needs of men and women, freed from the requirements of domination and exploration" (Marcuse, 1974b, p. 281). This "transcendence" involved a "negation of the exploiting and repressive values of patriarchal civilization" (*ibid.*), namely the capitalist performance principle, "the rule of functional rationality discriminating against emotions ..., condemnation to alienated and inhuman labor, and the will to power, the display of strength, virility" (*ibid.*, p. 282). I believe that this new consciousness, Trask's feminist Eros, could help release a more vital feminist consciousness from the murky waters of "feminisms." Feminist ideology could be determined by whether a particular idea or practice supports or impedes Marcusean liberation.

One of the great debates among feminists concerns pornography. Pro-sex or sex-positive feminists argue that pornography is liberating to women, allowing men and women to engage in cathartic fantasies (McElroy, 2002), while anti-porn feminists claim pornography perpetrates racism (Mayall & Russell, 1993) and is degrading to women, and therefore antagonistic to women's liberation. The debate has raged for decades. However, a Marcusean-informed feminism would argue against pornography on the basis of liberation. Marcuse has claimed that "exchange society comes to completion with the commercialization of sex: the female body not only a commodity, but also a vital factor in the realization of surplus value" (Marcuse, 1974b, p. 285). Defense of pornography, particularly on the basis of free speech, is antagonistic to liberation since the support of commercialization and commodification of (women's) sexuality is a defense of the Performance Principle. Pornography, and its equally incendiary counterpart – prostitution – legitimize a capitalist system that equates human value with monetary value, happiness with accumulation. One can see this trend not only in defenses of pornography but also in arguments supporting the legalization of prostitution – that women should be allowed to choose how they want to enter the marketplace, with no questioning of why there should be a system in which people must sell their labor and their time to survive. By not challenging a world of alienated labor, feminists that support prostitution and pornography perpetuate a society where individuals "have innumerable choices, innumerable gadgets which are all of the same sort and keep them occupied and divert their attention from the real issue—which is the awareness that they could both work less and determine their own needs and satisfactions" (Marcuse, 1974a, p. 100). At present, "sexual liberty is harmonized with profitable conformity" (*ibid.*, p. 94) and, as such, "the existing liberties and the existing gratifications are tied to the requirements of domination; they themselves become instruments of repression" (*ibid.*, p. 92). If feminism is about women's liberation and free choice, feminists should work toward a society in which people are allowed to decide what they want, not just choosing what they want from a pre-selected variety of products and services between which they can choose. These consumer choices do little more than reinforce the idea that a repressive society is giving us all that we really want.

Marcuse is clear: as "the established order of domination … [c]ivilization has to defend itself against the specter of a world which could be free" (Marcuse, 1974a, p. 93). The choices we are given in this society are merely illusions, which support the idea that we are all seeking and getting what we really want. By seeking gratification in socially constructed fantasies, we perpetuate a system that deflects our attention from what we really want. In celebrating performance sexuality – "raunch culture" as Ariel Levy would describe it – we are not being daring or progressive, but instead falling into one of the oldest repressive traps: limiting the libidinal energies to mere sexuality.

If fundamental differences between Marcuse and feminism exist, feminists would benefit from addressing these head-on and working, as Trask has done, on making explicit and developing further the feminist dimension of *Eros and Civilization*. Ignoring the philosophy that Marcuse has to offer does a disservice not only to feminism but also to efforts directed at overcoming a world of poisoned Eros. In its inception, the women's movement was essentially and eminently radical. Its aims were egalitarian and anti-materialistic. The women's movement was not focused on merely pointing out the ways in which a patriarchal system devalued women and their experiences, but it sought the end of patriarchy. Nothing less than a complete re-visioning of society would suffice – capitalism, racism, and all other systems of oppression had to be actively eliminated. As qualitative changes within the system of repression were achieved, this focus shifted to creating equality within the system. In the classes I took, we were encouraged to understand different forms of oppression; yet, no one was really talking about how to end this oppression. We talked about giving more people more rights, changing people's attitudes, but radical re-visionings of society were few, far between, and relegated to the realm of fantasy.

Marcuse states in *Eros and Civilization* that "those who die before they must and want to die, those who die in agony and pain, are the great indictment against civilization" (Marcuse, 1974a, p. 235). The initial radicalism of the women's movement was its attempt to end this cruelty – to end suffering and to create a world that was genuinely and qualitatively free. By advocating change within the system, and by not challenging the repressive apparatus of the rationality of domination and the performance principle, feminists participate in this continuing crime against humanity and nature. Feminists who debate pornography, domestic violence, or any other issue, without attacking the repressive system, distract from the real issue – that we live in a society where our energies are constantly channeled for the purpose of furthering actually existing social goals, as they represent and perpetuate what Marcuse called civilization as "the established order of domination," instead of enabling individuals as members of social groups to direct their energies, efforts, and time toward actual personal fulfillment – toward personal actualization, as the necessary precondition of society's self-actualization and liberation. Yet, feminism has been co-opted to the point of pop culture consumerist sexuality being paraded as "liberation." Feminism was initially founded in challenging structures of domination by raising consciousness. We owe it to all those who have come before and lived unnecessarily in want and in pain to continue in that fight – of creating a world in which everyone can be truly free, not just free to buy and buy in.

REFERENCES

Allison, D. (1994). *Skin: Talking about sex, class, and literature*. New York: Firebrand.

Anderson, K. L. (1997). Gender status and domestic violence: An integration of feminist and family violence approaches. *Journal of Marriage and the Family*, *59*, 655–669.

Archer, J. (2000). Sex differences in aggression between heterosexual partners: A meta-analytic review. *Psychological Bulletin*, *126*, 651–680.

Campbell, A. (1993). *Men, women, and aggression*. New York: Basic Books.

Combahee River Collective. (2003). A black feminist statement. In: C. McCann & S.-K. Kim (Eds), *Feminist theory reader* (pp. 353–361). New York: Routledge.

Dworkin, A. (1988). *Letters from a war zone*. New York: Dutton.

Johnson, M. P., & Ferraro, K. J. (2000). Research on domestic violence in the 1990s: Making distinctions. *Journal of Marriage and the Family*, *62*, 948–963.

Lerner, G. (1986). *The creation of patriarchy*. New York: Oxford University Press.

Levy, A. (2005). *Female chauvinist pigs: The rise of raunch culture*. New York: Free Press.

Marcuse, H. (1974a). *Eros and civilization: A philosophical inquiry into Freud*. Boston: Beacon Press.

Marcuse, H. (1974b). Marxism and feminism. *Women's Studies*, *2*, 279–288.

Mayall, A., & Russell, D. E. H. (1993). Racism in pornography. In: D. E. H. Russell (Ed.), *Making violence sexy: Feminist views on pornography*. (pp. 167–178).

McElroy, W. (2002). A feminist overview of pornography. Available at http://www.zetetics.com/mac/freeinqu.htm. Accessed on October 22.

Mohanty, C. T. (2003). Under western eyes: Femminist scholarship and colonial discourse. In: R. Lews & S. Mills (Eds), *Feminist postcolonial theory* (pp. 49–73). New York: Routledge.

Morgan, R. (1977). *Going too far: The personal chronicle of a feminist*. New York: Random House.

Narayan, U. (1997). *Dislocating cultures: Identities, traditions, and third-world feminisms*. New York: Routledge.

Rich, A. (1976). *Of woman born*. New York: Norton.

Rich, A. (1986). *Blood, bread, and poetry*. New York: Norton.

Trask, H.-K. (1986). *Eros and power: The promise of feminist theory*. Philadelphia: University of Pennsylvania Press.

West, C., & Zimmerman, D. H. (1987). Doing gender. *Gender & Society*, *1*, 125–151.

Whedon, C. (1997). *Feminist practice and postsructuralist theory*. London: Blackwell.

Zinn, M. B., & Dill, B. T. (2003). Theorizing difference from multiracial feminism. In: C. McCann & S.-K. Kim (Eds), *Feminist theory reader* (pp. 353–361). New York: Routledge.

AFTER POST-MODERNISM: TOWARD THE RECOVERY OF THEORY

James Block

The fateful question for our time is what comes after post-modernism, what comes after the after? Building upon the insight of Thomas Mann in *Dr. Faustus*, the great fictional study of the post-modern composer Adrian Leverkuhn, and upon insights of some of the post-modernists themselves, this after hovers as the lingering but ultimately expunged final (dissonant) chord whose uncanny presence in absence turns cosmic emptiness into the cult of memory, ritualized attentiveness to the faded chord that connects us back to a departed world of meaning by a gossamer thread. A vision of this ritual is acted out in the greatest of all works of deconstruction, *The Recovery of Lost Time*, in which Marcel Proust guards this attenuating thread with his life like the Wichita lineman, for that is his life. The traveler stranded at the beginning of the epic novel can no longer go forward – there is no more track to lay and no destination to lay it toward – only the obscured recesses of a lost world (long ago when the future still existed), now as fugitive as the years.

But what about those of us for whom the after the after is not enough? Those of us for whom post-modernism played a critical role in clearing away exploitative master narratives and crumbling theoretical universals through its radical interrogation of philosophical/cultural presumptions and political/conceptual privileges, yet left us in Derrida's term "late to the end

Nature, Knowledge and Negation
Current Perspectives in Social Theory, Volume 26, 339–345

ISSN: 0278-1204/doi:10.1108/S0278-1204(2009)0000026017

of history?"[1] Those of us who demand a future? What can future mean after the after?

It is clearly in some sense to begin anew, but this is insufficient. It is not just putting another quarter into the pinball machine for a new game or drawing upon a blank slate. For with the industrial strength dissolvents of post-modernism spread upon everything, some stronger substance is required as we head out in the face of gale-force winds. In fact, and perhaps unexpectedly, the uncomforting nostrums of dissolution achieved the capacity to quiet the savage beast, to console us for our failure to make history, or to even try. In its radical interrogation of all claims of possibility, ultimately dissolving things into unconnected monads (without God), it sharpened the scalpels to use on all efforts at possibility and connectivity. Connectivity and possibility, we now all-too-easily believe, are radical illusion, bad faith, the effort to impose an ordered middle term or narrative on discrete, disparate, and ultimately inert phenomena, to forge a union or project out of the imperial urge to colonize.

One way out, of course, is to turn in revulsion from the impossibility of living this radical faithlessness. In *Dr. Faustus*, the purity of Adrian's quest to see into the world's meaninglessness, which rendered even God's omniscient eye a place of wish fulfillment (after all, who would have more at stake in a project than its own creator), is ultimately undone by infantile needs that turn him – and Germany (whom he represents) – into a monster of pseudo-innocent primal hunger for recognition and ultimately domination. For Proust, memory becomes an infinite loop of fraying cohesion and coherence, perpetually papered over by the romantic will to reinstate in a mythic unity. Derrida's case is equally poignant: asking late in life, "I would like to learn how to live finally;" haunted by the ghosts of time past produced by trauma and mourning, he inquires about "seers" and "visionaries," and their power of "conjuration," of effecting a "promised return" of the "presence to come," or what he will call elsewhere simply, candidly, at "night-fall," the "desire of resurrection."[2]

In my understanding, proposing the naked and impossible need for a Savior as the answer to how to live finally betrays the final, desperate, and tragic result of the radical interrogation of possibility. As Derrida himself admits, it was the culture of inflated, all-enveloping narratives of possibility, Hegel, Marx, and their ilk, that led his generation to the work of decomposition, in Mann's terms, revenge against the self-righteous totalizing faiths of the masters, a work in which they were themselves ultimately consumed by the need, futile, to be saved from themselves.

How do we, then, begin after the after without succumbing to the wish for a *deus ex machina*? Here, I propose that with the only all-enveloping narrative remaining being that of post-modernism, rather than submit to its own radical interrogation once again we turn its project back upon itself – that we radically interrogate its presumption of impossibility. Is it the case that there can be no narratives, authors, connections, gifts, progress, or selves? The post-modern animus is directed against the Western assumption that God, History, or Nature mandated these, but its extreme rhetoric – very much the legacy of the totalizing ancestors it opposed – appeared to rule them out *tout court*. But what if learning how to live and all of its achievements – story, authority, mutual bonds, generosity, growth, and authenticity – are not, it is true, written in Stone and on display in the Temple, but rather developmental achievements possible for those serious about undertaking the quest for how to live.

In this context, the radical interrogation of impossibility would ask, "under what conditions and in what way is absence a repression, a suppression, of the human potential, my potential, for reaching beyond the bare given." Two quick examples to make the point. Objectivity, a.k.a. God's perspective, we now know, is as such impossible. Yet, we also intuit that some uncoverings reveal and some accountings encompass more of their worlds than others. They enable us to experience more deeply and to inhabit a space more comprehendingly. We make such judgments about what we are given in terms of what Robert Pirsig calls "quality" all the time. What lies at the base of these judgments? This is a complicated matter, worthy of further inquiry. One lead may lie in J. S. Mill's reflections on Tocqueville's *Democracy in America*. How did Tocqueville, Mill asks, a latter-day French aristocrat, produce such an insightful, even revelatory account of early American democracy, 180° removed from the world he knew? Perhaps, because he was not simply an aristocrat, but a dispossessed aristocrat belonging to a class no longer with a stake in or promises to be gained from history. So removed, he could witness the parade of other players with a detachment that is rare but perhaps not impossible, and which may under certain conditions allow for unusual levels of clarity.

In J. M. Coetzee's novel *Age of Iron*, Mrs. Curren is a dying South African white liberal in the final days of apartheid. Always an intellectual opponent of the regime, she finds her cynical detachment and condemnation equally of brutal whites and uncouth blacks challenged by the regime's barbarism in her backyard, which she can no longer deflect. Her sudden, shattering awakening to the question of how to live, finally, leads her first to propose a quixotic act of terrorism – a wish to end her life in heroic flames in a plea for instant resurrection. But, she quickly realizes that this will not teach her how to live but only how to die. Learning how to live is a much lengthier and ultimately demanding process, involving a surmounting of the defense that the world is meaningless, after which all the questions conclusively deferred by heroic death have to be faced. But she addresses them and comes to her end profoundly changed from when we first meet her, able to die with at least an inkling of having lived.

In the final interrogation of impossibility, I want to consider the grounding for the other forms of possibility, the subject, the author, the self, and the humans as the makers of history. Three of the four greatest modern novels, the aforementioned *Dr. Faustus*, *The Recovery of Lost Time*, and *Moby-Dick*, address the impossibility of self-wholeness (excepting only George Eliot's *Middlemarch*, which stands alone as a story of this-worldly transcendence). These novels strangely locate that impossibility, or perhaps the symbolic moment of recognition of the impossibility, in a single traumatic incident – Adrian's infatuation with the prostitute Esmeralda, Marcel's "black night" as a boy in which his mother spent the night in his room, and Ahab's loss of his leg to the great white whale. This "necessary" and yet "impossible" wish (to quote Derrida; *Specters of Marx*, p. xviii) for full self-realization sets up all that follows as a futile quest, in which the journey toward narcissistic wholeness, the I am I, is replaced by endlessly repeated impulses and acts of compensatory narcissism, wish fulfillments that substitute fantasized completeness for the now recognizably unobtainable reality. For the novelists, this self-consciously compensatory narcissism by individuals who know their quest is doomed defining the late modern era helps us account for the psychic roots of America's imperialism, twentieth-century fascism, empty Romanticism, and ultimately the despair of all forms of possibility constituting post-modernism.

Derrida explores this terrain in his article "Autoimmunity" on the 9/11 attack on the World Trade Center. In brief summary, he argues that such an attack, involving the collapse of national fantasies of primitive narcissistic – and imperial – omnipotence, triggered an irreparable trauma rendering the subject incapable of carrying on toward the future. As with Proust, the future is now either the space for the ritual enactment of such delusions ending inevitably in failure or it is not, that is, it is nothing. Once shattered, the subject is consigned as in an Ionesco play to repetition compulsion, the endless reimagination and wished-for repair in a fantasy of the impossible. And yet, the impossible, narcissistic development, is the inexpungable core of the psyche, leading to the result of the permanent condition of mourning, loss, rage, aggression, and futility.

What would the radical interrogation of this primal dynamic regarding primitive narcissism entail? In fact, this is the strenuously contested ground separating Freudian psychoanalysis and transformative psychodynamic accounts from Rousseau to Marcuse, Fromm, and beyond. This is the terrain on which the struggle for a new beginning must be waged. This is the future of a liberatory psychopolitics and transformative social vision, the development of new narratives, authorities, selves, and communities.

Freud is the master psychologist of post-modernism in that he posits a traumatic break – developed by Lacan and bequeathed to the post-modernists – with primitive narcissism as the passageway into maturation and social adaptation. But it is a form of self-sacrifice. Maturation is not development, because authentic internal growth ceases at that point of rupture, replaced by projective idealizations, introjected authors, substitute gratification, wish fulfillment, and sublimation – all of these forms of substitution, as-if-development in a false world of simulated objects and desires, the only fulfillment we are allowed. In this world, we have the spectacle of those who do not know how to live (and who would under these conditions?) and can barely enunciate the wish to find out, cut off from authenticity like Mrs. Curren who realizes in her 70s that she was raised to be a hollow doll, the protagonist in *Surfacing* whose head is radically disconnected from her body, the narrator in *Zen and the Art of Motorcycle Maintenance*, an inauthentic, unknowing not-self who carries his displaced real self deep within under layers of denial and amnesia, in Rousseau's terms driven by vanity, *amour proper*, the wish to adapt in place of self-love, *amour de soi* – condemned to live in the Cave, best by the shadows of things whose reality cannot be discerned. In Marcuse's terms, the repressed and repressively desublimated false and socially constructed id (or not-I) has replaced the self in this process of accommodation.

The ultimate question, then, is whether the narcissistic development of the self is possible, or whether the collapse of primitive feelings of narcissistic omnipotence portend the collapse of the self into Oedipal submission – that is, a self-willing of self-defeat even without further social repression, from being convinced of the sheer impossibility of narcissistic wholeness. Perhaps, this scenario of defeat occurs because the subject has already internalized the repressive matrix and its demands for self-repression. Perhaps, it is the obedient child only too anxious to save its parents the trouble – and pain – of sacrificing it by willing its self-sacrifice, thus excusing the repressers by making self-loss a natural and thus unavoidable matter like death that authority must all too patiently endure. Tragedy indeed.

A different scenario, suggested by Rousseau, Emerson, Whitman, Bourne, Pirsig, Fromm, Marge Piercy, and others (Marcuse to an extent), offers that the collapse of infantile omnipotence is the gateway to healthy narcissism. There are two pathways here. The first reframes the historical revisions that have undone the inflated fantasies of the human role: Copernicus, Darwin, and Freud. In my class, we have rethought this in terms of Mrs Curren's decision to live – a lost woman in a lost society. Because we mediated, she needs "to set things right." "What?," we asked. "To not die having fooled oneself perpetually." Fooling others, perhaps. To have lied systematically to oneself – a narcissistic blow from which one, I, who "knew what I was doing all along," will not recover. Suddenly the specter of *Specters of Marx* revisited. It begins with Hamlet's plaint "The time is out of joint" and "I must set it right," I must set myself aright, and all illusions of omnipotence constitute false, self-defeating detours.

Such a narcissistic revelation, secondly, leads to a new relation to one's time. Rather than our Western fetish-object God, whom we have constructed as the ultimate narcissist – lonely, living alone, no companions because of an insistence that the creation and its management is his, all his – we must imagine how to live on the far side of realizing the existence of other selves, other wills, other purposes, and other dreams than our own. Perhaps, our depression in this era is our sympathetic empathy with our God's agony at the loss of his primitive fantasies, his World Trade Center, but our task is to go on from there, together in a web of realized and realizing selves.

If so, we would be further along – because a politics of wholeness presupposes a model of human development, which enables each of us to become selves, authors, with narratives and meaning systems, and connectivity. Philosophy and political theory after post-modernism must attend not to the knowable, but to what has been made and what can be makeable. This is not sheer instrumentalism ... it is not simply "how." Rather, lest we forget, post-modernism arose after the 19th-century realization that *all is permitted.* Once it was discovered that though not that all can be accomplished, there is no *inherent* reason for not trying, a limitless world was opened. This in effect collapsed claims of essential meaning, and very nearly meaning itself.

In recasting meaning after the disappearance of transcendental boundaries, the issue is not what is permitted in the absolute sense, but what possibilities not yet there can be realized – and which are worth realizing. This imposes upon us and us alone the task of both shaping and clarifying viable forms of community, freedom, equality, social justice, and the like.

The clearing away of absolutes in the 20th century thus gives to political thought a renewed life.

A shared and meaningful world can of course be radically interrogated out of existence and then out of even being possible. For it is never yet completed. But part of the developmental process is to come to terms with our own doubts about our journey and the future, our capacity for grounding, for wholeness and for connection, and to discover our radical capacity to discover how to live. After post-modernism, where there are no absolute scripts, there are only the bonds and promises we achieve not as compensation but through the struggle for completeness, finally the achievement of a fit world for our healthiest visions of self and society.

NOTES

1. Jacques Derrida, *Specters of Marx* (New York: 1994), p. 15.
2. Derrida, *Specters*, pp. xvii, 36, 47, 99, 101.